Large Sample Basin Experiments for Hydrological Model Parameterization:

Results of the Model Parameter Experiment – MOPEX

Edited by

VAZKEN ANDRÉASSIAN
Hydrology group, Cemagref Antony, France

ALAN HALL
Water Resources Application Project/GEWEX, Cooma, Australia

NANÉE CHAHINIAN
Institut de Recherche pour le Développement (IRD), UMR HydroSciences–Université Montpellier II, France

JOHN SCHAAKE
Office of Hydrologic Development, NOAA/National Weather Service, USA

Financial support for production of this volume was provided by Cemagref, the French Agricultural and Environmental Research Centre

IAHS Publication 307
in the IAHS Series of Proceedings and Reports

Published by the International Association of Hydrological Sciences 2006

IAHS Publication 307
ISBN 978-1-901502-73-2

British Library Cataloguing-in-Publication Data.
A catalogue record for this book is available from the British Library.

The papers included in this volume have been reviewed and some were extensively revised before publication.

IAHS is indebted to the employers of the Editors for the invaluable support and services provided that enabled them to carry out their task effectively and efficiently.

Publications in the series of Proceedings and Reports are available from:
IAHS Press, Centre for Ecology and Hydrology, Wallingford, Oxfordshire OX10 8BB, UK
tel.: +44 1491 692442; fax: +44 1491 692448; e-mail: jilly@iahs.demon.co.uk

Printed in The Netherlands by Krips BV, Meppel.

Contents

4 Regionalization and parameterization studies based on large basin samples

5 Perspectives

Introduction and Synthesis: Why should hydrologists work on a large number of basin data sets?

VAZKEN ANDRÉASSIAN[1], ALAN HALL[2], NANÉE CHAHINIAN[3] & JOHN SCHAAKE[4]

1 *Cemagref, Hydrosystems and Bioprocesses Research Unit, Antony, France*
vazken.andreassian@cemagref.fr

2 *Water Resources Application Project/GEWEX, Cooma, Australia*

3 *Agrocampus Rennes, Agricultural Engineering Laboratory, Rennes, France*

4 *Office of Hydrologic Development, NOAA/National Weather Service, Silver Spring, USA*

"*Because almost any model with sufficient free parameters can yield good results when applied to a short sample from a single catchment, effective testing requires that models be tried on many catchments of widely differing characteristics, and that each trial cover a period of many years*" (Linsley, 1982).

INTRODUCTION

There is now such a wealth of publications in hydrology, that it seems unlikely that any hydrologist can find the time to read even a decent part of it. Therefore, we have prepared this rapid introduction and summary of the papers that constitute this volume. Our aim is to convince you, the reader, to keep reading in order to discover the original contributions, which we believe are really worth the time you will spend on them. This volume contains 25 papers, many of which follow on from the presentations made at the last two MOPEX workshops (held in July 2004 at ENGREF, Paris, France, and in April 2005 at the IAHS Assembly in Foz do Iguaçu, Brazil. One of the objectives of this volume is to show how valuable it is to work on large data sets in hydrological modelling. The contributions are organized into five sections:

- The first section provides an introduction to the goals of the MOPEX project, and presents the databases that were used by the participants, a part of which is made available to the hydrological community on the DVD accompanying this volume.
- The second section groups four review papers, which were not presented during the MOPEX workshops, and which have been solicited especially for this volume, in order to provide alternative views on the use of large sample basin experiments in hydrology.
- The third and the fourth sections present model parameterization experiments based on samples of a large number of basins: in the third section, the focus is on the databases that were gathered specifically for the MOPEX program, while the fourth section presents regionalization and parameterization studies based on other large hydrometeorological databases.

– Lastly, the fifth section presents a compilation of the most recent results of the project, and discusses its perspectives.

But let us now examine the justifications for the type of research presented here. After all, why is it so important to consider the results of experiments considering a large number of basin data sets, while the trend of the last decade has been towards more and more extremely detailed studies of a single basin or even a single hillslope?

WHY WE BELIEVE THAT HYDROLOGICAL MODELLING RESEARCH SHOULD FOCUS ON A LARGE NUMBER OF BASIN DATA SETS

In the early days of hydrological modelling, computation power was a limiting factor and hydrologists could generally only afford to work on a few flood events on a single basin. Forty years later, this problem of computation power has almost become a detail. However, the work on a single basin or a limited number of basins still remains the rule in most hydrological modelling studies. Instead, we believe that hydrologists can and should take advantage of working on large sets of basins. We detail here a few reasons why this should allow some progress in hydrological research.

Reason 1: Model intercomparisons can definitely be useful … provided they are based on large data sets

An approach proposed over the last three decades to improve basin models is model intercomparison. The successive international intercomparisons organized by WMO since the end of the 1960s (Askew, 1989; WMO, 1975, 1986, 1992) have been very efficient in promoting a sound competitive spirit among research teams and in forcing modellers to question some of their preconceptions. The same applies for intercomparisons organized by single groups (Vandewiele *et al.*, 1992; Perrin *et al.*, 2001).

Some authors have been rather critical of intercomparisons and their possibility to identify guidelines for model selection (Wheater *et al.*, 1993; Woolhiser, 1996). Our opinion is that, as long as the number of basins included in the comparison is limited (as was the case in most of the comparisons published up to now), their conclusions may well be a matter of luck, and so the intercomparison exercise loses most of its interest. What is needed is a statistically significant number of basins (such as in some of the work presented in this volume), to get a robust model assessment, even if it must be acknowledged that such comparisons are still the exception rather than the rule. However, an increasing number of studies based on large data sets have been published recently (Perrin *et al.*, 2001; Merz & Blöschl, 2004; Oudin, 2004; Mouelhi *et al.*, 2006), and this volume will still add some more: for example, the paper by Folton & Lavabre (this issue) based on 880 basins, and that by Rojas Serna *et al.* based on a sample of 1111 basins, are current records (however, that may not last very long).

When based on several hundreds of basins, intercomparisons can be extremely instructive and they can definitely help to improve models and assess their generality (Perrin *et al.*, 2003). However, these intercomparisons can only be implemented when models can be set-up (i.e. structured, parameterized) in a reasonably automated way. Thus, this does not apply to most of the so-called "physically-based" models, since running such models on just one basin usually requires several months of work.

Reason 2: Only large data sets can allow us to move from climate/region-specific towards general catchment models

Some modellers insist on the fact that catchment models should be climate- or region-specific. This is in line with the prescriptions of the "conceptual" school, which advocate keeping in a catchment model only those "driving processes" that the modeller believes to be important in a given basin. As the "driving processes" may vary depending on hydro-climatic zones, it then seems natural to recommend a climate-specific modelling structure.

However, some of the founding fathers of hydrological modelling recommended looking for models with a certain ambition of generality. Ray Linsley (1982), while alluding to the great variety of driving processes which may affect the rainfall–runoff relationship, wrote that "*these differences do not mean that a single model cannot be applied in all cases. The model must represent the various processes with sufficient fidelity so that irrelevant processes can be 'shut off' or will simply not function*". And Linsley concluded that "*it is no longer necessary for each hydrologist to develop his or her own model for each basin, since [...] a new model for every application eliminates the opportunity for learning that comes with repeated applications of the same model.*" The MOPEX approach is in line with such a statement.

Reason 3: The application of models on ungauged basins requires parameter estimation methods or laws that must be elaborated and/or calibrated on a wide range of conditions

Ungauged basins pose a formidable challenge to hydrology, as the methods proposed to estimate their parameters are still extremely uncertain. If we are to move forward on this topic during the PUB decade, existing and coming methods should be tested and validated over many basins, representative of a range of climates. This is because any unknown function f can, on a set of basins with similar climate conditions, be reduced to a linear first-order series expansion $y = f(x_0) + (x - x_0)f'(x_0)$. Conversely, if a test set encompasses semi-arid as well as humid basins, x will vary over a wide range of values and the proper structure of function f will be tested effectively.

WHAT ARE THE MAIN CONCLUSIONS OF THIS VOLUME?

This volume is the result of enthusiastic (but sometimes contradictory) exchanges held within a diversified group of modellers. This means that not all of the papers necessarily agree with each other. However, they all look with confidence towards data sets which have a large number of basins and intercomparison studies to guide their future work on model parameterization and regionalization. Let us here try to synthesize the main lessons of each section:

- **Data sets of a large number of basins are becoming widely available (Section 1)** We believe that the data set that comes along with this volume provides a wonderful opportunity for hydrologists all over the world to test their own methods and models. We hope in the future to be able to extend this data set and make it available through ftp.

- **Methods for regionalizing catchment models are still in their infancy (Section 2)** A lot remains to be done for simple lumped models and complex distributed ones as well. The explanation of the model parameter values representative of basin behaviour by basin characteristics remains unsolved. As long as this cannot be determined, it is unlikely we can hope for any progress in model regionalization.
- **Parallel work on the same data set of a large number of basins is extremely instructive (Section 3)** For the individual hydrologist, it is always instructive to read about the experience of his colleagues published in the scientific literature, but working on the same data set is a wonderful opportunity to better understand what others actually do, what assumptions they really make, and what approach or model structure is actually superior to the other.
- **Alternative approaches are emerging that may change the way we look at model regionalization (Section 4)** Though they have to repeatedly face the failure of their past efforts to deal with ungauged basins, hydrologists are not at a loss. Innovative approaches are proposed and tested, either to make the ungauged basins less ungauged, or to make progress with the methods of model parameterization.
- **Large samples of basins, as promoted by MOPEX, have much to contribute towards the success of the PUB decade (Section 5)** There is no consensus yet on the fact that experiments based on data sets of a large number of basins are needed in hydrology, and this volume should be seen as a very partial effort to promote the idea of using large data sets for hydrological studies (science does not necessarily need consensus to move forward!) It is often argued that as the basin sample increases in size, it becomes impossible to perform a detailed validation of the raw input data time series, and this will make the data set unusable. But in a comparative setting, there is no reason why a model would be less sensitive to bad input data than another: poor quality data will undoubtedly equally disadvantage all models. Thus, we believe that it is fallacious to object to the use of a large number of basin data sets on the grounds of the difficulty to control quality. For us, one thing is sure: a large sample of basin data sets and well organized intercomparisons may not be sufficient to ensure progress in hydrology, but at least, they are a necessary condition.

REFERENCES

Askew, A. (1989) Real-time intercomparison of hydrological models. In: *New Directions for Surface Water Modelling* (ed. by M. L. Kavvas), (Baltimore, USA), 125–132. IAHS Publ. 181. IAHS Press, Wallingford, UK.

Linsley, R. K. (1982) Rainfall–runoff models-an overview. In: *Proc. Int. Symp. on Rainfall-Runoff Modelling* (ed. by V. P. Singh), 3–22. Water Resources Publications, Littleton, Colorado, USA.

Merz, R. & Blöschl, G. (2004) Regionalization of catchment model parameters. *J. Hydrol.* **287**(1–4), 95–123.

Mouelhi, S., Michel, C., Perrin, C. & Andréassian, V. (2006) Linking stream flow to rainfall at the annual time step: the Manabe bucket model revisited. *J. Hydrol.* **328**(1–2), 283–296.

Oudin, L. (2004) Recherche d'un modèle d'évaporation potentielle pertinent comme entrée d'un modèle pluie-débit global. PhD Thesis, ENGREF, Paris, France.

Perrin, C., Michel, C. & Andréassian, V. (2001) Does a large number of parameters enhance model performance? Comparative assessment of common catchment model structures on 429 catchments. *J. Hydrol.* **242**, 275–301.

Perrin, C., Michel, C. & Andréassian, V. (2003) Improvement of a parsimonious model for streamflow simulation. *J. Hydrol.* **279**, 275–289.

Vandewiele, G. L., Xu, C. Y. & Win, N. L. (1992) Methodology and comparative study of monthly models in Belgium, China and Burma. *J. Hydrol.* **134**, 315–347.

Wheater, H. S., Jakeman, A. J. & Beven, K. J. (1993) Progress and directions in rainfall–runoff modelling, Chapter 5. In:, *Modelling Change in Environmental Systems* (ed. by A. J. Jakeman, M. B. Beck & M. J. McAleer), 101–132. John Wiley & Sons, Chichester, UK.

WMO (1975) Intercomparison of conceptual models used in operational hydrological forecasting. *Operational Hydrology Report no. 7*. World Meteorological Organisation, Geneva, Switzerland.

WMO (1986) Intercomparison of models of snowmelt runoff. *Operational Hydrology Report no. 23*. World Meteorological Organisation, Geneva, Switzerland.

WMO (1992) Simulated real-time intercomparison of hydrological models. *Operational Hydrology Report no. 38*. World Meteorological Organization, Geneva, Switzerland.

Woolhiser, D. A. (1996) Search for physically based runoff model—a hydrologic El Dorado? *J. Hydraul. Engng* **122**(3), 122–129.

1 Introduction

The US MOPEX Data Set

JOHN SCHAAKE[1], SHUZHENG CONG[1] & QINGYUN DUAN[2]

1 *Office of Hydrologic Development, NOAA National Weather Service, 1325 East–West Avenue, Silver Spring, Maryland 20910, USA*
john.schaake@noaa.gov

2 *Lawrence Livermore National Laboratory, Energy and Environment Directorate, 7000 East Avenue, Livermore, California 94550, USA*

Abstract A key step in applying land surface parameterization schemes is to estimate model parameters that vary spatially and are unique to each computational element. Improved methods for parameter estimation (especially for parameters important to runoff response) are needed and require data from a wide range of climate regimes throughout the world. Accordingly, the GEWEX Hydrometeorology Panel (GHP) endorsed the concept of an international Model Parameter Estimation Project (MOPEX) at its Toronto meeting, August 1996. Phase I of MOPEX was funded by NOAA in financial year (FY) 1997, Phase II in FY 2000 and Phase III in FY 2003. MOPEX was adopted as a project of the IAHS/WMO Working Group on GEWEX and of the WMO Commission on Hydrology (CHy), and is now a contributor to the Combined Enhanced Observing Period (CEOP) of the World Climate Research Program (WCRP). In 2004, MOPEX became a Working Group of the IAHS Prediction in Ungauged Basins (PUB) Initiative. MOPEX is also expected to contribute to the work of the Hydrologic Ensemble Prediction Experiment (HEPEX). The primary goal of MOPEX is to develop techniques for the *a priori* estimation of the parameters used in land surface parameterization schemes of atmospheric models and in hydrological models. A major early effort of MOPEX has been to assemble a large number of high-quality historical hydrometeorological and river basin characteristics data sets for a wide range of river basins (500–10 000 km^2) throughout the world. MOPEX data sets are available via the Internet (ftp://hydrology.nws.noaa.gov/pub/gcip/mopex/US_Data/). This paper documents the development of data sets for river basins in the USA. Several highly successful parameter estimation workshops have been organized by MOPEX. The first was held as part of the IAHS General Assembly in Birmingham, UK, in July 1999. The second workshop was hosted in April 2002 in Tucson, Arizona, USA, by the SAHRA/University of Arizona. The third MOPEX workshop was held as part of the IAHS General Assembly in Sapporo, Japan, July 2003. The fourth, in Paris, France, July 2005 was organized by the CEMAGREF in collaboration with the ENGREF, Météo France, National Weather Service and the SAHRA/University of Arizona. The fifth workshop was held as part of the IAHS Scientific Assembly, February 2005, Foz do Iguacu, Brazil. The purpose of the future phases of the project is to: (a) continue collecting additional international data sets; update data from the USA by adding recent years, including data for elevation zones in mountainous areas and refining energy forcing data; (b) continue to conduct international MOPEX workshops; (c) provide leadership to develop a better scientific understanding of how to improve procedures for *a priori* parameter estimation; (d) make a significant hydrological contribution to CEOP and PUB; and (e) demonstrate transferability of MOPEX results. The basic data collection strategy being used in MOPEX is to seek the most readily available and highest quality data first. During the next three years, analyses of the available MOPEX data sets by the international scientific community will be emphasized.

Key words *a priori*; calibration; hydrological models; parameters; regionalization; transferability; uncertainty

INTRODUCTION

Background

A critical step in applying a hydrological model to a basin or a land surface parameterization scheme (LSPS) of an atmospheric model to a specific grid element is to estimate the coefficients or constants (i.e. parameters) in the model. Parameters are inherent in all models. In general, they vary spatially so they are unique to each basin or grid point. Some may also vary seasonally. Moreover, some parameters may be space-time scale dependent.

A common approach in the hydrological modelling community to parameter estimation is to calibrate hydrological models to historical observations by tuning model parameters. For ungauged basins and for LSPS applications, it is difficult to obtain adequate data needed for model calibration. A further complication is that LSPSs are typically applied to large spatial scales and involve many grid elements. To estimate model parameters in those cases, it is necessary to assign model parameters *a priori*.

A priori parameter estimation procedures are available for many hydrological models and LSPSs. But these procedures have not been fully validated through rigorous testing using retrospective hydrometeorological data and corresponding land surface characteristics data. This is partly because the necessary database needed for such testing has not been available until recently. Moreover, there is a gap in our understanding of the links between model parameters and the land surface characteristics. Generally available information about soils (e.g. texture) and vegetation (e.g. type or vegetation index) only indirectly relates to model parameters such as the hydraulic properties of soils and rooting depths of vegetation. Also, it is not clear how heterogeneity associated with spatial land surface characteristics data affects those characteristics at the scale of a basin or a grid cell. Consequently, there is a considerable degree of uncertainty associated with the parameters given by existing *a priori* procedures. It is necessary to develop enhanced *a priori* parameter estimation methodologies for hydrological models and LSPSs. Toward this goal, a project known as the Model Parameter Estimation Experiment (MOPEX) was initiated in 1996. The MOPEX project has been truly an international collaborative endeavour, with the involvement of international scientists and hydrological data assembled from different countries.

MOPEX Science Strategy

The MOPEX Science Strategy involves three major steps, as illustrated in Fig. 1. The first step is to develop the necessary data sets. The next is to use these data to develop *a priori* parameter estimation methodology. Step three is to demonstrate that new *a priori* techniques produce better model results than existing *a priori* techniques for basins not used to develop the new *a priori* techniques.

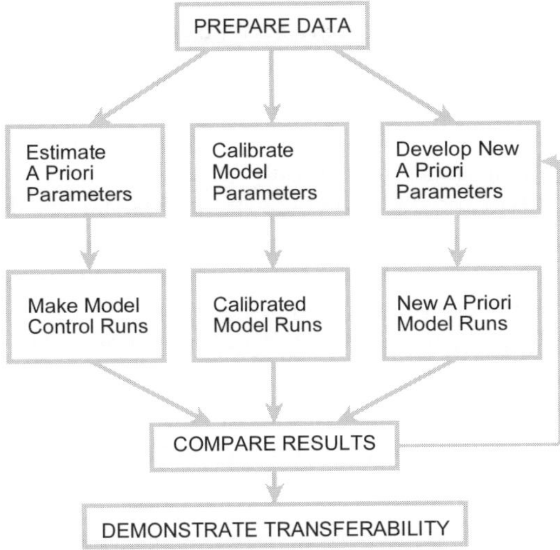

Fig. 1 The MOPEX Science Strategy.

Step two is accomplished using a three-path strategy. The first path is to make reference runs with model parameters estimated by using existing *a priori* parameter estimation procedures. The second path is to make model runs using calibrated or tuned values of selected model parameters. Then, the calibrated parameters are analysed to improve the relationships between model parameters and basin characteristics including climate, soils, vegetation and topographic features. The new relationships are then used to estimate the new *a priori* parameters. The third path is to make new model runs using the new *a priori* parameter estimates. The success of step two is measured by how much improvement in model performance is achieved when the model is operated using new *a priori* parameters as compared to the reference runs.

DATA REQUIREMENTS

The data required for MOPEX can be grouped into four categories:

(a) basic required observations for development and testing;
(b) required physical characteristics;
(c) desirable additional observations;
(d) observations for detailed testing and evaluation.

These are explained below. In each case, the minimum data required are believed to be readily available for the basins to be used. The desired level of data are also believed to be available in many basins.

Basic required observations for development and testing

Historical/retrospective data are needed for many years (as long as possible, e.g. in the USA the period 1948–to date) for at least several hundred test basins which have the minimum observations and basin physical characteristics data and which cover a wide range of climate, soils and vegetation characteristics. (Where available, basins which have the additional data discussed in the following section will be selected to cover the range of characteristics.) The main types of required historical data are: hourly and daily gauged precipitation; daily maximum, minimum and average temperature; surface meteorological observations and daily average stream discharges. These minimum data requirements for MOPEX are actually quite modest, although a higher level of data would be desirable. Since the desirable level of data is unachievable for all basins, the most important requirement is the minimum level. The data requirements are summarized in Table 1.

The most basic minimum requirement is to have daily precipitation and streamflow with climatological monthly mean statistics of the following surface meteorological variables: air temperature; relative humidity; wind speed; and cloud cover. The surface observation statistics would be used to estimate potential evaporation for some schemes and radiative forcing for others. Experience in hydrological modelling is that good parameter estimates can be made with climatological statistics to estimate energy forcing.

Table 1 Summary of minimum basin required observations.

Description requirement	Minimum	Desired
Precipitation	Daily	Hourly
Streamflow	Daily	Hourly
Surface meteorology observations	Monthly statistics	Daily/Hourly

Basins from a wide range of climate regimes are required. Basins must be free of upstream flow regulation. Basins must have sufficient hydrometeorological observations (precipitation, temperature and streamflow). Some basins should have a strongly dominant soil type and a strongly dominant vegetation type. Data for a large number of basins are required.

Data for several hundred basins for a wide range of climate, soils, and vegetation regimes are believed to be needed to meet the MOPEX goal. Although it should be possible to do this globally, it will take many years to have a truly worldwide set of basins. To make the most of limited resources and to demonstrate results quickly, the MOPEX strategy for acquiring basin data is first to seek data from places where it is most readily available. By far the best single source of data is the USA, which covers a wide range of climate regimes, where there are high quality data, and where there are no restrictions on data distribution. Accordingly, more than 400 USA basins were identified that met the basin selection criteria. Then, data from additional basins, globally, can be used to test whether results from the USA basins are transferable to other basins throughout the world and to evaluate how much new information may be contained in data sets from other parts of the world. At the same time, steps have been and are being taken, to encourage countries and scientists throughout the world to contribute to the MOPEX database.

Required observations

Required observations include daily values of mean areal precipitation, mean areal maximum and minimum temperature, streamflow and climatological mean potential evaporation. Additional observations are desirable as discussed below.

A critical aspect of data set preparation to meet MOPEX objectives is to have research quality estimates of mean areal precipitation. A practical estimate of gauge density requirements was made by Schaake (1981) and Schaake *et al.* (2000) for river forecasting applications. The required number of gauges for a basin of area, A (km^2), is:

$$N = 0.6A^{0.3} \tag{1}$$

The exponent 0.3 implies that the required number of gauges doubles as the basin size increases by a factor of 10. The number of gauges given by this equation should give mean areal precipitation estimates for each time step that are accurate to within 20%, 80% of the time during thunderstorm rainfall events (in the 20 000 km^2 Muskingum River, Ohio basin). Equation (1) is reasonable to apply for basins between 200 and 20 000 km^2. The required number of gauges for basins of different size, according to equation (1), are given in Table 2.

Table 2 Desired minimum number of raingauges per basin.

Area (km^2)	Number of gauges
1	1
10	2
100	3
1000	6

Required basin characteristics

Supporting basin boundary, stream and land characteristics data relating to topography, soils and vegetation are also needed (Table 3). Some of these supporting data are available on an ISLSCP CD-ROM, but additional data and refinement of the ISLSCP data to a scale greater than 0.5° are required.

Table 3 Required basin physical characteristics.

Description requirement	Minimum	Desired
Elevation	5 km/5 m contours	1 km/1 m
Basin boundaries	10 km/location	1 km
Streams	10 km/location	1 km
Soils—texture, hydraulic properties, etc.	20 km	1 km
Vegetation—type, rooting depth, phenology, etc.	20 km/monthly	1 km/weekly
Geology	10 km	1 km

Desirable additional observations

Having actual measurements of meteorological surface variables at daily or hourly steps will improve the simulations of land surface schemes. If diurnal fluctuations of surface fluxes are to be simulated, detailed measurements of energy forcing variables are needed. These data are not critical to estimate those parameters that can be extracted from long periods of precipitation and runoff, although they might contribute to the development of improved parameter estimation techniques and to testing the techniques developed only with the minimum required data. Table 4 lists the desired additional observations.

Table 4 Summary of desired additional observations.

Description requirement	Minimum	Desired
Snow cover—satellite product	Seasonal statistics	Daily/1 km
Snow water equivalent	Seasonal statistics	Daily
Pan evaporation	Seasonal statistics	Daily
Clouds	Daily	3-hourly
Short wave radiation	Daily	Hourly
Long wave radiation	Daily	Hourly
Soil moisture	Weekly	Daily

DATA SET DEVELOPMENT FOR US BASINS

Streamgauges

The streamgauge data were selected from a subset of the US Geological Survey (USGS) streamgauge network. This subset includes most of the gauges in the USGS hydro-climatic data network (HCDN) (Slack *et al.*, 1992) or in a similar network selected by Wallis *et al.* (1991). Both of these networks include only gauges believed to be unaffected by upstream regulation and with long enough data records to be suitable for climate studies.

Basin boundaries

Basin boundaries were developed for each of the potential streamgauges from the subset of USGS gauges explained above. These boundaries were based on a DEM derived by the NOHRSC (National Operational Hydrologic Remote Sensing Center) (http://www.nohrsc.noaa.gov/). The NOHRSC provides and maintains the NWS Integrated Hydrologic Automated Basin Boundary System (IHABBS) GIS database to support river and flood forecasting throughout the nation. The IHABBS system uses a 15-arc second DEM that has been processed to have a hydrologically consistent connectivity of neighbouring grid points. The IHABBS connectivity files were used to create a hydrological connectivity file upstream from each of the potential stream gauges. Then basin boundaries were generated from the basin connectivity files. The location of each streamgauge in the DEM was adjusted slightly to get the best possible match between the digital boundary area and the total gauged area specified by the USGS. If the areas matched to within 10% the basin was accepted. Boundaries for several small basins could not be found. In the end a set of 1861 streamgauges were selected for potential use by MOPEX. The locations of these streamgauges are shown in Fig. 2.

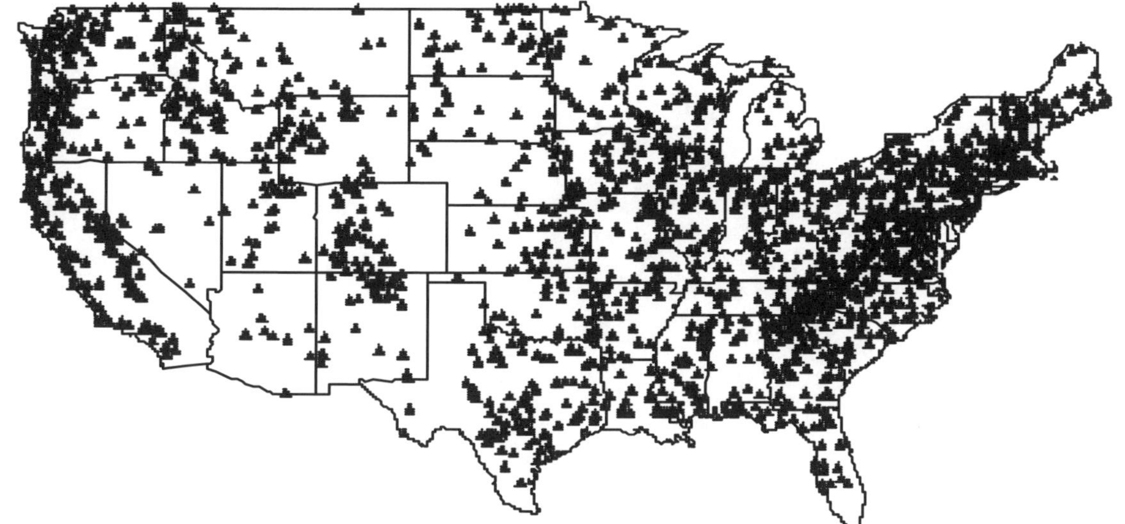

Fig. 2 Location of the 1861 potential MOPEX basins.

Precipitation observations

Hourly and daily precipitation data sets from 1948 to 2003 were assembled for the USA. Data sources included daily and hourly data sets from the National Climate Data Center (NCDC) (http://www.ncdc.noaa.gov/) and daily data from the Natural Resources Conservation Service (NRCS) SNOTEL network (http://www.wcc.nrcs.gov/factpub/-sntlfct1.html). Details about how these data were processed to create mean areal precipitation estimates are presented below.

Basin selection

Figure 3 compares the number of available gauges for each basin with the required number of gauges according to equation (1). Points that lie above the solid curve representing equation (1) have potentially sufficient data. Only 23% (438) of all basins have at least 80% of the required number of gauges. Figure 4 shows the locations of these 438 gauges. Subsequent analysis showed that adequate data for only 431 of the 438 stations were available.

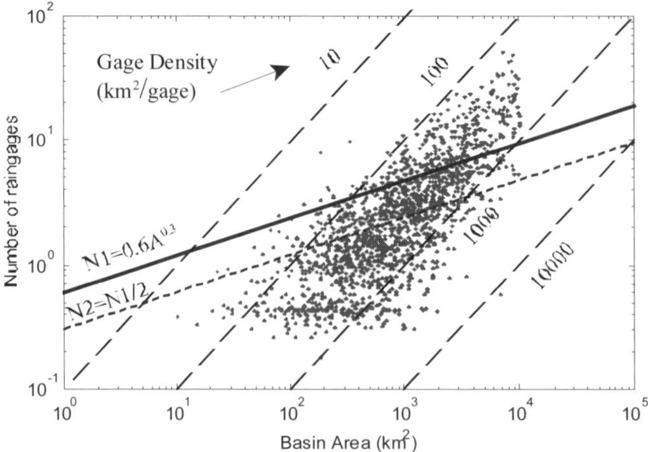

Fig. 3 Comparison of number of available gauges to number of required gauges at 1861 potential MOPEX basins.

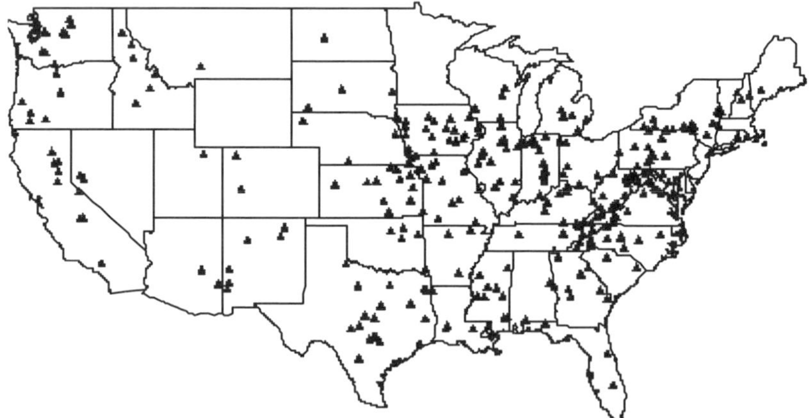

Fig. 4 Location of the 431 basins with an adequate number of precipitation gauges.

Basin characteristics

Gridded values of climate, soils and vegetation characteristics were used to derive basin characteristics for each basin. A useful variable to characterize the climate of each basin is the ratio of mean annual precipitation, P, to mean annual potential evaporation, EP. P/EP for each basin was estimated from gridded values of P from the PRISM project (http://www.ocs.orst.edu/prism_new.html) (Daly *et al.*, 1994) and gridded values of EP from the NOAA Evaporation Atlas (Farnsworth *et al.*, 1982). The distributions of P/EP values for basins within the Mississippi River basin are shown in Fig. 5. A number of approaches have been developed to classify vegetation and gridded files of these were processed to identify the vegetation distributions in each basin. These can be used to identify basins with very large fractions (say 80%) of each vegetation type. Soil hydraulic properties for each were developed from 1-km

STATSGO soil data provided by the Penn State Earth System Science Center (Miller *et al.*, 1999). STATSGO provides soil texture information. Hydraulic properties are derived from soil texture computed using several different empirical relationships (Clapp *et al.*, 1978; Cosby *et al.*, 1983). Figure 6 shows the distribution of basin average saturated hydraulic conductivity values derived for 39 basins in the Arkansas-Red River basin.

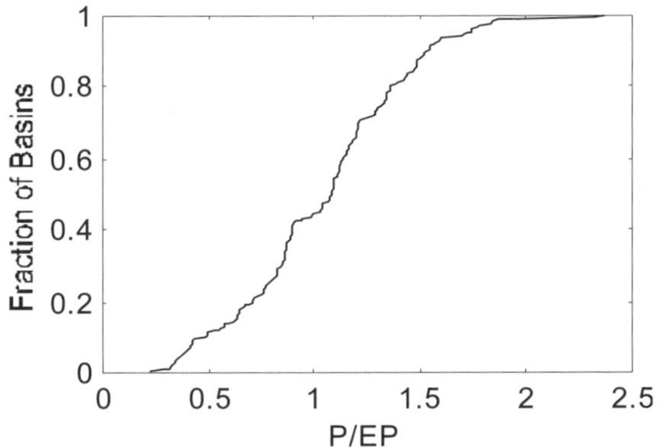

Fig. 5 Distribution of the ratio of mean annual precipitation to mean annual potential evaporation in the Arkansas-Red River basin.

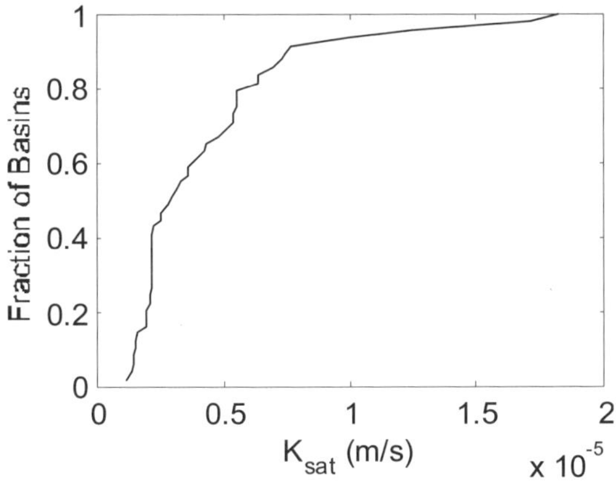

Fig. 6 Distribution of basin average saturated hydraulic conductivity in the Arkansas–Red River basin.

Soil texture The fractional spatial coverage of each of the 16 US Department of Agriculture (USDA) soil types (12 soil, 4 other category) was compiled for each basin. The analysis used 1-km gridded maps of USDA soil texture class for each of 11 soil layers. The 1-km gridded data sets were produced by Miller (1999) based on STATSCO polygon representation of soil texture. Table 5 provides the USDA soil classification definitions.

Table 5 The soil texture classification definitions.

1	S	Sand
2	LS	Loamy sand
3	SL	Sandy loam
4	SIL	Silt loam
5	SI	Silt
6	L	Loam
7	SCL	Sandy clay loam
8	SICL	Silty clay loam
9	CL	Clay loam
10	SC	Sandy clay
11	SIC	Silty clay
12	C	Clay
13	OM	Organic materials
14	W	Water
15	BR	Bedrock
16	O	Other

Table 6 The IGBP vegetation classification definitions.

1	Evergreen needleleaf forest
2	Evergreen broadleaf forest
3	Deciduous needleleaf forest
4	Deciduous broadleaf forest
5	Mixed forest
6	Closed shrublands
7	Open shrublands
8	Woody savannah
9	Savannahs
10	Grasslands
11	Permanent wetlands
12	Croplands
13	Urban and built-up
14	Cropland / natural vegetation mosaic
15	Snow and ice
16	Barren or sparsely vegetated
17	Water bodies

Table 7 The University of Maryland vegetation classification definitions.

0	Water (And Goode's Interrupted Space)
1	Evergreen needleleaf forest
2	Evergreen broadleaf forest
3	Deciduous needleleaf forest
4	Deciduous broadleaf forest
5	Mixed cover
6	Woodland
7	Wooded grassland
8	Closed shrubland
9	Open shrubland
10	Grassland
11	Cropland
12	Bare ground
13	Urban and built-up

Vegetation type Fractional coverage of each vegetation type according to each of two vegetation classification systems was compiled for each basin. The two vegetation classification systems were the International Geosphere-Biosphere Programme (IGBP) and University of Maryland (UMD). Table 6 gives the IGBP vegetation classes. Table 7 gives the UMD vegetation classes.

Greeness fraction Average monthly values of the fractional area coverage of vegetation were derived from NDVI data (Gutman & Iganatov, 1998). These data were originally compiled for most of North America Land Data Assimilation (NLDAS) project. Basin average values for each month were derived from the NLDAS grids for each of the 438 MOPEX basins.

Potential evaporation

Alternatives Experience in the calibration of hydrological models for river forecasting is that the year-to-year variability in the seasonal pattern of potential evaporation is not critical for model parameter estimation, although the effect of this can be seen in model performance and there is some influence on model parameters. One effect is that the active range of total water storage is about 10% greater if interannual variability of potential evaporation is accounted for and this affects the tuning of some model parameters. Some models do not use potential evaporation as an input and require the energy forcing data (e.g. SSIB and BATS). Energy forcing also is required for snow and frozen ground processes in some models. Energy forcing data sets for the US NLDAS domain on a 1/8 degree grid that were compiled by the University of Washington using empirical relationships between surface temperature and energy forcing variables (wind data were taken from the global reanalysis) were processed to produce basin-average energy forcing data sets. Comparisons between the 1/8 degree energy variable estimates and surface meteorological observations at 80 stations throughout the USA show that the temperature values are highly correlated but that other variables, especially wind, are not.

Climatology Climatological estimates of mean daily potential evaporation were made for each basin. The basis for these estimates is data from the NOAA Evaporation Atlas (Farnsworth *et al.*, 1982). This atlas contains maps of average annual and May–October free water surface potential evaporation. The NOAA Evaporation Atlas maps were digitised on a 1/6 degree grid. The NOAA Evaporation Atlas maps were derived by analysis of evaporation pan data. Pan evaporation was converted to free water surface evaporation using pan coefficients derived from studies of lake evaporation at several locations in the USA.

The annual cycle of mean potential evaporation was developed using average monthly pan data taken from the NOAA Evaporation Atlas. A Fourier series with only an annual cycle was fitted to evaporation pan monthly averages. This single frequency component accounted for almost all of the variance of the annual cycle of monthly average pan evaporation. Values of the maximum amplitude and phase angle were gridded. For each basin the amplitude and phase angle values were used together with values of the mean annual potential evaporation and mean May–October potential evaporation to estimate the basin average annual cycle of potential evaporation. Daily values of mean potential evaporation were produced by this analysis.

Temperature index Several methods exist to estimate potential evaporation from air temperature. One of these by Hargreaves (Jensen *et al.*, 1990), used air temperature as an index to energy budget terms in a combination equation (e.g. Penman). The Hargreaves equation uses daily maximum and minimum air temperature data. Therefore, daily mean areal maximum and minimum air temperature data were produced for each basin. The analysis technique is described below.

PRECIPITATION ANALYSIS

Priority was given to creating the highest possible quality basin-average hourly precipitation estimates. This required analysis of both hourly and daily precipitation data from NCDC since there are more than four times more daily precipitation gauges. The nearest hourly gauge was used to disaggregate the daily data to hourly estimates. This required daily gauge observation times to be used. About 1/3 of the observation times were missing, so a highly reliable correlation technique to estimate missing daily observation times was developed. Hourly mean areal precipitation estimates were made using the observed hourly and disaggregated hourly data. Also, a daily mean areal precipitation data set was created for a day ending at midnight so that the daily precipitation data are synchronized with the USGS daily streamflow data.

In the USA, MOPEX basins located west of longitude 100W are likely to be affected by orographic processes. This region includes 80 of the 438 total number of MOPEX basins. It also includes 35 of the 188 basins selected for streamflow verification as part of the NLDAS project. PRISM precipitation climatology data were used to assure that the estimated areal average precipitation estimates in the west preserved the PRISM climatology.

Observation times at NCDC daily COOP stations

Precipitation and maximum and minimum temperature data at NCDC daily COOP (co-operative network) stations are observed once per day at a specified time, e.g. 07:00 h. Different stations may have different observation times. Even for one station, the observation time may change from time to time during the period of record. The frequency at which different observations times occur for stations in Ohio are given in Table 8.

The total number of stations in Table 8 is 169. There are 90 stations where there is no observation time available at all. The major observation times are 7, 8, 18 and 24 hours.

It is essential to account for the effect of observation time in the computation of mean areal precipitation for the MOPEX basins. The strategy used to do this is to disaggregate the daily data to hourly using the time distribution of hourly precipitation at the nearest hourly gauge and using the observation time at the daily gauge.

It also is essential to consider on which calendar day the maximum and minimum temperatures occurred. Since minimum temperatures usually occur in the early morning, it is assumed that the minimum temperature occurred on the day of observation. But maximum temperatures typically occur in the afternoon. In that case it is likely that the maximum temperature for stations with AM observation times actually

Table 8 Frequency of Observation times of daily precipitation in Ohio.

Observation time (hours)	Duration in station-months	Frequency (%)
5	126	0.1
6	767	0.9
7	33635	37.7
8	30286	34.0
9	1073	1.2
10	109	0.1
11	435	0.5
13	20	0.0
16	195	0.2
17	3412	3.8
18	6539	7.3
19	2186	2.5
20	226	0.3
21	680	0.8
22	16	0.0
23	195	0.2
24	9256	10.4

occurred on the previous day. Therefore the station observation times were used to create a "local 24 hour" maximum daily temperature data set that was used to compute mean areal maximum temperature values.

Estimation of missing observation times at NCDC daily COOP stations

About one-third of the station observations times for NCDC daily COOP stations are missing from the digital NCDC station history records. These observation times may have a paper record at NCDC but there is no digital record of them. Because observation time is important to the MOPEX analyses, a method was developed to estimate missing observation times at precipitation stations and the same observation time was assumed to apply if the station also reported maximum and minimum temperature.

Table 9 Classification of the observation times of daily precipitation.

Class	Definition	Number of stations
I	percent of AM duration ≥ 80%	78
II	50%≤ percent of AM duration ≤ 79%	46
III	percent of PM duration ≥ 80%	14
IV	50%≤ percent of PM duration ≤ 79%	5
V	percent of 24 PM duration ≥ 80%	10
VI	50%≤ percent of 24 PM duration ≤ 79%	4
No observation time available at all		90
The rest		12
Total stations		259
Subtotal of stations in I to VI		157

Analysis of the daily COOP precipitation data for Ohio showed that the correlation coefficient for daily precipitation between two stations a given distance apart was strongest if the two stations had the same observation time. The correlation distances were found to be greatest during winter months.

The procedure to estimate missing observation times was first to classify station observation times as shown in Table 9.

Next, exponential distance decorrelation functions were developed for stations in each observation class and depending on the azimuth (eight compass points) of the line connecting the pair of stations. A set of these exponential decorrelation functions were developed for 5-degree latitude–longitude grid boxes for the contiguous USA.

It might be expected that the observation time for a given station could be inferred by comparing the correlation coefficients between the given station and its neighbouring stations with the expected correlation coefficient if the two stations had the same observations time class. The statistic used to make this comparison is the root-mean-square (RMS) difference for each observation time class between the correlation coefficients for the given station and the corresponding expected correlation coefficient for neighbouring stations in that class. The observation time for the given station was assumed to be given by the class with the minimum RMS difference.

Analysis of data for Ohio showed that the correlation coefficients in December and January were higher than that in other months. So January and December were selected for use. The analysis also showed that the averages of the correlation coefficients in azimuth sectors 5 and 6 were highest followed by averages in sectors 4 and 7. Therefore, sectors 5 and 6, and 4 and 7 were selected for use.

The reliability of the estimation method was tested for stations in Ohio. Observation times for stations with known observation times were estimated with the method using data from neighbouring stations. The results are given in Table 10. The rows in Table 10 correspond to the true observation time class. The columns correspond to the estimated observation class. The numbers in the table are the number of stations in a given class (row) that were classified in each of the other classes (column). If the technique was able to detect the correct class for every station, all of the numbers (>0) in Table 10 would fall on the diagonal.

The data in Table 10 suggest the technique is very reliable at detecting AM *vs* PM stations but not that reliable at determining the exact AM or PM time. Therefore, a set of three observation time classes was created by combining classes 1–2, 3–4 and 5–6. The results are in Table 11. Only a few values are off the diagonal so the technique seems to be highly reliable at detecting AM *vs* PM stations.

Table 10 Test of the observation class procedure using Sectors 5 and 6 in December for six classes.

	Estimation						
	59	18	1	0	0	0	78
T	27	13	2	4	0	0	46
r	0	1	10	2	0	1	14
u	0	0	2	2	0	1	5
e	0	0	2	0	5	3	10
	0	0	2	0	0	2	4
	86	32	19	8	5	7	157

Table 11 Test of the observation class procedure using Sectors 5 and 6 in December for three classes.

	Estimation			
T	117	7	0	124
r	1	16	2	19
u	0	4	10	14
e	118	27	12	157

Table 12 Test of the observation class procedure using Sectors 4 and 7, and 5 and 6 in January and December for three classes.

	Estimation			
T	120	4	0	124
r	0	17	2	19
u	0	2	12	14
e	120	23	14	157

By using inter-station correlations for more winter months and for more sectors, it might be possible to improve the reliability of the procedure. Accordingly the best results are given in Table 12 and the procedure used for Table 12 was adopted to estimate missing observation times.

It is not likely that stations with observations times at 24:00 h have missing observation times. Table 12 suggests that the procedure has a 95% reliability to detect AM *vs* PM observations times.

Construction of station data for different time periods

Several processed data sets were created to make the raw NCDC and SNOTEL data easier to use with hydrological models. These include:

(a) disaggregated daily to hourly NCDC COOP precipitation data;
(b) disaggregated daily to hourly SNOTEL precipitation data;
(c) daily 12z to 12z precipitation data;
(d) local 24 h daily precipitation data;
(e) local 6 h precipitation data;
(f) local 24 h maximum temperature for NCDC COOP stations.

Mean areal precipitation analysis

The approach used to estimate mean areal precipitation (MAP) for each of the MOPEX basins is essentially the same as used for gauge-only MAP estimation in the National Weather Service River Forecast System (NWSRFS) (www.nws.noaa.gov/oh/hrl/-nwsrfs/users_manual). The underlying interpolation procedure uses an inverse distance algorithm. The MAP procedure has the following steps:

(1) Get the basin boundary coordinates.

(2) Create an n by n analysis grid within the latitude/longitude window containing the basin. The value used for n is 30.
(3) Create a list of the N grid points within the basin boundaries.
(4) For each month of the year get the PRISM (Daly *et al.*, 1994) climatological mean precipitation value for each grid point in the basin (for the period 1961–1990).
(5) Create a station selection window with latitude/longitude limits a distance d_w outside of the latitude/longitude limits of the basin. The value used for d_w in this analysis is 50 km.
(6) For each time step that a MAP value is needed:
   (i) Get gauge precipitation data for each gauge in the station selection window.
   (ii) Get the PRISM climatological mean precipitation value for each gauge.
   (iii) Compute the ratio, f, of gauge precipitation to PRISM average precipitation for the current month.
(7) For each grid point:
   (i) Select the stations to be used to estimate precipitation at that grid point. Select the two nearest stations in each quadrant.
   (ii) Compute station weights to estimate precipitation at each grid point using inverse distance weighting with exponent m. The value used for m in this analysis is two.
   (iii) Apply the station weights for that grid point to gauge values of f to estimate f at the grid point.
   (iv) Multiply the grid point value of PRISM mean precipitation by the grid point estimate of f, to get the grid point estimate of precipitation.
   (v) Sum the grid point precipitation estimates and divide the sum by N.

This MAP procedure can be used for both daily and hourly time steps. This assures that daily totals of hourly MAP estimates are the same as estimated from an analysis of daily data if data for the same stations are used and the daily data at each station are equal to the sum of the hourly data.

The MAP program code was optimized to avoid computing station weights and getting PRISM values if the station list did not change or if the PRISM values had already been obtained. Also, the interpolation procedure was optimized by computing an effective station weight (implied by the procedure described above) that could be applied to each station to estimate MAP directly as a linear combination of gauge precipitation values. These weights were re-computed at the beginning of each month and whenever the station list changed within a given month.

The MAP procedure used for MOPEX is slightly different than the NWSRFS MAP procedure. The difference is that the value of f at the precipitation gauge is computed as the ratio of the gauge precipitation value to the PRISM mean value for the PRISM grid element where the gauge is located. The NWSRFS procedure uses the historical *station mean* (also called station normal in NWSRFS terminology) for the gauge, not the PRISM mean. This difference has the following implications:

(a) There must be enough data at a gauge to estimate the station normals for each month to use the NWSRFS procedure.
(b) The station normal may not be for the same climatological period as the PRISM mean value unless the period of record is the same or some method is used to account for the difference between the period gauged and the PRISM period.

(c) If the station normal is for the same period as the PRISM data, the NWSRFS estimation procedure would produce a mean precipitation estimate at a location very near the gauge that would equal the PRISM value, not the gauge mean value.
(d) If the station normal is not for the same period as the PRISM data, the NWSRFS estimation procedure would produce a mean precipitation estimate for the period of the gauge record at a location very near the gauge that would equal the PRISM value for gauge period, not the PRISM period.

The procedure used here to process the PRISM data has the following advantages:

(a) The MAP procedure used here can use precipitation data for any station without regard for its period of record.
(b) The MAP procedure used here will produce a mean precipitation estimate at a location very near a gauge that is equal to the gauge value.
(c) The mean value of estimated precipitation anomalies (relative to the PRISM grid) near gauges tend to be driven by the gauge anomaly (relative to the PRISM grid).
(d) The current procedure is less sensitive to the particular period of time used to estimate the PRISM climatology. It does not actually use the magnitude of the PRISM grid directly. It only uses the ratio of the PRISM value at one point to the PRISM value at another point. Although this ratio depends on the types of precipitation events that occurred over a period of time, it should not be very sensitive to differences in mean precipitation climatology over different periods of time. In other words, it might be expected that spatial ratios would be much less sensitive to the period chosen than the magnitude of the precipitation. As a result it is more important to update the PRISM grids if the PRISM analysis method improves than if a more recent period of data are used in the PRISM analysis.
(e) The existing NWSRFS procedure requires manual construction of precipitation isolines.

Because the quality of the gauge data used in the MAP analysis is very important and the quality of data for gauges that have operated for only short periods is problematic, only gauges with at least five years data are used in this analysis.

Another difference between the MAP procedure used here and the MAP procedure used by the NWSRFS is that the current procedure does not estimate missing data for a station. The NWSRFS procedure first estimates missing station data so there is a data value for every station for the entire period of the analysis. The NWSRFS procedure can be shown not to modify the effective weights applied to the non-missing data to get the MAP value. The NWSRFS procedure is easier to program than the current procedure and is only slightly more computationally efficient.

TEMPERATURE ANALYSIS

Temperature data sources

Data sources for temperature analysis were daily maximum and minimum temperatures at NCDC COOP stations and at SNOTEL sites.

Estimation of missing temperature observation times

The observation times estimated for missing temperature observation times were assumed to be the same as for the corresponding precipitation observation.

Mean areal temperature analysis

The procedure to estimate mean areal temperature involved first estimating mean areal maximum and minimum daily temperature. Then, if temperature values during the day are needed, the procedure suggested by Parton & Logan (1981) to estimate hourly temperatures from the daily maximum and minimum was used.

US DATA SETS

Several different kinds of data sets were produced for the USA by the MOPEX project. These are available by anonymous ftp at ftp://hydrology.nws.noaa.gov/pub/gcip/-mopex/US_Data/. Gauge data were obtained for the period of record for each station for the period 1948 to the present. A general description of the data available at this site follows. More detailed documentation is available on the ftp site.

Mean areal precipitation and temperature data are organized as time series data for each basin. A composite daily time series data set was also produced for each basin that contains daily mean areal values for precipitation, potential evaporation, stream-flow, maximum temperature and minimum temperature.

The gauge-based data sets used to generate the mean areal time series are organized in a "random-spatial" format. Daily gauge-based data are organized so there is a data file for each month. This file contains a record for each station with data for at least one day of that month. All stations for the contiguous USA are in that file. The monthly files are organized into directories for the decades 1940, 1950, ..., 2000. Hourly gauge-based data are organized so there is a file for each day with a record for each station containing 24 hourly values for that day. The daily data files are organized into monthly subdirectories.

hydrology.nws.noaa.gov/pub/gcip/mopex/US_Data/US_438_Daily
This directory contains a time series daily data file of each of the MOPEX basins that met the minimum precipitation gauge density requirements. There are data for 431 basins in this directory.

hydrology.nws.noaa.gov/pub/gcip/mopex/US_Data/Basin_Characteristics
This directory contains files and subdirectories with basin characteristics data for each of the 431 basins with data subdirectory US_438_Daily.

hydrology.nws.noaa.gov/pub/gcip/mopex/US_Data/Basin_Boundaries
This directory contains basin boundary data files for each of the 431 basins with data subdirectory US_438_Daily.

hydrology.nws.noaa.gov/pub/gcip/mopex/US_Data/Hourly_MAP
This directory contains hourly MAP files for each of the 431 basins with data subdirectory US_438_Daily.

hydrology.nws.noaa.gov/pub/gcip/mopex/US_Data/6hr_MAP
This directory contains 6-hourly MAP files for each of the 431 basins with data subdirectory US_438_Daily.

hydrology.nws.noaa.gov/pub/gcip/mopex/US_Data/6hr_MAT
This directory contains 6-hourly MAT files for each of the 431 basins with data subdirectory US_438_Daily.

hydrology.nws.noaa.gov/pub/gcip/mopex/US_Data/Daily_Q_1800
This directory contains time series files of daily streamflow for each of the 1862 gauges that were potential US MOPEX basins.

hydrology.nws.noaa.gov/pub/gcip/mopex/US_Data/Ameriflux_Data
This subdirectory contains information about the Ameriflux network, including data sets for some stations.

hydrology.nws.noaa.gov/pub/gcip/station_data/precipitation/dpndata/
This directory contains subdirectories for the following gauge-based daily precipitation data sets:

(a) ncdc_24lcl—NCDC daily station data processed using hourly data to be on a local 24-hour clock. This subdirectory contains separate subdirectories for data from daily and hourly stations. Each of these subdirectories contains a subdirectory for each decade from 1940 to 2000.
(b) ncdc_ 12z—NCDC daily station data processed using hourly data to be on a daily 12z to 12z clock. The date corresponds to the day at the end of the 12z valid observation period. This subdirectory contains separate subdirectories for data from daily and hourly stations. Each of these subdirectories contains a subdirectory for each decade from 1940 to 2000.
(c) snotel_24lcl—original daily SNOTEL data on a local 24-hour clock. This subdirectory contains subdirectories for each decade 1970 to 2000.
(d) snotel_12z—SNOTEL daily station data processed using hourly data to be on a daily 12z to 12z clock. The date corresponds to the day at the end of the 12z valid observation period. This subdirectory contains subdirectories for each decade 1970 to 2000.

hydrology.nws.noaa.gov/pub/gcip/station_data/precipitation/hpndata/
This directory contains subdirectories for the following gauge-based hourly precipitation data sets:

(a) ncdcz—NCDC hourly data on a UTC 0–24z clock. These data are from hourly gauges.
(b) ncdc_ disagg—disaggregated NCDC daily data to hourly on a UTC 0–24z clock.

(c) snotel_disagg—disaggregated hourly snotel daily data on a UTC 0–24z clock.

hydrology.nws.noaa.gov/pub/gcip/station_data/temperature/
This directory contains subdirectories for the following gauge-based daily station data sets:

(a) snoteldtmn—SNOTEL daily minimum temperature.
(b) snoteldtmx—SNOTEL daily maximum temperature.
(c) tmax_24lcl—NCDC daily maximum temperature (observation time adjusted).
(d) tmin—NCDC daily minimum temperature.

Acknowledgements The NOAA Office of Global Programs provided financial support under the GEWEX Americas Prediction Project (GAPP) program for development of the US basin data set for the MOPEX database.

REFERENCES

Clapp, R. B. & Hornberger, G. M. (1978) Empirical equations for some soil hydraulic properties. *Water Resour. Res.* **14**(4), 601–604.

Cosby, B. J., Hornberger, G. M. Clapp, R. B. & Ginn, T. R. (1984) A statistical relationship of soil moisture characteristics to the physical properties of soils. *Water Resour. Res.* **20**, 682–690.

Daly, C., Neilson, R. P. & Phillips, D. L. (1994) A statistical-topographic model for mapping climatological precipitation over mountainous terrain. *J. Appl. Met.* **33**, 140–158.

Farnsworth, R. K., Thompson, E. S. & Peck, E. L. (1982) Evaporation Atlas for the contiguous 48 United States. *NOAA Technical Report, NWS 33, Washington, DC, USA.*

Franz, K., Ajami, N., Schaake, J. & Buizza, R. (2005) Hydrologic Ensemble prediction experiment focuses on reliable forecasts. *Eos, Trans. AGU* **86**(25), 239.

Gutman, A. & Iganatov, A. (1998) The derivation of the green vegetation fraction from NOAA/AVHRR data for use in numerical weather prediction models. *Int. J. Remote Sens.* **19**(8), 1533–1543.

Hogue, T., Wagener, T., Schaake, J., Duan, Q., Hall, A., Gupta, H. V., Leavesley, G. & Andreassian, V. (2004) A new phase of the Model Parameter Estimation Experiment (MOPEX). *Eos, Trans. AGU* **85**(22), 217–218.

Jensen, M. E., Burman, R. D. & Allen, R. G. (1990) Evapotranspiration and irrigation water requirements. *ASCE Manual 70*, 232.

Miller, D. A. & White, R. A. (1999) A conterminous United States multi-layer soil characteristics data set for regional climate and hydrology modeling. *Earth Interactions 2* (available at http://EarthInteractions.org).

Parton, W. J. & Logan, J. A. (1981) A model for diurnal variation in soil and air temperature. *Agric. Met.* **23**, 205–216.

Schaake, J. C. (1981) Summary of river forecasting raingauge network density requirements (unpublished). Available at http://www.nws.noaa.gov/oh/mopex/raingauge%20density%20requirement.htm.

Schaake, J. C., Duan, Q., Smith, M. & Koren, V. (2000) Criteria to select basins for hydrologic model development and testing. Preprints in: *15th Conf. On Hydrology* (Long Beach, California, USA, Am. Met. Soc., 10–14 January 2000), Paper P1.8.

Slack, J. R. & Landwehr, J. M. (1992) Hydro-climatic data network (HCDN): A US Geological Survey streamflow data set for the United States for the study of climate variations, 1874–1988. *USGS Open-file Report 92-129, Reston, Viginia, USA.*

Wallis, J. R., Lettenmaier, D. P. & Wood, E. F. (1991) A daily hydroclimatological data set for the continental United States. *Water Resour. Res.* **27**(7): 1657–1663.

The MOPEX 2004 French database: main hydrological and morphological characteristics

N. CHAHINIAN[1,2], T. MATHEVET[1,3], F. HABETS[4] & V. ANDRÉASSIAN[1]

1 *Hydrology group, Cemagref Antony, Parc Tourvoie BP 44, F-92163 Antony cedex, France*
nanee.chahinian@agrocampus-rennes.fr

2 *Now at: Institut de Recherche pour le Développement(IRD), UMR HydroSciences–Université Montpellier II, Case Courrier MSE, Place Eugène Bataillon,F- 34095 Montpellier cedex 5, France*

3 *Now at: EDF-DTG, BP 41, F-38040 Grenoble cedex, France*

4 *Météo-France, 42, avenue G. Coriolis, F-31057 Toulouse cedex 1, France*

Abstract The MOPEX 2004 workshop was held in Paris in July 2004. As for former workshops, the participants were asked to test their models and explore new parameter estimation strategies using the common databases. However, for this workshop a new series of 40 French catchments was added to the existing MOPEX database. The new data consists of hourly estimates of hydrometric data as well as information on land surface characteristics. This paper presents and analyses the contents of the Paris workshop database.

Key words database; MOPEX; parameter estimation; rainfall–runoff modelling

INTRODUCTION

Since its launch in 1996, the Model Parameter Estimation Experiment, MOPEX, was aimed at the development of a comprehensive database containing many years of historical hydrometeorological time series and land surface characteristics data. The database thus compiled is available to the hydrological community and serves as a common base for model and parameterization inter-comparison experiments.

In the previous experiments (Birmingham, 1999; Arizona 2002; Sapporo, 2003), the database included daily measurements of precipitation and runoff (Duan *et al.*, 2006). For the 2004 workshop a new selection of 40 French catchments, with both daily and hourly meteorological data, was added to the MOPEX database pool. As in previous workshops, the participants were asked to run their models and simulate runoff series. However, as part of the IAHS's PUB decade (Predictions on Ungauged Basins) participants were asked to further explore alternative ways to estimate parameters without using runoff data for model calibration. This paper is a synthesis of the French database information.

The database

The database was put together especially for the MOPEX 2004 workshop. All the selected catchments satisfy the number of raingauge *vs* area criterion set by Schaake *et al.* (2000), i.e. a minimum of two raingauges for catchments with an area <50 km^2 and an increasing number of gauges with area, to reach nine raingauges for catchments with an area >6000 km^2.

In addition to hourly and daily hydrometric data, the database contains information on the catchments' elevation, land use and soil type. The files are available in digital format both as ASCII files and as ArcView coverages.

Catchment location

The database is composed of 40 French catchments scattered all around continental France (Fig. 1) covering the full range of climatic conditions encountered in France. Special care was taken to avoid catchments with significant snow cover/contribution and flow regulation structures. The catchment contours and the river network were digitized from the 1/25 000 topographic analogue maps produced by the National Geographic Institute. The digitized river network hence corresponds to the "blue lines" found on these maps.

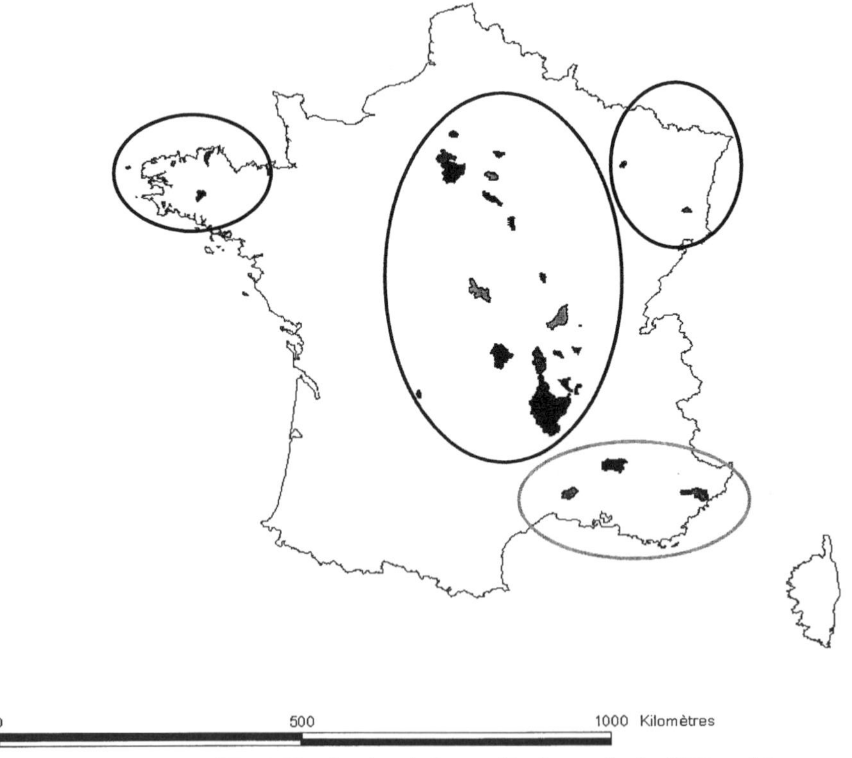

Fig. 1 Simulated and observed hydrographs for VIC model.

Two of the selected catchments (A1522020, A5723010) are located in the Alsace region of eastern France with a semi-continental climate, i.e. with harsh winters and hot summers. Four catchments (J2034010, J3024010, J4124420, J4712010) are located in Brittany, western France under oceanic-humid climatic conditions. Seven

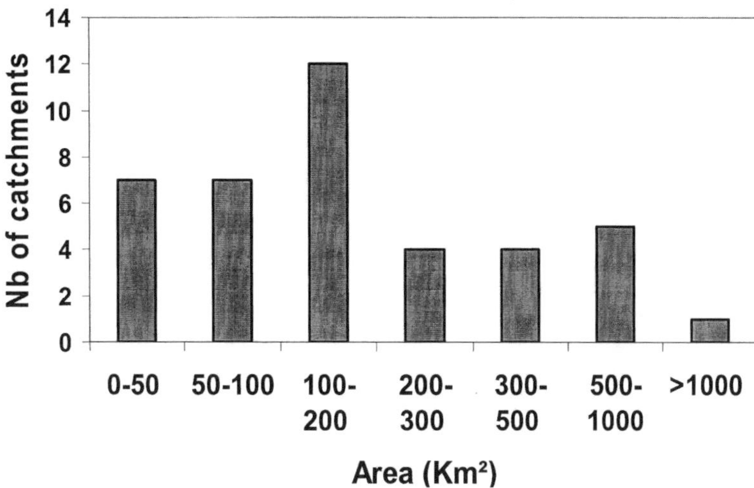

Fig. 2 Catchment area distribution.

catchments (V6035010, V6052010, X2414030, Y3514020, Y5615010, Y5615030 and Y5625020) are located in southern France, near the Mediterranean Sea where the winters are mild and the summers hot. The remaining 27 catchments are sited in the Parisian area and in central France i.e. they have an intermediate climate with cold winters and hot summers.

Catchment size varies between 11 km^2 for the smallest (La Denante at Davayé) and 3234 km^2 (La Loire at Bas en Basset). The mean area is 297 km^2 (Fig. 2), with 30 catchments having an area smaller than 291 km^2.

Elevation

The elevation data was derived from a Digital Elevation Model with 75-m resolution (Table 2). The minimal elevation values range between 5 m (J2034010) and 1010 m (X2414030) while maximum elevation ranges between 121 m (J3024010) and 1900 m (V6052010). The catchment with the highest span is the Loup at Villeneuve-Loubet (5615030) with 1750 m. Six out of the seven Mediterranean catchments have a higher span than the median, whereas all four Breton catchments have a lower span than the median. These two groups have very contrasted characteristics and performances and it would be interesting to see how the model parameters will try to account for and translate this variability.

Hydrological data

The hydrological data consisted of climatic data provided by Météo-France and runoff data provided through the Banque Hydro database. The climatic data consists of hourly estimates of evapotranspiration, downward solar and infrared radiation, specific air

Table 1 Catchment list and characteristics.

Level	Code	Name	Area (km²)	Instantaneous peak discharge-maximum value ever recorded ($m^3 s^{-1}$)*	Median of annual discharge ($m^3 s^{-1}$)*	Mean annual rainfall 1995–2002 (mm year^{-1})	Mean annual evapo-transpiration 1995–2002 (mm year^{-1})
3C	J3024010	Le Guillec à Trézilidé	43	12	0.67	1014	686
	V6035010	Le Toulourenc à Malaucène	150	81	1.32	1059	1081
	Y5615030	Le Loup à Villeneuve-Loubet	279	228	4.47	1159	1120
12C	A1522020	La Lauch à Guebwiller	68	41	1.65	1665	735
	H2001020	L'Yonne à Corancy	98	45	2.89	1299	742
	H3613020	Le Lunain à Épisy	252	12	0.74	808	736
	H5723011	L'Orgeval à Boissy-le-Châtel	104	33	0.63	804	753
	J2034010	Le Guindy à Plouguiel	125	27	1.21	960	710
	J4124420	La Rivière de Pont-l'Abbé à Plonéour-Lanvern	32	4	0.53	1236	719
	K0744010	L'Anzon à Débats-Rivière-d'Orpra	181	72	2.54	980	727
	K0753210	Le Lignon du Forez à Boën	371	285	5.70	1012	727
	Y3514020	Le Vistre à Bernis	291	43	2.11	847	1161
40C	A5723010	L'Ingressin à Toul	54	10	0.44	881	666
	H2513110	Le Tholon à Champvallon	131	18	0.86	824	761
	H3613010	Le Lunain à Paley	163	18	0.56	818	736
	H3923010	Le ru d'Ancoeur à Blandy	181	24	0.59	780	753
	H4252010	L'Orge à Morsang-sur-Orge	922	41	3.96	720	707
	H7853010	Le Sausseron à Nesles-la-Vallée	101	3	0.56	763	726
	H7913030	La Mauldre à Aulnay-sur-Mauldre	369	29	2.15	711	685
	J4712010	L'Éllé au Faouët	142	59	2.75	1192	729
	K0100020	La Loire à Goudet	432	1600	5.70	1395	773
	K0253020	La Borne occidentale à Espaly-Saint-Marcel	375	261	3.66	840	773
	K0550010	La Loire à Bas-en-Basset	3234	3500	38.60	979	784
	K0614010	Le Furan à Andrézieux-Bouthéon	178	142	2.54	849	773
	K0813020	L'Aix à Saint-Germain-Laval	193	195	3.02	988	727
	K0974010	Le Gand à Neaux	85	58	0.90	798	727
	K1173210	L'Arconce à Montceaux-l'Étoile	599	147	5.77	890	794
	K2724210	L'Artière à Clermont-Ferrand	49	9	0.27	955	825
	K2783010	La Morge à Maringues	713	103	4.29	862	825
	K5623010	L'Auron au Pondy	199	30	0.98	783	733
	K5653010	L'Auron à Bourges	585	84	3.77	801	786
	P3245010	Le Mayne à Saint-Cyr-la-Roche	49	23	0.70	1166	771
	U4305410	La Denante à Davayé	11	8	0.13	872	792
	U4525210	Le Morgon à Villefranche-sur-Saône	68	18	0.49	828	749
	V3315010	La Valencize à Chavanay	36	17	0.36	851	735
	V3517010	Le Ternay à Savas	25	16	0.34	867	735
	V6052010	L'Ouvèze à Vaison-la-Romaine	585	1000	6.07	990	1081
	X2414030	L'Artuby à la Bastide	91	104	1.04	1241	1267
	Y5615010	Le Loup à Tourrettes-sur-Loup	206	147	3.67	1226	1120
	Y5625020	La Cagne à Cagnes-sur-Mer	95	160	0.82	1059	1120

(*) Record length is variable. The longest time series available in "Banque Hydro" is retained.

Table 2 Synthesis of elevation data.

Catchment	Elevation (m)				
	Minimum	Maximum	Mean	1st quantile	3rd quantile
A1522020	299	1408	782	547	1014
A5723010	213	421	303	246	358
H2001020	335	900	594	506	672
H2513110	90	322	176	137	212
H3613010	86	201	149	135	159
H3613020	55	201	134	124	152
H3923010	67	146	115	107	124
H4252010	40	180	132	102	160
H5723011	77	185	148	140	157
H7853010	40	216	107	88	126
H7913030	24	187	124	104	149
J2034010	5	300	83	61	98
J3024010	33	121	87	77	97
J4124420	20	156	85	62	105
J4712010	79	302	197	180	212
K0100020	765	1602	1174	1101	1255
K0253020	631	1281	951	878	1024
K0550010	440	1720	967	843	1098
K0614010	362	1307	661	513	752
K0744010	405	1344	758	654	841
K0753210	387	1628	866	678	1039
K0813020	372	1183	745	634	848
K0974010	355	886	569	483	647
K1173210	245	760	357	320	389
K2724210	341	1018	613	433	788
K2783010	291	1458	526	336	674
K5623010	168	313	212	196	226
K5653010	133	313	191	170	210
P3245010	117	445	279	196	360
U4305410	195	483	314	249	375
U4525210	169	797	327	257	374
V3315010	191	1355	639	446	806
V3517010	505	1391	893	736	1043
V6035010	315	1876	838	652	989
V6052010	180	1900	695	456	884
X2414030	1010	1640	1192	1095	1262
Y3514020	20	213	81	48	108
Y5615010	130	1760	1050	890	1242
Y5615030	10	1760	827	349	1180
Y5625020	10	1688	597	240	939

humidity, air temperature and wind speed. They are extrapolated over each catchment with Météo-France's SAFRAN code (Durand *et al.*, 1993), hence a single set of average climatic variables are given for each catchment. The hydrometric data consists of hourly and daily estimates of runoff depth at each catchment outlet.

Table 1 and Figs 3 and 4 indicate that the database is indeed representative of the hydrological regimes encountered in France. Figure 3 shows a rough linear trend

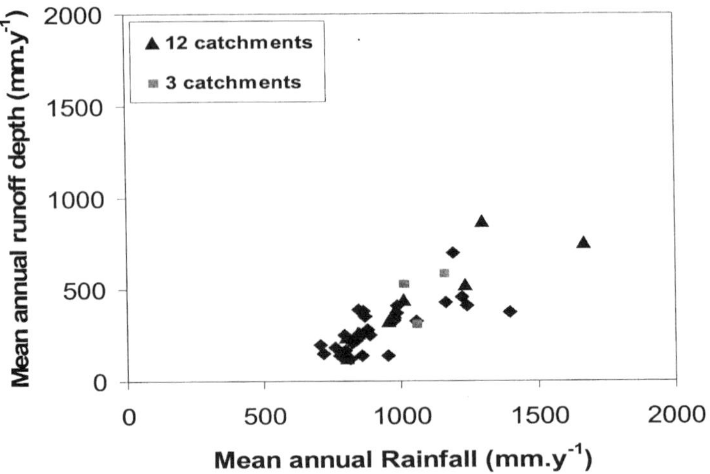

Fig. 3 Mean yearly rainfall *vs* runoff.

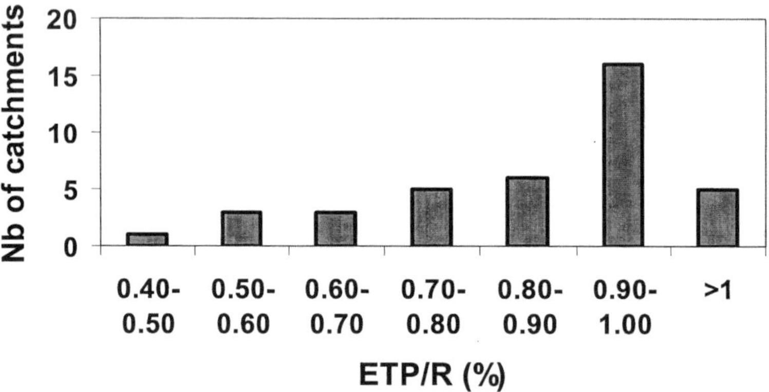

Fig. 4 Mean yearly aridity index distribution.

between the average annual rainfall and runoff depth. Analysis of the catchments' rainfall and runoff data over the 1995–2002 period indicates that the catchment receiving the least amount of rain is the Mauldre at Aulnay-sur-Mauldre (H7913030) with 711 mm year^{-1} whereas the wettest catchment is the Lauch at Guebwiller (A1522020) with 1665 mm year^{-1}. The mean annual evapotranspiration varies between 666 mm year^{-1} for the Ingressin at Toul (A5723010) and 1267 mm year^{-1} for the Artuby at La Bastide (X2414030). Not surprizingly, the highest evapotranspiration values are reported for the Mediterranean catchments.

Table 1 also indicates that the catchment producing the highest amount of runoff is the Yonne at Corancy (H2001020). The driest catchment in terms of runoff depth is the Lunain at Paley (H3613010). The mean runoff yield over the study period varies between 14% and 67%, with a mean value of 33%. Thus the selected catchments have moderate runoff coefficients with only 11 catchments transforming more than 40% of the mean rainfall input into runoff (Fig. 5).

Table 1 and Fig. 6 also present the long-term behaviour of these catchments, as reported in the Banque Hydro database. Figure 6 indicates that the highest value of peak discharge recorded for the majority of catchments does not exceed 25 $m^3 s^{-1}$. The highest values of peak discharge were recorded for the Loire at Bas en Basset (K0550010) and Goudet (K0100020). The third highest value is reported for the Ouvèze at Vaison la Romaine (V6052010). The Ouvèze is a tributary of the Rhone, the fastest of French rivers, and the largest tributary of the Mediterranean after the Nile. The peak discharge value reported corresponds to the 1992 flood.

Land use

Land-use percentages were calculated using the Corine Land Cover inventory (EEA, 1995), which is a homogeneous European land use database. The information is based

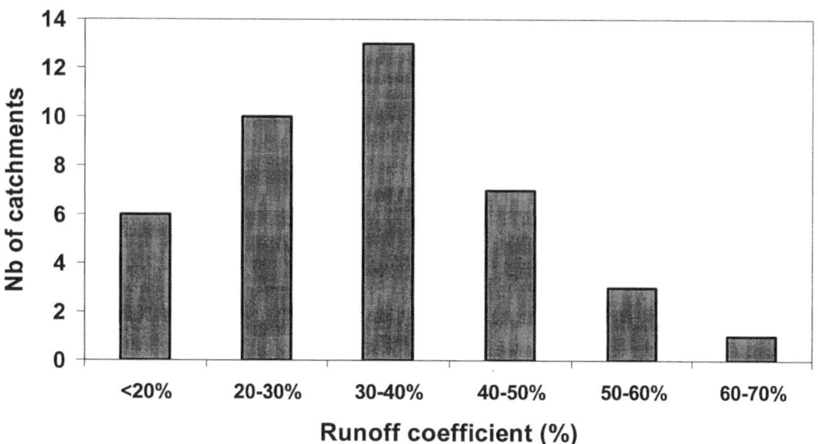

Fig. 5 Mean yearly runoff coefficient.

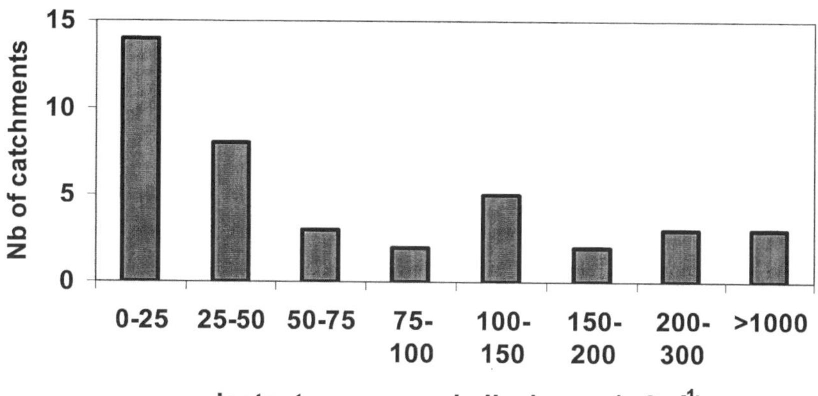

Fig. 6 Peak discharge distribution.

Table 3 Synthesis of main land use classes.

Catchment code	Urban fabric (%)	Industrial or commercial units (%)	Non irrigated arable land (%)	Vineyards and olive groves (%)	Fruit trees and berry plantations (%)	Pastures	Complex cultivation patterns (%)	Land principally occupied by agriculture, with significant area of natural vegetation (%)	Broad leaved forest (%)	Coniferous forest (%)	Mixed forest (%)	Natural grassland (%)	Moors and heath lands (%)	Sclero-phylous vegetation (%)	Trans-itional wood land-scrub (%)	Total (%)
A1522020	4	0	1	0	0	5	2	0	17	25	35	5	5	0	0	99
A5723010	4	3	11	0	0	9	7	7	51	2	4	0	1	0	0	100
H2001020	1	0	0	0	0	27	5	4	18	36	9	0	0	0	0	100
H2513110	1	0	59	0	0	2	11	5	21	0	1	0	0	0	0	100
H3613010	1	0	74	0	0	1	2	4	18	0	0	0	0	0	0	100
H3613020	1	0	70	0	0	1	2	3	22	0	1	0	0	0	0	100
H3923010	2	1	63	0	0	0	1	0	31	0	0	0	0	0	0	99
H4252010	14	2	49	0	0	0	0	0	28	3	1	0	0	0	0	97
H5723011	0	0	81	0	0	0	2	0	17	0	0	0	0	0	0	100
H7853010	3	0	75	0	0	0	1	1	21	0	0	0	0	0	0	100
H7913030	15	2	54	0	1	1	0	0	21	0	2	0	0	0	1	96
J2034010	1	0	43	0	0	1	35	17	3	0	0	0	0	0	0	100
J3024010	4	0	38	0	0	9	35	13	1	0	0	0	1	0	0	100
J4124420	1	0	16	0	0	10	51	15	7	0	0	0	0	0	0	100
J4712010	1	0	10	0	0	15	36	19	9	1	1	1	4	0	1	98
K0100020	0	0	0	0	0	26	8	5	8	25	5	15	5	0	2	99
K0253020	1	0	3	0	0	32	30	8	1	22	2	0	0	0	0	100
K0550010	1	0	1	0	0	33	15	8	2	26	7	3	2	0	1	99

Table 3 *(cont.)*

K0614010	14	9	0	0	0	30	10	8	8	12	3	0	0	0	0	95
K0744010	1	0	0	0	0	36	1	7	4	33	14	1	1	0	4	100
K0753210	1	0	0	0	0	31	1	6	4	33	10	4	4	0	5	100
K0813020	1	0	0	0	0	45	3	5	3	31	10	0	1	0	1	100
K0974010	3	0	1	0	0	54	16	10	4	5	6	0	0	0	0	100
K1173210	0	0	0	0	0	70	8	4	9	7	2	0	0	0	0	100
K2724210	23	3	1	0	0	18	14	16	4	1	10	0	6	0	1	98
K2783010	7	1	30	0	0	14	18	8	13	3	2	0	2	0	0	99
K5623010	0	0	32	0	0	53	2	4	9	0	0	0	0	0	0	100
K5653010	1	0	46	0	0	26	1	3	22	0	1	0	0	0	0	100
P3245010	1	0	1	0	4	23	38	10	21	3	1	0	0	0	0	100
U4305410	1	0	0	40	0	8	34	7	6	0	1	0	0	0	4	100
U4525210	11	1	0	36	0	6	30	6	9	0	0	0	0	0	0	100
V3315010	3	0	0	2	1	12	19	8	23	23	7	0	1	0	0	100
V3517010	1	0	0	0	0	32	5	5	6	37	11	0	1	0	2	100
V6035010	0	0	3	0	0	0	4	5	36	3	12	3	19	1	9	95
V6052010	1	0	2	6	0	0	10	6	26	13	11	1	12	4	5	97
X2414030	0	0	10	0	0	12	2	2	0	57	0	11	0	0	2	98
Y3514020	5	1	5	46	7	0	1	0	9	0	0	0	3	5	13	95
Y5615010	1	0	0	0	0	2	1	2	16	13	1	51	0	1	8	96
Y5615030	7	0	1	0	0	1	1	1	14	16	6	38	1	4	6	97
Y5625020	24	0	2	0	0	1	1	1	14	2	12	40	0	0	0	97

on Landsat MS and Spot XS satellite images which were later combined with aerial photographs, and topographic and forest maps by the French National Institute for the Environment (IFN) and the National Geographical Institute (IGN). The end product is a set of 1/100 000 scale digital maps with a 3-level nomenclature. The three levels are aggregated such that there are 5 classes in the first level, 15 in the second and 44 in the last. This level has the most detailed information. Table 3 summarizes the main land use classes, i.e. those occupying >5% of total area of the catchment, encountered on the 40 catchments selected for this study. The catchments retained for this study are not affected directly by urban activities. Pastures, non-irrigated arable land and coniferous forest cover the greatest areas when considering all the catchments (Fig. 7).

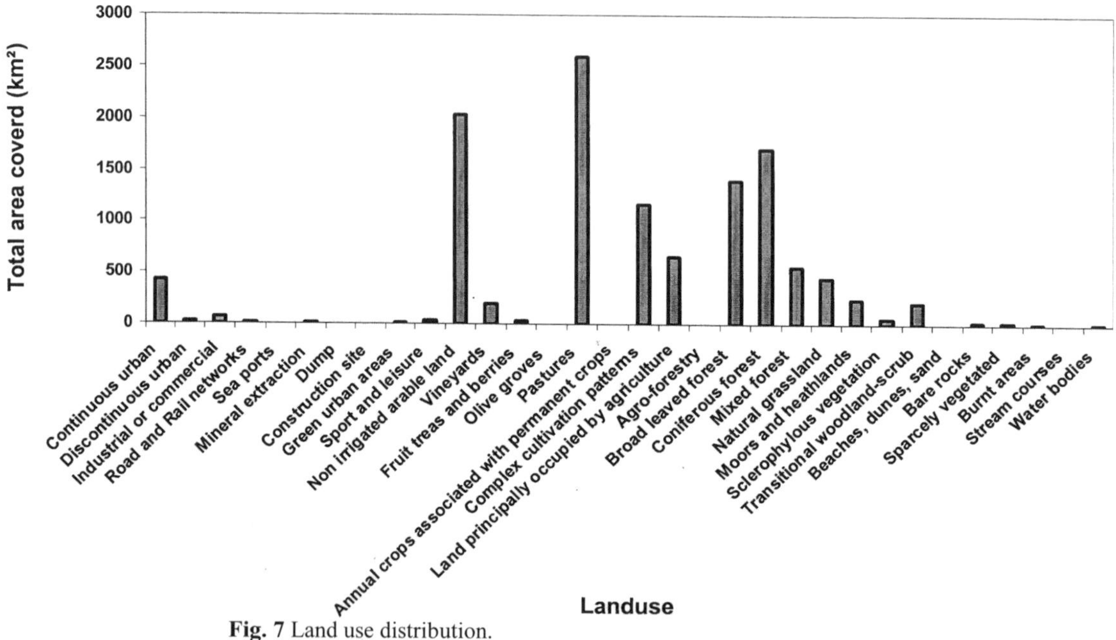

Fig. 7 Land use distribution.

Soil

The soil data consist of texture data derived from the 1/1 000 000 digital soil map compiled by the French National Institute for Agricultural Research (Dupuis, 1967; INRA, 2005). The maps are in vector form and correspond to five broadly defined texture classes. Although the data represented at this scale is rather crude, it is the only scale for which homogeneous and continuous soil information is available for the whole of continental France.

Each soil unit is organized into Soil Cartographic Units (SCU); this is the smallest geographic entity represented for a given scale. A SCU often groups several Soil Typological Units (STU). A STU is the smallest semantic entity according to a predefined nomenclature. At the 1/1 000 000 scale it is impossible to locate and delimit them. However, the SCUs are entities that can be localized in space and are composed of well-identified STUs.

Table 4 Minimum, maximum and median of land-use percentages.

Land use	Minimum percentage covered (%)	Maximum percentage of catchment area covered (%)	Median (%)
Urban fabric	0.1	24.1	1.2
Industrial or commercial units	0.0	9.1	0.0
Non irrigated arable land	0.0	80.6	3.0
Vineyards and olive groves	0.0	45.7	0.0
Fruit trees and berry plantations	0.0	6.9	0.0
Pastures	0.0	69.6	12.0
Complex cultivation patterns	0.1	50.8	4.9
Land principally occupied by agriculture, with significant area of natural vegetation	0.0	19.3	5.1
Broad leaved forest	0.0	51.5	9.1
Coniferous forest	0.0	57.0	3.1
Mixed forest	0.0	34.7	1.6
Natural grassland	0.0	50.9	0.0
Moors and heathlands	0.0	18.6	0.1
Sclerophylous vegetation	0.0	5.3	0.0
Transitional woodland-scrub	0.0	13.2	0.3

Table 5 Soil texture classification (Dupuis, 1967).

Texture	Composition
Coarse	Clay < 18% and sand > 65%
Moderate	18% < Clay < 35% and Sand > 15% or Clay < 18% and 15% < Sand < 65%
Moderately fine	Clay < 35% and Sand < 15%
Fine	35% < Clay < 60%
Very fine	Clay > 60 %

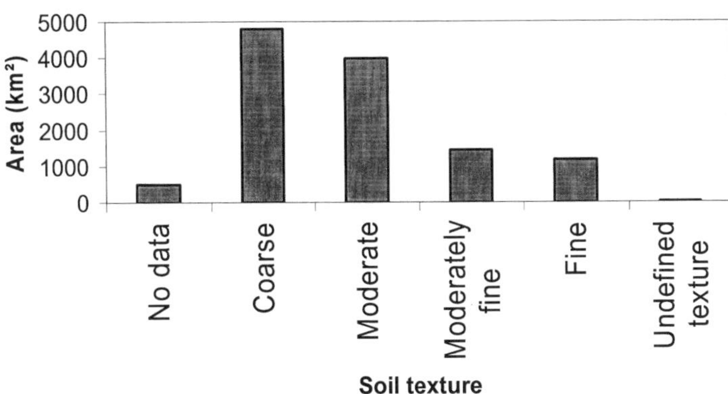

Fig. 8 Soil texture distribution.

Hence, the authors advise caution while using these maps, given the fact that "*the 1/100 000 scale limits the representation of the variability... The database should be used for projects effectively corresponding to the 1/1 000 000 scale i.e. for regional or national projects requiring a global reasoning on large landscape units*" (INRA,

2005). The texture defined on the maps corresponds to five classes, coarse, moderate, moderately fine, fine and very fine. Given the spatial resolution of the maps, the class limits are not determined precisely (Table 5). Determining a soil type or inferring soil hydrodynamic properties based on the data provided is not a straightforward process.

The two most represented texture classes are the coarse and moderate textures. The "moderately fine" class is the least represented while the "very fine" class is totally absent (Fig. 8).

CONCLUSION

A specific database containing both hourly and daily hydrometric and morphological data was compiled specially for the MOPEX 2004 workshop in Paris.

Although every effort has been made to ensure data quality, in some instances we were forced to use coarse resolution maps to produce a homogenous database. With the growing efforts into soil mapping, these problems should be easily overcome.

Acknowledgements The MOPEX workshop organization funds were provided by Cemagref's International Relations' department. The authors would like to thank Météo-France, the French Ministry of the Environment (Banque Hydro), Cemagref and ENGREF for hosting the 2004 workshop.

REFERENCES

Banque HYDRO (2004) Banque de données pour l'hydrométrie et l'hydrologie. http://hydro.rnde.tm.fr/.

Duan, Q., Schaake, J., Andréssian, V. *et al.* (2006) Model Parameter Estimation Experiment (MOPEX): An overview of science strategy and major results from the second and third workshops. *J. Hydrol.* **320,** 3–17.

Dupuis J. (1967) *Carte pédologique de la France* à 1/1 000 000. Editions INRA, France.

Durand, Y., Brun, E., Mérindol, L., Guyomarc'h, G., Lesaffre, B. & Martin E. (1993) A meteorological estimation of relevant parameters for snow schemes used with atmospheric models. *Ann. Glaciol.* **18,** 65–71.

European Environmental Agency (1995) *Corine Landcover*. EEA Report. Commission of the European Communities.

INRA (2005) http://gissol.orleans.inra.fr/programme/bdgsf/tarifs.php.

Schaake, J., Duan, Q., Smith, M. & Koren, V. (2000) Criteria to select basins for hydrologic model development and testing. Conference on Hydrology, AMS, Long Beach, California, USA.

Catalogue of the models used in MOPEX 2004/2005

V. ANDREASSIAN, S. BERGSTRÖM, N. CHAHINIAN, Q. DUAN, Y. M. GUSEV, I. LITTLEWOOD, T. MATHEVET, C. MICHEL, A. MONTANARI, G. MORETTI, R. MOUSSA, O. N. NASONOVA, K. O'CONNOR, E. PAQUET, C. PERRIN, A. ROUSSEAU, J. SCHAAKE, T. WAGENER & Z. XIE

First author: Cemagref, Parc de Tourvoie, BP 44, F-92163 Antony cedex, France
vazken.andreassian@cemagref.fr

Abstract A description of each of the 15 models used in MOPEX (the Model Parameter Experiment), as reported at the MOPEX workshops in 2004 and 2005, is provided. The following models are included: AFFDEF, GR4H, GR4J, HBV, HYDROTEL, IHACRES, MODSPA, MORDOR, NOAH, RRMT, SAC-SMA, SMAR, SWAP, SWB and VIC. Each is described systematically by the original author(s) with details of where the model was first published and of its subsequent use.

Key words AFFDEF; GR4H; GR4J; HBV; HYDROTEL; IHACRES; MODSPA; MORDOR; NOAH; RRMT; SAC-SMA; SMAR; SWAP; SWB; VIC

INTRODUCTION

It is common for most scientific publications to present the materials and methods used to reach a certain conclusion. Evidently, a special issue on a model intercomparison exercise should abide by the same rule. Given the large number of contributions and the relatively short length of the articles, the editors sought for a standardized presentation of all the models, fifteen in total. Our main objective was to help readers make their way through the "jungle" of hydrological models used throughout this publication. An electronic form was sent out to all the authors. The form had three sections: (1) general information about the model (name, acronym, creation date …); (2) model description both in terms of structure and parameterization; (3) references.

The fifteen models presented in this catalogue are: AFFDEF, GR4H, GR4J, HBV, HYDROTEL, IHACRES, MODSPA, MORDOR, NOAH, RRMT, SAC-SMA, SMAR, SWAP, SWB and VIC. Each model structure is described by its authors to ensure the accuracy of the information and avoid all misinterpretation possibilities. For the same reason, the forms are reproduced without modification or editing. Readers who require additional information about a given model are advised to consult the corresponding references or to contact the model developers directly.

AFFDEF

GENERAL INFORMATION

Model acronym: AFFDEF
Model full name: —
Authors first publication: Brath, A., Montanari, A. & Moretti, G. (2002) On the use of simulation techniques for the estimation of peak river flows. Proceedings of the International Conference on Flood Estimation, Berna, 6–8 March 2002, Rep II-17, 587–599, CHR/KHR, International Commission for the Hydrology of the Rhine basin, Lelystad, The Netherlands.
Original application domaine: Flow simulation and applications; assessment of the effect of the anthropogenic influence on the hydrological cycle; hydrological applications where long simulation runs of river flows are needed at different locations of the catchment; prediction in ungauged basin.
Type: Spatially-distributed, continuously (in time) simulating rainfall–runoff model
Contact: Greta Moretti
Ingenieurbüro Winkler und Partner GmbH, Schlossstrasse 59a, D-70176 Stuttgart, Germany
Email: moretti@iwp-online.de
Alberto Montanari
Faculty of Engineering, University of Bologna, Viale Risorgimento 2, I-40136 Bologna, Italy
Email: alberto.montanari@mail.ing.unibo.it
web site: www.costruzioni idrauliche.ing.unibo.it/people/alberto/affdef.html

MODEL DESCRIPTION

Brief model description The main characteristic of AFFDEF is that long simulation runs can be performed with short time steps in limited computational times. The model is robust and thus applicable to a wide spectrum of real world case studies.

AFFDEF is raster-based. It takes as input the Digital Elevation Model (DEM) of the basin in raster form, as a rectangular matrix covering the whole basin. The cells of the DEM can be of any size. It also needs input rainfall and temperature data collected at an arbitrary number of thermometers and raingauges. Many of the hydrological processes involved in the rainfall–runoff transformation have been schematized by using conceptual approaches. The simulation can be carried out for a single event as well as in continuous time. In the first case, the Curve Number (CN) method is used to separate between surface and sub surface flows. In the case of continuous simulations, a more complex schematization of the rainfall–runoff transformation has been implemented, which also accounts for the interception and evapotranspiration. The model computes the local contribution to the surface runoff by applying a modified CN method. In order to compute the soil storativity, one must provide the matrix of the CN numbers for any given DEM cell. The local contribution to the surface runoff and the groundwater flows are transferred to the basin outlet by using a Muskingum-Cunge model with variable parameters, which are determined on the basis of the "matched diffusivity" concept. Distinction between the hillslope and network channel is based on the concept of the constant critical support area. Some of the model parameters have a well defined physical meaning and can be estimated on the basis of *in situ* surveys; the remainder have to be optimized by calibration on the basis of some historical hydro-metereological records.

Since a number of conceptual schemes were used in modelling the rainfall–runoff transformation, AFFDEF cannot be considered physically-based in the strict sense. Although it can be used for any kind of basin, it should be noted that AFFDEF only uses a simplified solution to model the contribution of groundwater flows. Therefore it is best suited for basins where the runoff production is mainly due to infiltration excess.

The model code, written in Fortran programming language, provides a user friendly and ready to use tool that runs on a personal computer, on a classic DOS platform. A routine for performing automatic calibration that makes use of the SCE-UA algorithm is included in the code.

Main hydrological processes Interception of rainfall by the vegetation cover; computation of the local rainfall (Thiessen polygons or the inverse squared distance interpolation method); evapotranspiration of the intercepted precipitation and from the soil; distinction between surface and sub-surface flow and computation of the local runoff response; and infiltration and formation of groundwater flow.

Rainfall–runoff module The distinction between surface and sub-surface flow is based on a modified version of the CN method. Conceptual schemes are used to model the interaction between soil, vegetation and atmosphere.

Transfer function Surface and groundwater flow are propagated towards the basin outlet by applying the Muskingum-Cunge model with variable parameters.

Groundwater/percolation module The formation of the groundwater flow is based on the excess infiltration scheme.

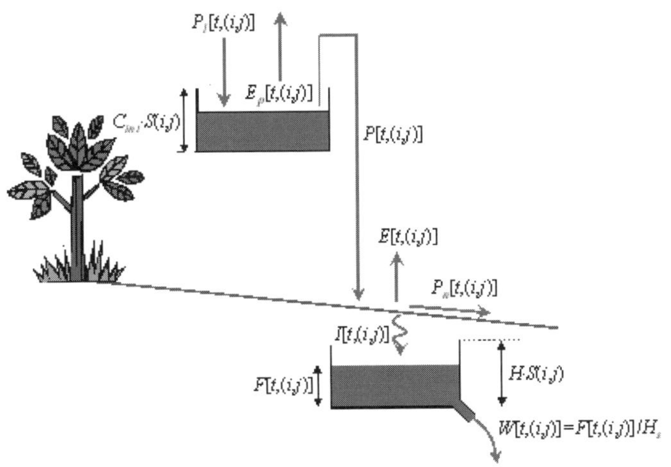

Fig. 1 Schematic representation of AFFDEF model structure.

Additional components None

Model applications In Brath *et al.* (2004), AFFDEF was applied to the Reno River Basin in Italy to assess its reliability when calibrated using data sets of increasing length. Overall, the results indicate that the best out-of-sample performances are obtained by calibrating the model with minimum periods of three months. Brath *et al.* (2002) used the model for estimating the flood frequency distribution of the Samoggia River basin, located in northern Italy. The model proved to be robust in the simulation of the observed flood frequency distribution, even if only short historical rainfall, temperature and river flow records were available for model calibration. An example of the success of the model in investigating the effects of land use change on flood flows may be found in Brath *et al.* (2003, 2006).

AFFDEF was applied for the prediction in ungauged basins to the Upper Neckar catchment, located in western Germany (Das *et al.*, 2006), and to the Riarbero Torrent (north of Italy, Moretti & Montanari, 2003).

Schematic representation of model structure (Fig. 1)

$P_l[t,(i,j)]$: rainfall depth at time t for the cell of coordinates (i,j)

$P[t,(i,j)]$: rainfall that reaches the ground at time t for the cell of coordinates (i,j)

$P_n[t,(i,j)]$: surface runoff at time t for the cell of coordinates (i,j)

$I[t,(i,j)]$: infiltrated water at time t for the cell of coordinates (i,j)

$F[t,(i,j)]$: water content of the infiltration reservoir at time t located in the cell of coordinates (i,j)

$W[t,(i,j)]$: outflow from the infiltrated reservoir at time t located in the cell of coordinates (i,j)

$E[t,(i,j)]$: effective evapotranspiration from the soil at time t for the cell of coordinates (i,j)

$E_P[t,(i,j)]$: evapotranspiration from the intercepted water at time t for the cell of coordinates (i,j)

$S[t,(i,j)]$: soil storativity according to the CN method for the cell of coordinates (i,j)

C_{int}: multiplying parameter for the interception reservoir capacity

H: multiplying parameter for the infiltration reservoir capacity

H_S: bottom discharge parameter for the infiltration reservoir capacity

DATA REQUIREMENTS; PARAMETERIZATION

Input data The input meteorological data consists of both observed precipitation (rainfall depths) and air temperature, at the same time step. The input topography data are required as a grid based Digital Elevation Model (DEM), which is given in raster format.

To characterize the spatial pattern of the infiltration capacity the Curve Number (CN) parameters associated to each DEM cell must be provided in input. Finally, a matrix is used to represent the spatial variability of the roughness on the hillslope for the overland flow. According to land use, different classes of roughness may be determined and a value for the Strickler coefficient is assigned to each class.

Overall model parameters

Parameter	Dimension and symbol	Method of estimation
Channel width/height ratio for the hillslope	w_v (dimensionless)	Calibrated
Strickler coefficients for the N-classes of roughness on the hillslope	$k_{sv}(\mathrm{i})$, i=1,N ($\mathrm{m}^{1/3}\mathrm{s}^{-1}$)	Estimated
Channel width/height ratio for the channel network	w_r (dimensionless)	Estimated
Maximum and minimum Strickler roughness for the channel network	k^0_{sr} , k^1_{sr} ($\mathrm{m}^{1/3}\mathrm{s}^{-1}$)	Estimated
Value of the Curve Number for each cell	CN (dimensionless)	Estimated
Constant critical source area	A_0 (km^2)	Estimated
Saturated hydraulic conductivity	K_{sat} ($\mathrm{m\ s}^{-1}$)	Calibrated
Width of the rectangular cross section of the sub-surface water flow	B_P^{sub} (m)	Calibrated
Bottom discharge parameter for the infiltration reservoir capacity	H_S (s)	Calibrated
Multiplying parameter for the infiltration reservoir capacity	H (dimensionless)	Calibrated
Multiplying parameter for the interception reservoir capacity	C_{int} (dimensionless)	Calibrated

The values of some parameters, namely A_0, k^0_{sr}, k^1_{sr} and w_r, and the values of the Strickler coefficients for the different classes of roughness on the hillslope, can be estimated by physical reasoning or *in situ* measurements ("estimated" in the Table above). In particular, the value of A_0 is usually identified by comparing the river network determined by the model with a topography map of the catchment showing the natural flow paths. As an initial estimate one may set $A_0 = 0.5\ \mathrm{km}^2$. A first trial value for k^0_{sr}, k^1_{sr} and w_r along the river network can be derived from the analysis of the river network geometry. The parameter C_{int} is not well correlated with the other parameters, but can be calibrated separately by comparing observed values of the runoff coefficient with estimates derived by simulating sufficiently long records of synthetic river flows. The remaining parameters ("calibrated" in the Table above) can be calibrated manually with a trial and error procedure by comparing observed and simulated hydrographs.

Sensitivity analysis results In the sensitivity analysis carried out in Brath *et al.* (2004) it was found that certain input parameters had a more direct effect on the model outputs than others did. For example, the parameters of the surface Muskingum model (w_v, k_{sv}, w_r, k^0_{sr}, k^1_{sr}) mainly affect peak flow timing and somewhat affect hydrograph shape and peak flow magnitude. The parameters that refer to hillslopes (w_v, k_{sv}) have a more significant influence on the simulated river flows than the river network parameters, especially if A_0 is not small. The infiltration reservoir parameters H and H_s have a large effect on peak flow magnitude. Finally, the parameter H_s and those of the sub-surface Muskingum model (K_{sat}, B_P^{sub}) have an effect on the recessing limb of the hydrograph.

Calibrated parameters (See Table above)

Calibration procedure-algorithm Manual calibration based on the trial and error method; automatic calibration by means of the Shuffled Complex Evolution global optimization algorithm (implemented in AFFDEF code). The user must select the parameters to be automatically calibrated, and define their lower and upper bounds. The automatic calibration procedure, using a least square objective function, results in a better agreement between observed and simulated flows.

REFERENCES

Brath, A. & Montanari, A. (2000) Effects of the spatial variability of soil infiltration capacity in distributed rainfall runoff modelling. *Hydrol. Processes* **14**(5), 2779–2794.

Brath, A., Montanari, A. & Toth, E. (2001) Comparing the calibration requirements and the simulation performances of lumped and distributed hydrological models: an Italian case study. *Eos. Trans. AGU 82, Spring Meet. Suppl.*, Abstract: H31D-04.

Brath, A., Montanari, A. & Moretti, G. (2002) On the use of simulation techniques for the estimation of peak river flows. In: *Proceedings of the International Conference on Flood Estimation* (Berna, March 2002), Rep II-17, 587–599, CHR/KHR, International Commission for the Hydrology of the Rhine basin, Lelystad, The Netherlands.

Brath, A., Montanari, A. & Moretti, G. (2003) Assessing the effects on flood risk of the land-use changes in the last five decades: An Italian case study. In: *Hydrology in the Mediterranean and Semiarid Regions* (ed. by E. Servat, W. Najem, C. Leduc & A. Shakeel), 435–441. IAHS Publ. 278. IAHS Press, Wallingford, UK.

Brath, A., Montanari, A. & Toth, E. (2004) Analysis of the effects of different scenarios of historical data availability on the calibration of a spatially-distributed hydrological model. *J. Hydrol.* **291**, 272–288.

Brath, A., Montanari, A. & Moretti, G. (2006) Assessing the effect on flood frequency of land use change via hydrological simulation (with uncertainty). *J. Hydrol.* (in press).

Das, T., Moretti, G., Bárdossy, A. & Montanari, A. (2006) Assessing the predictive ability of the spatially distributed conceptual AFFDEF model for a meso scale catchment. In: *Prediction in Ungauged Basins: Promises and Progress* (ed. by M. Sivapalan, T. Wagener, S. Uhlenbrook, E. Zehe, V. Lakshmi, X. Liang, Y. Tachikawa & P. Kumar), 351–359. IAHS Publ. 303. IAHS Press, Wallingford, UK.

Montanari, A. & Brath, A. (2004) A stochastic approach for assessing the uncertainty of rainfall–runoff simulations. *Water Resour. Res.* **40**(1) W01106 10.1029/2003WR002540.

Moretti, G. & Montanari, A. (2003) Estimation of the peak river flow for an ungauged mountain creek using a distributed rainfall–runoff model. In: *Proc. ESF LESC Exploratory Workshop* (24–25 October 2003, Bologna, Italy).

Moretti, G. & Montanari, A. (2006) AFFDEF: a spatially distributed grid based rainfall–runoff model for continuous time simulations of river discharge. *Environ. Modelling Software* (in press).

GR4H

GENERAL INFORMATION

Model acronym: GR4H
Model full name: modèle du Génie Rural à 4 paramètres Horaire
Authors-first publication: Mathevet (2005) Which rainfall–runoff model at the hourly time-step? Empirical development and intercomparison of rainfall–runoff models on a large sample of watersheds, PhD thesis, ENGREF, Paris, France (in French).
Original application domain: Flow simulation and application such as flood estimation, flood forecasting for headwater basins.
Type: Lumped
Contact: Thibault Mathevet
eDF–DTG, Département Surveillance - Service CADE, 21, Avenue de l'Europe, BP41, F-38040 Grenoble cedex 9, France
Fax.: +33 (0) 476 202045
Email: thibault.mathevet@edf.fr

MODEL DESCRIPTION

Brief model description The GR4H model is an hourly lumped rainfall–runoff model, derived from the GR4J model (Perrin *et al.*, 2003). Its structure is similar to that of many conceptual type models (i.e. based on interconnected storages). However it was developed following an empirical approach, i.e. without *a priori* ideas on the rainfall–runoff transformation at the watershed scale. Moreover, the structure was developed by trying to find the more efficient formulation on a large sample of watersheds with hourly data, covering a wide range of hydro-climatic conditions. The model has two storages, four parameters to calibrate: two for the production function and two for the routing function. Full mathematical details are provided by Mathevet (2005). The GR4H structure is close to the GR4J structure, but is typically more efficient at the hourly time-step and thus dedicated to reactive watersheds. It has also been shown that GR4H was at least as efficient as GR4J at the daily time-step.

Main hydrological processes The model has no *a priori* physical underpinning. It includes a production function and a routing function.

Rainfall–runoff module The production function is based on: an interception phase using an interception store with zero capacity (potential evapotranspiration directly acts on input rainfall); a soil moisture accounting store (SMA) to determine (i) the part of raw rainfall that will become effective rainfall and (ii) the actual evapotranspiration; and a water-exchange function that can simulate import or export from/to the subterranean outside of the catchment. It acts on the two flow components simulated by the routing module.

The difference between GR4J and GR4H in the production function is the percolation rate.

Transfer function The routing function is based on: a percolation from the SMA store; a constant volumetric split of effective rainfall into direct flow component (10%) and an indirect flow component (90%); one unit hydrograph (UH) for the direct and indirect flow components; a nonlinear routing store that transfers the indirect flow component; the difference between GR4J and GR4H in the routing function is the use of only one UH, with a smooth shape.

Groundwater/percolation module Interactions with groundwater are accounted for via the two model stores and the water exchange function.

Model applications The model was applied on more than 300 watersheds worldwide (Mathevet, 2005), covering a wide range of hydro-climatic conditions (semiarid, Mediterranean, oceanic, temperate, mountainous and continental) and of watershed area (from 0.5 km^2 to 5000 km^2). Watersheds were located in France, the USA, Australia, Slovenia and Spain.

Schematic representation of model structure

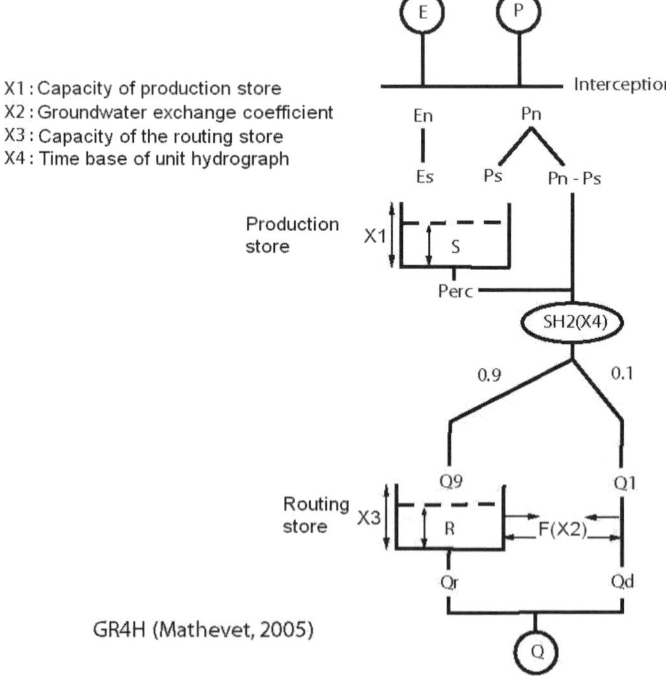

DATA REQUIREMENTS; PARAMETERIZATION

Input data The only inputs are: hourly potential evapotranspiration (PE) time series, derived from the disaggregation of the long-term average regime curve (i.e. the same PE curve is used every year); hourly rainfall time series. Rainfall is an estimate of area rainfall, e.g. calculated each day as an average on all available raingauges.

The model simulates flow time series. Observed hourly flow time series are required for model calibration and evaluation.

Overall model parameters The model has four free parameters:
$x1$: maximum capacity of the production store,
$x2$: groundwater exchange coefficient,
$x3$: one-day-ahead maximum capacity of the routing store,
$x4$: time base of unit hydrograph UH2.

Sensitivity analysis results An empirical sensitivity analysis of the model structure was performed by Mathevet (2005).

Calibrated parameters All four parameters are calibrated.

Calibration procedure-algorithm Given the low number of parameters, model errors, climatic variability, input and output variables uncertainties, the model can be calibrated by a simple direct search algorithm. The "step-by-step" optimization algorithm developed at Cemagref was found to be effective and efficient enough to successfully calibrate the model. It is a local search procedure (Edijatno *et al.*, 1999), that starts from a default parameter set, that is usually the media parameter of the parameter distribution obtained after the optimization of the model over a large sample of watersheds. Then, the optimization search step-by-step, or by trial and error, in the parameter space establishes the direction that improves the objective function the most. During the search, the search step is progressively reduced to refine the location of the optimum. The use of a progressively detailed search step allows the algorithm to locate the region of the optimum, with a low probability of getting trapped in a local optimum. Then, once the region of the optimum is located and no objective function improvement is achieved, the search step is reduced. The search stops when the search step is below a given threshold, i.e. when one considers that the optimum was located precisely enough. This method was tested in several studies and used to optimize parameter sets of thousands of watersheds. A comparative study with global search algorithms (SCE-UA and a Genetic Algorithm) showed that this method was able to optimise four to ten free parameters models, as successfully as global search algorithms.

REFERENCES

Edijatno, N., Nascimento, O., Yang, X., Makhlouf, Z. & Michel, C. (1999) GR3J: a daily watershed model with three free parameters. *Hydrol. Sci. J.* **44**(2), 263–277.

Mathevet, T. (2005) Which rainfall–runoff model at the hourly time-step? Empirical development and intercomparison of rainfall–runoff models on a large sample of watersheds, PhD Thesis, ENGREF, Paris, France (in French).

Perrin, C., Michel, C. & Andréassian, V. (2003) Improvement of a parsimonious model for streamflow simulation. *J. Hydrol.* **279**(1–4), 275–289.

GR4J

GENERAL INFORMATION

Model acronym: GR4J
Model full name: modèle du Génie Rural à 4 paramètres Journalier
Authors first publication: Edijatno *et al.* (1999) Mise au point d'un modèle élémentaire pluie-débit au pas de temps journalier. PhD Thesis, University Louis Pasteur/ENGEES, Strasbourg, France.
Original application domain: Flow simulation and applications such as flood estimation, flood forecasting (Yang, 1993; Yang & Michel, 2000; Tangara, 2005), drought forecasting, design of water regulation structures, detection of anthropogenic influence over the hydrological cycle.
Type: Lumped
Contact: Charles Perrin, Claude Michel and Vazken Andréassian
Cemagref, Parc de Tourvoie, BP 44, F-92163 Antony cedex, France
Fax: +33 (0)1 40 96 61 99
Email: charles.perrin@cemagref.fr
Web site: http://www.cemagref.fr/webgr/

MODEL DESCRIPTION

Brief model description The GR4J model is a daily lumped continuous rainfall–runoff model. Its structure is similar to that of many conceptual type models (i.e. built using storages). However, it was developed following an empirical approach, i.e. without *a priori* ideas on the rainfall–runoff transformation, but trying to find the model structure that performs best on a large set of hydro-climatic conditions. The model has four parameters to calibrate. Full mathematical details are provided by Perrin *et al.* (2003).

Main hydrological processes The model has no *a priori* physical underpinning. It includes a production module and a routing module.

Rainfall–runoff module The model production module is based on:

(a) an interception phase using an interception store with zero capacity (potential evapotranspiration directly acts on input rainfall);
(b) a soil moisture accounting (SMA) store to determine: (i) the part of raw rainfall that will become effective rainfall; and (ii) the actual evapotranspiration;
(c) a water-exchange function that can simulate import or export of water from/to the subterranean outside of the catchment. It acts on the two flow components simulated by the transfer module.

Transfer function The transfer production module is based on:

(a) a percolation from the SMA store;
(b) a constant volumetric split of effective rainfall into a direct flow component (10%) and an indirect flow component (90%);

(c) two unit hydrographs (UH), each one acting on one flow component;
(d) a nonlinear routing store that routes the indirect flow component.

Groundwater/percolation module Interactions with groundwater are accounted for via the two model stores and the water-exchange term.

Additional components An optional snowmelt module was proposed by Makhlouf (1994).

Model applications The model was applied on more than 1000 catchments worldwide. The model was intensively tested and compared to other models in: France (Edijatno *et al.*, 1999; Perrin, 2000; Oudin, 2004; Mathevet, 2005), the UK (Perrin & Littlewood, 2000), Slovenia (Mathevet, 2005), the USA (Perrin, 2000; Oudin, 2004; Mathevet, 2005), Australia (Perrin, 2000; Oudin, 2004; Mathevet, 2005), Mexico (Rojas-Serna, 2005), Brazil (Perrin, 2000; Oudin, 2004; Mathevet, 2005), the Ivory Coast (Servat & Dezetter, 1991, 1992; Perrin, 2000).

Schematic representation of model structure

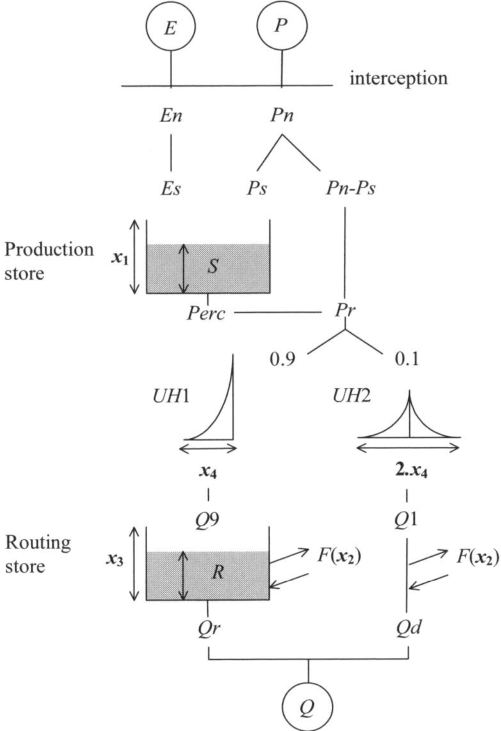

DATA REQUIREMENTS; PARAMETERIZATION

Input data The only inputs are daily potential evapotranspiration (PE) time series PE can be a long-term average regime curve, i.e. the same PE curve is used every year; and daily rainfall time series. Rainfall is an estimate of areal rainfall, e.g. calculated each day as an average on all the available raingauges.

The model simulates daily flow time series. Observed daily flow time series are required for model calibration and evaluation.

Overall model parameters The model has four free parameters:

$x1$ maximum capacity of the production store (mm),

$x2$ groundwater exchange coefficient (mm),

$x3$ one-day-ahead maximum capacity of the routing store (mm),

$x4$ time base of unit hydrograph UH1 (days).

Sensitivity analysis results Several structural sensitivity analyses were performed during PhD research projects (Edijatno, 1991; Nascimento, 1995; Perrin, 2000; Mathevet, 2005). Sensitivity analyses to model inputs were performed by Oudin (2004), Oudin *et al.* (2004, 2005) and Andréassian *et al.* (2001, 2004a,b). Uncertainty analyses were also carried out by Yang & Parent (1996) and Kuczera & Parent (1998).

Calibrated parameters All four model parameters are calibrated.

Calibration procedure-algorithm Given the low number of parameters, the model can be calibrated using any calibration approach. The step-by-step optimization algorithm developed at Cemagref was found to be effective and efficient to calibrate the model.

It is a local gradient search procedure (Edijatno *et al.*, 1999). The optimization starts from a default parameter set that is the average of parameter sets obtained from a large number of gauged watersheds. Then the optimization procedure searches step by step in the parameter space for the direction that improves the objective function the most. During the search, the search step progressively reduces to refine the location of the optimum. The search stops when the search step is below a given threshold, i.e. when one considers that the optimum was located precisely enough. This method was tested in several studies and showed good performances for models having up to eight parameters to calibrate (Nascimento, 1995; Perrin, 2000; Mathevet, 2005).

REFERENCES

Anctil, F., Michel, C., Perrin, C. & Andréassian, V. (2004). A soil moisture index as an auxiliary ANN input for stream flow forecasting. *J. Hydrol.* **286**(1–4), 155–167.

Andréassian, V. (2002) Impact de l'évolution du couvert forestier sur le comportement hydrologique des bassins versants. PhD Thesis, University Pierre et Marie Curie Paris VI, Cemagref (Antony), France.

Andréassian, V., Perrin, C., Michel, C., Usart-Sanchez, I. & Lavabre, J. (2001) Impact of imperfect rainfall knowledge on the efficiency and the parameters of watershed models. *J. Hydrol.* **250**, 206–223.

Andréassian, V., Oddos, A., Michel, C., Anctil, F., Perrin, C. & Loumagne, C. (2004a).Impact of spatial aggregation of inputs and parameters on the efficiency of rainfall–runoff models: A theoretical study using chimera watersheds. *Water Resour. Res.* **40**(5), W05209, doi:10.1029/2003WR002854.

Andréassian, V., Perrin, C. & Michel, C. (2004b) Impact of imperfect potential evapotranspiration knowledge on the efficiency and parameters of watershed models. *J. Hydrol.* **286**(1–4), 19–35.

Edijatno (1991) Mise au point d'un modèle élémentaire pluie-débit au pas de temps journalier. PhD Thesis, University Louis Pasteur/ENGEES, Strasbourg, France.

Edijatno & Michel, C. (1989) Un modèle pluie-débit journalier à trois paramètres. *La Houille Blanche* **2**, 113–121.

Edijatno, Nascimento, N. O., Yang, X., Makhlouf, Z. & Michel, C. (1999) GR3J: a daily watershed model with three free parameters. *Hydrol. Sci. J.* **44**(2), 263–277.

Kuczera, G. & Parent, E. (1998) Monte Carlo assessment of parameter uncertainty in conceptual catchment models: the Metropolis algorithm. *J. Hydrol.* **211**, 69–85.

Lavabre, J., Sempere Torres, D. & Cernesson, F. (1993). Changes in the hydrological response of a small Mediterranean basin a year after a wildfire. *J. Hydrol.* **142**, 273–299.

Makhlouf, Z. (1994) Compléments sur le modèle pluie-débit GR4J et essai d'estimation de ses paramètres. PhD Thesis, University Paris XI Orsay, France.

Mathevet, T. (2005) Quels modèles pluie-débit globaux pour le pas de temps horaire? Développement empirique et comparaison de modèles sur un large échantillon de bassins versants. PhD Thesis, ENGREF, Cemagref, Paris, France.

Michel, C. (1983) Que peut-on faire en hydrologie avec un modèle conceptuel à un seul paramètre? *La Houille Blanche* **1**, 39–44.

Nascimento, N. O. (1995) Appréciation à l'aide d'un modèle empirique des effets d'action anthropiques sur la relation pluie-débit à l'échelle du bassin versant. PhD Thesis, CERGRENE/ENPC, Paris, France.

Oudin, L. (2004) Recherche d'un modèle d'évapotranspiration potentielle pertinent comme entrée d'un modèle pluie-débit global. ENGREF (Paris) / Cemagref (Antony), France.

Oudin, L., Andréassian, V., Perrin, C. & Anctil, F. (2004). Locating the sources of low-pass behaviour within rainfall–runoff models. *Water Resour. Res.* **40**(11), doi:10.1029/2004WR003291.

Oudin, L., Hervieu, F., Michel, C., Perrin, C., Andréassian, V., Anctil, F. & Loumagne, C. (2005) Which potential evapotranspiration input for a rainfall–runoff model? Part 2 - Towards a simple and efficient PE model for rainfall–runoff modelling. *J. Hydrol.* **303**(1–4), 290–306.

Ouédraogo, M., Servat, E., Paturel, J. E., Lubès-Niel, H. & Masson, J. M. (1998) Caractérisation d'une modification éventuelle de la relation pluie-débit autour des années 1970 en Afrique de l'ouest et centrale non-sahélienne. In: *Water Resources Variability in Africa during the XXth Century* (ed. by E. Servat, D. Hughes, J. M. Fritsch & M. Hulme), 315–321. (Proc. of the Abidjan Conf., Ivory Coast). IAHS Publ. 252. IAHS Press, Wallingford, UK.

Perrin, C. & Littlewood, I. G. (2000) A comparative assessment of two rainfall–runoff modelling approaches: GR4J and IHACRES. In: *Proc. Liblice Conference* (22–24 September 1998), 191–201. IHP-V, Technical Documents in Hydrology no. 37, UNESCO, Paris, France.

Perrin, C., Michel, C. & Andréassian, V. (2003) Improvement of a parsimonious model for streamflow simulation. *J. Hydrol.* **279**(1–4), 275–289.

Rojas-Serna, C. (2005) Quelle connaissance hydrométrique minimale pour définir les paramètres d'un modèle pluie-débit? PhD Thesis, ENGREF, Paris, Cemagref, France.

Servat, E. & Dezetter, A. (1991) Selection of calibration objective functions in the context of rainfall–runoff modelling in a Sudanese savannah area. *Hydrol. Sci. J.* **36**(4), 307–331.

Servat, E. & Dezetter, A. (1992) Modélisation de la relation pluie-débit et estimation des apports en eau dans le nord-ouest de la Côte d'Ivoire. *Hydrologie Continentale* **7**(2), 129–142.

Servat, E. & Dezetter, A. (1993). Rainfall–runoff modelling and water resources assessment in northwestern Ivory Coast. Tentative extension to ungauged catchments. *J. Hydrol.* **148**, 231–248.

Tangara, M. (2005) Nouvelle méthode de prévision de crue utilisant un modèle pluie-débit global. PhD Thesis, EPHE, Paris, France.

Yang, X. (1993) Mise au point d'une méthode d'utilisation d'un modèle pluie-débit conceptuel pour la prévision des crues en temps réel. Thèse de Doctorat, ENPC/CERGRENE.

Yang, X. & Michel, C. (2000) Flood forecasting with a watershed model: a new method of parameter updating. *Hydrol. Sci. J.* **45**(4), 537–546.

Yang, X. & Parent, E. (1996) Analyse de fiabilité en modélisation hydrologique: concepts et applications au modèle pluies-débits GR3. *Revue des Sciences de l'Eau* **1**, 31–49.

Yang, X., Parent, E., Michel, C. & Roche, P.A. (1991). Gestion d'un réservoir pour la régularisation des débits. *La Houille Blanche*, 6, 433–440.

Yang, X., Parent, E., Michel, C. & Roche, P. A. (1995). Comparison of real-time reservoir-operation techniques. *J. Water Resources Planning and Management* **121**(5), 345–351.

HBV

GENERAL INFORMATION

Model acronym: HBV
Model full name: HBV hydrological model
Authors first publication: Bergström, S. & Forsman, A. (1973). Development of a conceptual deterministic rainfall–runoff model. *Nordic Hydrology* **4**, 147–170.
Original application domaine: streamflow simulation and hydrological forecasting
Type: Semi-distributed conceptual model
Contact: Sten Bergström
The Swedish Meteorological and Hydrological Institute, SE-601 76 Norrköping, Sweden
Email: sten.bergstrom@smhi.se

MODEL DESCRIPTION

Brief model description The HBV model consists of three main components; subroutines for snow accumulation and melt, subroutines for soil moisture accounting and response, and river routing subroutines. It uses sub-basins as primary hydrological units and an area-elevation distribution and a crude classification of land use. The sub-basin option is used in geographically or climatologically heterogeneous basins or in the presence of large lakes.

Main hydrological processes Snow accumulation and snowmelt; soil moisture accounting; response function; river and lake routing.

Rainfall/runoff module A variable source-area concept based on a soil moisture accounting routine.

Transfer function A set of boxes representing the saturated zone with variable drainage depending on storage. River routing and explicit routing through lakes based on storage discharge relationships.

Groundwater/percolation module All excess water from the soil moisture accounting routine will enter the saturated (groundwater) zones. Water will leave the upper saturated zone either as runoff via rivers or as deep percolation to a lower saturated zone. Water from the lower saturated zone will eventually drain into rivers as well.

Additional components The HBV model has been developed to meet the needs of the environmental sector. Initially acidification was the main focus, but later nonpoint source pollution and transport of nutrients from land to sea became a major field of application.

Model applications The HBV model is a standard tool for runoff simulations and flood forecasting in Sweden, Norway and Finland. The model has been applied in more than 50 countries all over the world. Some of these applications are made by modified versions of the model developed in, for example, Norway, Finland, Germany and Switzerland (Bergström, 1995). More recently, the HBV model has been used extensively for analyses related to the impacts of global warming on water resources and hydropower production (Andréasson *et al.*, 2004).

Schematic representation of model structure

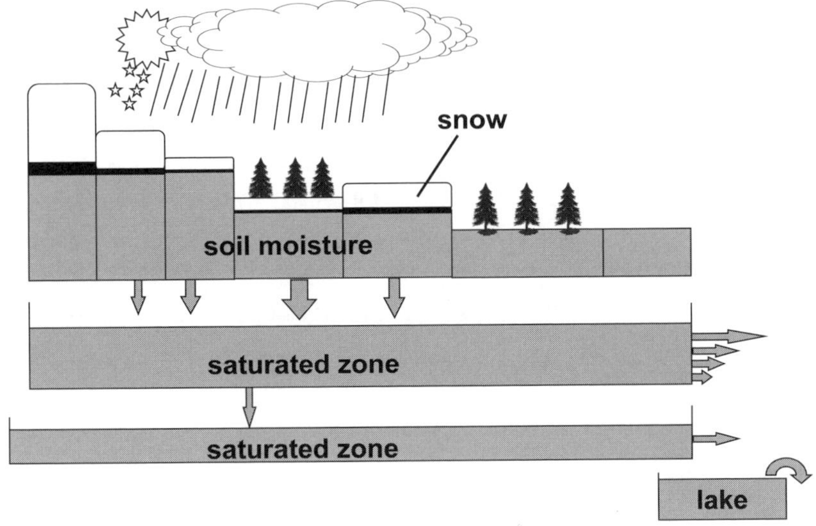

DATA REQUIREMENTS; PARAMETERIZATION

Input data Input variables to the HBV model are normally 24-hourly values of precipitation and air temperature and some estimate of potential evapotranspiration, which can either be daily or of lower resolution in time. A version of the model with hourly resolution in time is also available.

Overall model parameters Depending on the choice of the modeller and version of the model the number of parameters to calibrate are normally 2–4 in the snow routine, 3 in the soil moisture routine and 4–5 in the response function. In addition there are two general input correction factors, which should be used with some caution.

These are the most basic parameters used in calibration of the HBV-96 model version, which is standard for most Swedish applications (Lindström *et al.*, 1997):

General input corrections SFCF, snowfall correction factor; RFCF, rainfall correction factor.

Snow accumulation and snowmelt CFMAX, degree-day melt factor; TT, threshold temperature.

Soil moisture accounting FC, maximum soil moisture storage; LP, limit for potential evapotranspiration; BETA, exponent in the runoff generation equation.

Response function PERC, recharge of lower saturated zone; K, recession parameter; K4, recession parameter; ALFA, empirical coefficient.

Calibrated parameters Normally all the parameters above are the subject of calibration. However, the *general input corrections* are used restrictively.

Calibration procedure-algorithm Automatic calibration according to Lindström (1997).

REFERENCES

Andréasson, J., Bergström, S., Carlsson, B., Graham, L. P. & Lindström, G. (2004) Hydrological change—climate change impact simulations for Sweden. *Ambio*, **33**(4–5), 228–234.

Arheimer, B. (2005) Evaluation of water quantity and quality modelling in ungauged European basins. In: *Predictions in Ungauged Basins: Promises and Progress* (ed. by M. Sivapalan, T. Wagener, S. Uhlenbrook, E. Zehe, V. Lakshmi, X. Liang, Y. Tachikawa & P. Kumar), 99–107. IAHS Publ. 303. IAHS Press, Wallingford, UK.

Arheimer, B. & Brandt, M. (1998) Modelling nitrogen transport and retention in the catchments of southern Sweden. *Ambio* **27**(6), 471–480.

Bergström, S. (1975) The development of a snow routine for the HBV-2 model. *Nordic Hydrol.* **6**, 73–92.

Bergström, S. (1995) The HBV model. In: *Computer Models of Watershed Hydrology* (ed. by V. P. Singh), 443–476. Water Resources Publications. Colorado, USA.

Bergström, S. & Forsman, A. (1973) Development of a conceptual deterministic rainfall–runoff model. *Nordic Hydrol.* **4**, 147–170.

Bergström, S., Carlsson, B., Gardelin, M., Lindström, G., Pettersson, A. & Rummukainen, M. (2001) Climate change impacts on runoff in Sweden—assessments by global climate models, dynamical downscaling and hydrological modelling. *Climate Res.* **16**, 101–112.

Bergström, S. & Forsman, A. (1973) Development of a conceptual deterministic rainfall–runoff model. *Nordic Hydrol.* **4**, 147–170.

Brandt, M., Bergström, S. & Gardelin, M. (1988) Modelling the effects of clearcutting on runoff—examples from Central Sweden. *Ambio* **17**(5), 307–313.

Graham, L. P. (2004) Climate change effects on river flow to the Baltic Sea. *Ambio* **33**(4–5), 235–241.

Lindström, G. (1997) *A Simple Automatic Calibration Routine for the HBV Model. Nordic Hydrol.* **28**(3), 153–168.

Lindström, G., Johansson, B., Persson, M., Gardelin, M. & Bergström, S. (1997) Development and test of the distributed HBV-96 model. *J. Hydrol.* **201**, 272–288.

Lindström, G., Rosberg, J. & Arheimer, B. (2005). Parameter precision in the HBV-NP Model and impacts on Nitrogen scenario simulations in the Rönneå River, southern Sweden. *Ambio* **34**(7), 533–537.

HYDROTEL

GENERAL INFORMATION

Model acronym: HYDROTEL
Model full name: HYDROTEL
Authors first publication: Fortin *et al.* (1995) Fortin, J. P., Moussa, R., Bocquillon, C. & Villeneuve, J. P. (1995) HYDROTEL, un modèle hydrologique distribué pouvant bénéficier des données fournies par la télédétection et les systèmes d'information géographique. *Revue des sciences de l'eau* **8**(1), 97–124 (in French), Fortin *et al.* (2001a, 2001b) (in English).
Original application domain: Streamflow simulation
Type: Distributed
Contact: Alain Rousseau

Institut national de la recherche scientifique, Centre Eau, Terre & Environnement, 490 de la Couronne, Québec G1K 9A9, Canada
Tel: (418) 654-2621
Fax: (418) 654-2600
Email: alain_rousseau@ete.inrs.ca

MODEL DESCRIPTION

Brief model description HYDROTEL is a distributed hydrological model compatible with remote sensing and GIS data. For each subprocess of the water cycle, HYDROTEL offers the possibility of choosing among various sub-models depending on available data. Thus, when the necessary data are available, it is possible to choose more accurate sub-models based on physical processes. Otherwise, more conceptual sub-models compatible with the available data may have to be chosen. This allows application of HYDROTEL to a wide variety of problems.

With the exception of river routing, computations are performed independently on a number of relatively homogenous hydrological units (RHHU) chosen so as to take into account the spatial variability of topography, land use, soil types and meteorological variables within a basin. Runoff from each RHHU is used to estimate lateral flow conditions for a hydraulic model which takes care of river routing. Computations associated with river routing are performed on each modelled river reach, starting with the most upstream ones and going down in a cascade through the river network to the outlet of the basin. Consequently, HYDROTEL can provide simulated flows at the downstream end of each river reach and not only at the outlet of the basin.

Main hydrological processes Six hydrological processes are simulated by HYDROTEL: interpolation of meteorological data; accumulation and melt of snow cover; potential evapotranspiration; vertical water budget; surface and sub-surface runoff; river routing.

Rainfall/runoff module Overland flow can be triggered by infiltration excess and saturation excess. The vertical water budget is computed by solving the Richards equation on three tilted layers of soil to take into account the local slope.

Transfer function Runoff is routed to the stream network using a geomorphological unit hydrograph. Kinematic wave and diffusive wave methods are available for river routing with assumed or measured channel cross-section, slope and length.

Groundwater/percolation module Exponential decay controlled by a recession coefficient.

Additional components Snow accumulation and melt are simulated using an energy budget approach, with radiation estimated from temperature. Different algorithms are available to estimate evapotranspiration, all taking into account the leaf area index and rooting depth of the vegetation.

Model applications Used operationally for streamflow forecasting in Québec for snowmelt- and rainfall-driven events for basins having a temperate climate with snowy winters, relatively low elevations (up to 1000m), and sizes varying from a few hundred square kilometres to over twenty thousand square kilometres (Turcotte *et al.*, 2004).

Embedded within a decision-support system used for integrated watershed management, including which can be used to simulate land use and water use effects on water quantity and quality (Lavigne *et al.*, 2004).

Schematic representation of model structure

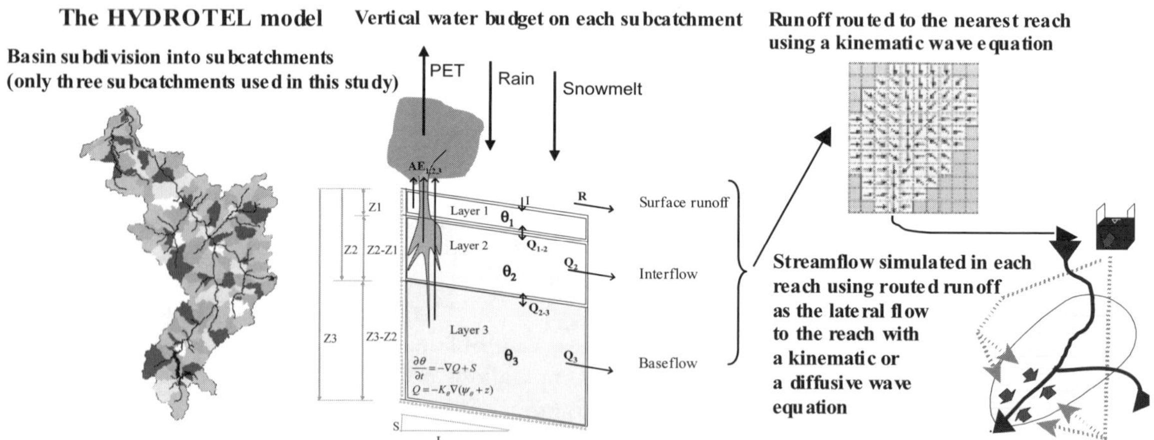

DATA REQUIREMENTS; AND PARAMETERIZATION

Input data

Physiographic information *Required*: digital elevation model, leaf area index map, root depth map, soil type map, soil properties for each soil type (hydraulic conductivity, matrix potential, and water content at saturation, field capacity and wilting point). *Suggested*: vectorized river network. *Optional*: length, slope and width of river reaches, vegetation height, albedo.

Hydrometeorological observations *Required*: temperature and precipitation observations (daily or hourly). *Optional*: streamflow, water level, snow water equivalent, snow depth, wind speed, relative humidity, hours of sunshine, snow water equivalent.

Overall model parameters HYDROTEL has a total of 24 parameters which are subject to calibration. They can either be constant over the watershed, constant over groups of computational units, or fully distributed.

Interpolation of meteorological data Three parameters: vertical gradient of precipitation; vertical gradient of temperature; temperature threshold for separating rain and snow.

Accumulation and melt of snow cover Nine parameters: snow-soil melt rate; maximum density of the snowpack; snowpack densification rate coefficient; snow melting rate at the snow–air interface in coniferous forests, deciduous forests and open areas (three parameters); temperature threshold for snowmelt in coniferous forests, deciduous forests and open areas (three parameters).

Potential evapotranspiration One parameter: multiplicative optimization coefficient for adjusting PET.

Vertical water budget Five parameters: depth of each soil layer (three parameters); coefficient affecting the efficiency of transpiration; recession coefficient for base flow.

Surface and sub-surface runoff Four parameters: Manning's N runoff coefficients for forested areas, unresolved lakes, and open areas (three parameters); reference precipitation excess for the estimation of the geomorphological unit hydrograph.

River routing Two parameters: Manning's N runoff coefficient for the channel bed; multiplicative optimization coefficient for the width of assumed cross-sections.

Sensitivity analysis results All parameters are sensitive, but on different time scales.

Sensitivity on an annual time-scale Seven parameters: vertical gradient of precipitation; multiplicative optimization coefficient for adjusting PET; depth of each soil layer (three parameters); coefficient affecting the efficiency of transpiration; recession coefficient for base flow.

Sensitivity on a seasonal time-scale 11 parameters: vertical gradient of temperature; temperature threshold for separating rain and snow; snow-soil melt rate; maximum density of the snowpack; snowpack densification rate coefficient; snow melting rate at the snow-air interface in coniferous forests, deciduous forests and open areas (three parameters); temperature threshold for snowmelt in coniferous forests, deciduous forests and open areas (three parameters).

Sensitivity on the scale of the concentration time of the basin Six parameters; Manning's N runoff coefficients for forested areas, unresolved lakes, and open areas (three parameters); reference precipitation excess for the estimation of the geomorphological unit hydrograph; Manning' N runoff coefficient for the channel bed; multiplicative optimization coefficient for the width of assumed cross-sections.

Calibrated parameters All parameters can potentially be calibrated, but 17 are typically calibrated.

Accumulation and melt of snow cover Seven parameters: snow-soil melt rate; snow melting rate at the snow-air interface in coniferous forests, deciduous forests and open areas (three parameters); temperature threshold for snowmelt in coniferous forests, deciduous forests and open areas (three parameters).

Calibration procedure-algorithm Snow accumulation and melt parameters can be calibrated using snow observations (Turcotte *et al.*, 2005). Calibration of the remaining parameters can be performed using a process-oriented, multiple-objective calibration strategy accounting for model structure (Turcotte *et al.*, 2003).

REFERENCES

Fortin, J. P., Moussa, R. Bocquillon, C. & Villeneuve, J. P. (1995) HYDROTEL, un modèle hydrologique distribué pouvant bénéficier des données fournies par la télédétection et les systèmes d'information géographique. *Revue des sciences de l'eau* **8**(1), 97–124.

Fortin, J. P., Turcotte, R., Massicotte, S., Moussa, R. & Fitzback, J. (2001a) A distributed watershed model compatible with remote sensing and GIS data, part 1: description of the model. *J. Hydrol. Engng ASCE* **6**(2), 91–99.

Fortin, J. P., Turcotte, R., Massicotte, S., Moussa, R. & Fitzback, J. (2001b) A distributed watershed model compatible with remote sensing and GIS data, Part 2: Application to the Chaudière watershed. *J. Hydrol. Engng ASCE* **6**(2), 100–108.

Lavigne, M. -P., Rousseau, A. N., Turcotte, R., Laroche, A. -M., Fortin, J. -P. & Villeneuve, J. -P. (2004) Validation and Use of a semidistributed hydrological modeling system to predict short-term effects of clear-cutting on a watershed hydrological regime. *Earth Interactions* **8**, 1–19.

Turcotte, R., Rousseau, A. N., Fortin, J. P. & Villeneuve, J. P. (2003) Development of a process-oriented, multiple-objective, hydrological calibration strategy accounting for model structure. In: *Calibration of Watershed Models* (ed. by Q. Duan, S. Sorooshian, H. Gupta, A. N Rousseau & R. Turcotte), 153–163. American Geophysical Union (AGU), Washington, USA.

Turcotte, R., Lacombe, P., Dimnik C. & Villeneuve J. -P. (2004). Prévision hydrologique distribuée pour la gestion des barrages publics du Québec. *Rev. can. génie civ./Can. J. Civ. Engng* **31**(2), 308–320.

Turcotte, R., Fortin, L.-G., Fortin, J.-P., Fortin, V. & Villeneuve, J. -P. (2005) Operational analysis of the spatial distribution and the temporal evolution of the snowpack water equivalent in southern Québec. *Nordic Hydrology* (submitted).

IHACRES

GENERAL INFORMATION

Model acronym: IHACRES (PC-IHACRES is downloadable from http://www.ceh.ac.uk/ and http://www.wmo.int/web/homs/projects/homsp1.html (WMO HOMS Component K22.2.11); IHACRES Classic Plus is downloadable from http://www.toolkit.net.au/ihacres.)

Model full name: Identification of unit hydrographs and component flows from rainfall, evaporation and streamflow data

Authors first publication: Jakeman, A. J., Littlewood, I. G. & Whitehead, P. G. (1990) Computation of the instantaneous unit hydrograph and identifiable component flows with application to two small upland catchments. *J. Hydrol.* **117**, 275–300.

Original application domain: Continuous flow simulation; unit hydrograph identification; hydrograph separation; catchment characterization; environmental change impact assessments.

Type: Spatially lumped

Contact: Ian Littlewood
Centre for Ecology and Hydrology, Wallingford, Oxfordshire OX10 8BB, UK
Email: igl@ceh.ac.uk
Barry Croke
Integrated Catchment Assessment and Management Centre, and Department of Mathematics, Australian National University, Canberra, Australia
Email: barry.croke@anu.edu.au.

MODEL DESCRIPTION

Brief model description A (nonlinear) loss module converts rainfall to effective rainfall, followed by a (linear) unit hydrograph module to convert effective rainfall to streamflow.

Main hydrological processes Conversion of catchment-scale rainfall to effective rainfall (i.e. the portion of rainfall that eventually leaves the catchment as streamflow) is based on a catchment wetness index; unit hydrograph routing of effective rainfall to streamflow at the catchment outlet.

Rainfall/runoff module Hybrid conceptual-metric.

Transfer function Unit hydrograph represented by a time-domain rational transfer function in the backwards shift operator, z^{-1}.

Groundwater/percolation module None (explicitly).

Additional components None (but additional snowmelt modules have been developed and applied for IHACRES modelling).

Model applications Many, please see References for a small selection.

Schematic representation of model structure (Fig. 1)

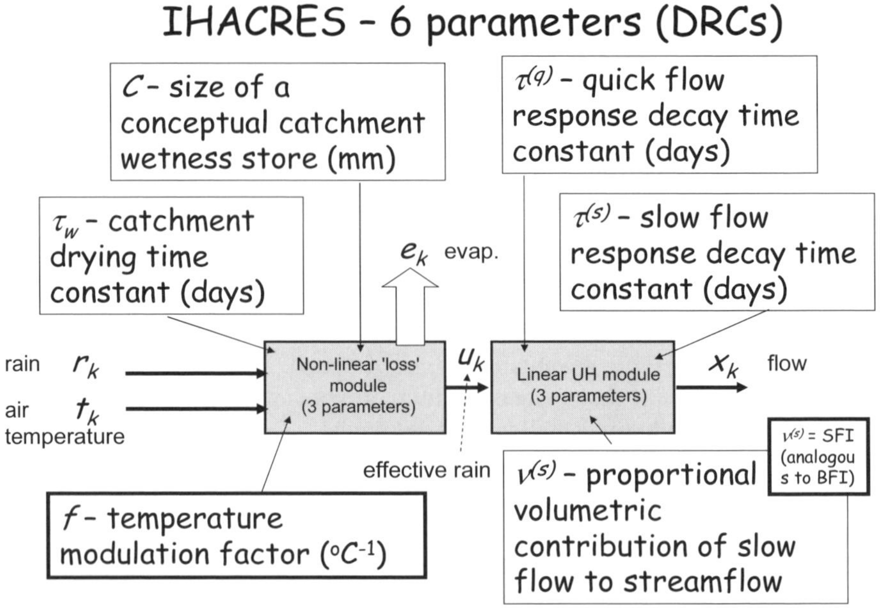

Fig. 1 Schematic representation of IHACRES model structure.

DATA REQUIREMENTS; PARAMETERIZATION

Input data Time series of rainfall, streamflow and air temperature. Apart from catchment area, no other data/information is required.

Overall model parameters See Fig. 1.

Sensitivity analysis results See literature.

Calibrated parameters See literature.

Calibration procedure-algorithm Usually six parameters. Loss module: Software-assisted, semi-automatic, grid-search for two of the three parameters in search of coincidentally high Nash-Sutcliffe efficiency for modelled streamflow and low "average relative parameter error" for the UH module. The third loss module parameter is calculated to give a water balance between effective rainfall and observed flow over the suitably chosen calibration period. The three UH module parameters are automatically identified using an advanced time series analysis technique. See Jakeman *et al.* (1990) for procedural details for initial calibration of the six parameters. The operator can use other model-fit statistics provided by IHACRES to help select a "best" model. In some cases, additional inspection of flow duration curves for observed and modelled flows can help the operator to adjust the loss module parameters in order to select a model "fit-for-purpose" (Littlewood *et al.*, 2003).

REFERENCES

Croke, B. F. W., Andrews, F., Jakeman, A. J., Cuddy, S. M. & Luddy, A. (2006) IHACRES Classic Plus: a redesign of the IHACRES rainfall–runoff model. *Environmental Modelling and Software* **21**, 426–427.

Jakeman, A. J., Littlewood, I. G. & Whitehead, P. G. (1990) Computation of the instantaneous unit hydrograph and identifiable component flows with application to two small upland catchments. *J. Hydrol.* **117**, 275–300.

Jakeman, A. J., Hornberger, G. M., Littlewood, I. G., Whitehead, P. G., Harvey, J. W. & Bencala, K. E. (1992) A systematic approach to modelling the dynamic linkage of climate, physical catchment descriptors and hydrologic response components. *Mathematics and Computers in Simulation* **33**, 359–366.

Jakeman A. J. & Hornberger, G. M. (1993) How much complexity is warranted in a rainfall–runoff model? *Water Resour. Res.* **29**(8), 2637–2649.

Littlewood, I. G. & Jakeman, A. J. (1992) Characterisation of quick and slow streamflow components by unit hydrographs for single- and multi-basin studies. In: *Fourth General Assembly of the European Network of Experimental and Representative Basins* (ed. by M. Robinson). (Proc., September 29–October 2, Oxford, UK) published as Institute of Hydrology Report 120.

Littlewood, I. G. & Jakeman, A. J. (1994) A new method of rainfall–runoff modelling and its applications in catchment hydrology. In: *Environmental Modelling*, vol. II (ed. by P. Zannetti), 143–171. Computational Mechanics Publications, Southampton, UK.

Littlewood, I. G., Down, K., Parker, J. R. & Post, D. A. (1997) *The PC version of IHACRES for catchment-scale rainfall–streamflow modelling: User Guide*. Institute of Hydrology Software report.

Littlewood, I. G. (2003) Improved unit hydrograph identification for seven Welsh rivers: implications for estimating continuous streamflow at ungauged sites. *Hydrol. Sci. J.* **48**(5), 743–762.

Littlewood, I. G., Clarke, R. T., Collischonn, W. & Croke, B. F. W. (2006) Hydrological characterisation of four Brazilian catchments using a simple rainfall–streamflow model. Summit on Environmental Modelling and Software: 3rd Biennial meeting of the International Environmental Modelling and Software Society, Burlington, USA, July 2006.

MODSPA

GENERAL INFORMATION

Model acronym: ModSpa
Model full name: Modèle Spatialisé
Authors first publication Moussa, R. (1993) Modélisation hydrologique spatialisée et système d'information géographique. *La Houille Blanche* **5**, 293–301.
Original application domain: Streamflow simulation, flood prediction, water budget simulation, water resources management.
Type: Distributed
Contact: Roger Moussa
Institut National de la Recherche Agronomique (INRA), Laboratoire d'étude des Interactions entre Sol, Agrosystème et Hydrosystème, UMR LISAH ENSA-INRA-IRD, 2 Place Pierre Viala, 34060 Montpellier Cedex 1, France
Tel : +33 (0)4 99 61 24 56
Fax : +33 (0)4 67 63 26 14
Email: moussa@ensam.inra.fr

MODEL DESCRIPTION

Brief model description In ModSpa, Digital Elevation Models are used in order to subdivide the catchment into right-banks, left-banks or source/head subcatchments and to extract the channel network (Fig. 1(a)). Each subcatchment is linked to only one reach of the tree-like channel network. Over each subcatchment, the vertical water budget is computed using a two-layer model (Fig. 1(b)). The first layer, denoted "soil-reservoir", represents the upper soil layer where surface runoff, infiltration, interflow, percolation and evapotranspiration occur. The second layer, named the "aquifer-reservoir", represents the aquifer where the base flow occurs. Three state variables are calculated as a function of the time: the regulating function f which separates rainfall into surface runoff and infiltration, the level S in the soil-reservoir and the level S_b in the aquifer-reservoir. Then, a transfer function, based on the diffusive wave equation, is used to route flows on each subcatchment (surface runoff, interflow and base flow) and then through the channel network.

Main hydrological processes Surface runoff, infiltration, interflow, percolation, evapotranspiration, base flow, transfer function on subcatchments, transfer function in the channel network.

Rainfall–runoff module Infiltration/runoff is modelled using a two-layer reservoir model.

Transfer function The kinematic wave or a unit hydrograph on subcatchments, the diffusive wave on the channel network.

Groundwater/percolation module A simple reservoir characterized by a recession curve.

Additional components No additional components.

Model applications Simulation of streamflow on each reach of the channel network, calculation of the terms of the water budget, applications on catchments in humid, temperate and arid regions.

Application on the Gardon d'Anduze basin located in the Cévennes Mediterranean mountains southern France (daily and hourly time steps).

Schematic representation of model structure (Fig. 1)

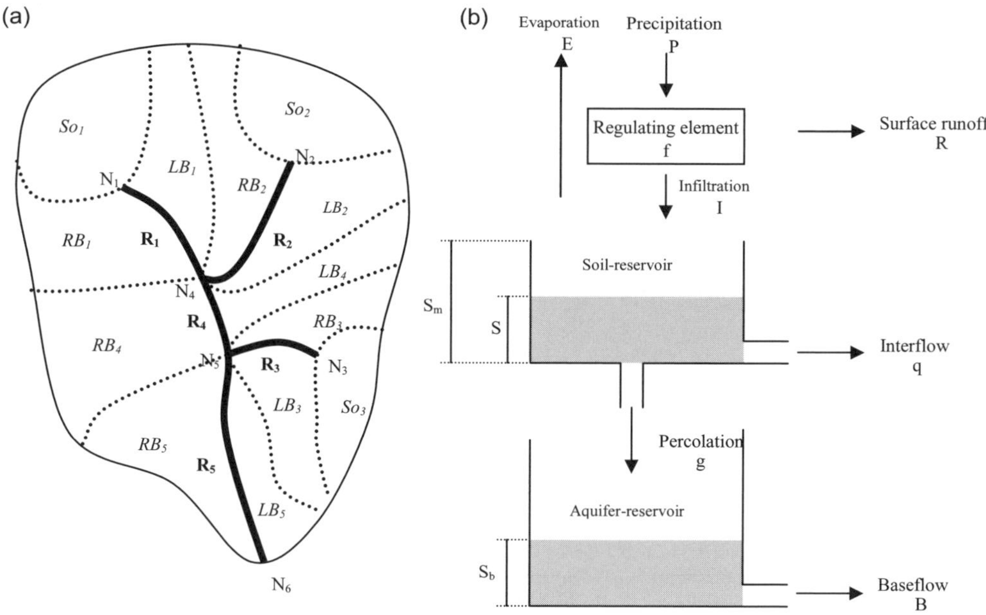

Fig. 1 ModSpa structure. (a) Representation of the basin topology in ModSpa: Channel network encoding: nodes (N_i), reaches (R_i in bold) and subcatchments (right-bank RB_i, left-bank LB_i and source basin So_i in italic). (b) Schematic structure of the vertical water budget in ModSpa.

DATA REQUIREMENTS; PARAMETERIZATION

Input data

(1) Topological structure of the basin calculated from the DEM: tree-like channel network, upstream and downstream nodes of each reach, links between subcatchments and reaches.

(2) On each subcatchment: (a) hydrometeorlogical data: rainfall and potential evapotranspiration; (b) geographical data from DEM: area, slope, distance to the reach; (c) soil hydrodynamic properties: hydraulic conductivity at natural saturation, soil-reservoir depth, the maximum value of the hydraulic conductivity; (d) leaf area index.

(3) On each reach: length, slope.

Overall model parameters

(1) On each subcatchment: the total storage of the soil-reservoir (S_m); the hydraulic conductivity at natural saturation of the soil (K_s); the maximum value of the hydraulic conductivity of the soil ($K_{max} = \alpha.K_s$ with α a parameter); the Leaf Area Index (LAI); a constant k representing the recession curve of the aquifer-reservoir; the celerity (C_u) on the subcatchment; the diffusivity (D_u) on each subcatchment.
(2) On each reach: the celerity (C_r); the diffusivity (D_r); sensitivity analysis results: the infiltration/runoff is sensitive to K_s, α, and S_m; the base flow is sensitive to k; the evapotranspiration is sensitive to LAI; the transfer function is sensitive to C_r and C_u.

Calibrated parameters Seven parameters are calibrated: α and k are considered similar for all subcatchments; C_u considered similar on all subcatchments; C_r is considered similar on all reaches; CK_s, CLAI and CS_m are three positive coefficients used as multiplication coefficients respectively of the *a priori* estimated K_s, LAI and S_m on each subcatchment.

Calibration procedure-algorithm Trial and error method.

REFERENCES

Moussa, R. (1991) Variabilité spatio-temporelle et modélisation hydrologique. Application au bassin versant du Gardon d'Anduze. PhD Thesis, University of Montpellier, France.

Moussa, R. (1993) Modélisation hydrologique spatialisée et système d'information géographique. *La Houille Blanche* **5**, 293–301.

Moussa, R. 1997. Geomorphological transfer function calculated from digital elevation models for distributed hydrological modelling. *Hydrol. Processes* **11**(5), 429–449.

Moussa, R., Chahinian, N. & Bocquillon, C. (2006) Distributed hydrological modelling of a Mediterranean mountainous catchment – model construction and multi-site validation. *J. Hydrol.* (accepted).

MORDOR

GENERAL INFORMATION

Model acronym: MORDOR
Model full name: Modèle à Réservoirs de Détermination Objective du Ruissellement
Authors first publication: Garçon, R. (1996) Prévision opérationnelle des apports de la Durance à Serre-Ponçon à l'aide du modèle MORDOR (*Operational inflow forecasting of the Durance River at Serre Ponçon using the MORDOR model*). *La Houille Blanche* **5**, 71–76.
Original application domain: Flow simulation and applications such as stochastic flood estimation, flood forecasting, long-term water resources forecasting, drought forecasting, reservoir inflow forecasting, snowpack water equivalent forecasting, design of water regulation structures, river temperature and sediment delivery forecasting.
Type: Lumped
Contact: E. Paquet, R. Garçon, J. Gailhard & T. Mathevet
EDF–DTG, Service CADE, 21, avenue de l'Europe, BP 41 38040 GRENOBLE 9, France
Email: thibault.mathevet@edf.fr

MODEL DESCRIPTION

Brief model description The MORDOR model is an hourly to daily lumped continuous rainfall–runoff model, developed since the 1990s by the French energy producer (EDF). The model development philosophy was close to the one of the HBV model. Its structure is similar to that of many conceptual type models (i.e. built using different interconnected storages) and composed of three main components: a snow accumulation and melt function, a production function (soil moisture accounting type) and a routing function. The whole model structure, with the snow component, is quite complex: it has five reservoirs and up to 23 free parameters to calibrate.

Main hydrological processes The model has no *a priori* physical underpinning. It includes a snow accumulation and melt function, a production function and a routing function.

Rainfall–runoff module The production function is based on: an evaporation function that determines the potential evaporation as a function of the actual temperature; a rainfall excess/soil moisture accounting store that determines: (a) the part of raw rainfall that will contribute to direct runoff; and (b) a part of the actual evapotranspiration; an evaporating reservoir, filled by a part of the indirect runoff component, that contributes to the actual evaporation.

Transfer function The transfer function is based on: an intermediate store that determines the split between (a) direct runoff, (b) indirect runoff and (c) percolation to a deep store; a deep store that determines a base flow component; an evaporating store, filled by a part of the indirect runoff component; a unit hydrograph (UH), based on a

Weibull law, that determines the routing of the total runoff (sum of the direct, indirect and baseflow components).

Additional components An optional snow accumulation and melt module is used. This module uses the hypsometric function of the basin to determine the actual part of the rainfall that is accumulated as snow. Then a melt function, based on a refined degree-day formulation, determines the part of the accumulated snow that contributes to (a) direct runoff or (b) baseflow.

Model applications The main application of the MORDOR model is in the hydro-meteorological operational forecasting centres of EDF-DTG. Models are mainly used to perform short term (sub-daily to daily) and mid term (weekly) streamflow and reservoir inflow forecasting, up to long-term snowmelt and reservoir inflow probabilistic forecasting. Models are coupled with quantitative forecasts and historical meteorological database. MORDOR models are currently used for operational forecasting on about 30 French watersheds (Fig. 1). MORDOR models are also used for general hydrological studies and for extreme flood probability studies.

MORDOR models have been developed and tested under various hydro-meteorological conditions in different countries over the world (Bolivia, Gabon, French Guyana, Laos, Honduras) for engineering applications.

A model intercomparison study (Mathevet, 2005), based on the assessment of 20 rainfall–runoff models, tested on a sample of 313 watersheds at the daily and hourly time-step, have shown that versions of the MORDOR model (with six and ten free parameters) were among the more efficient and robust rainfall–runoff model structures.

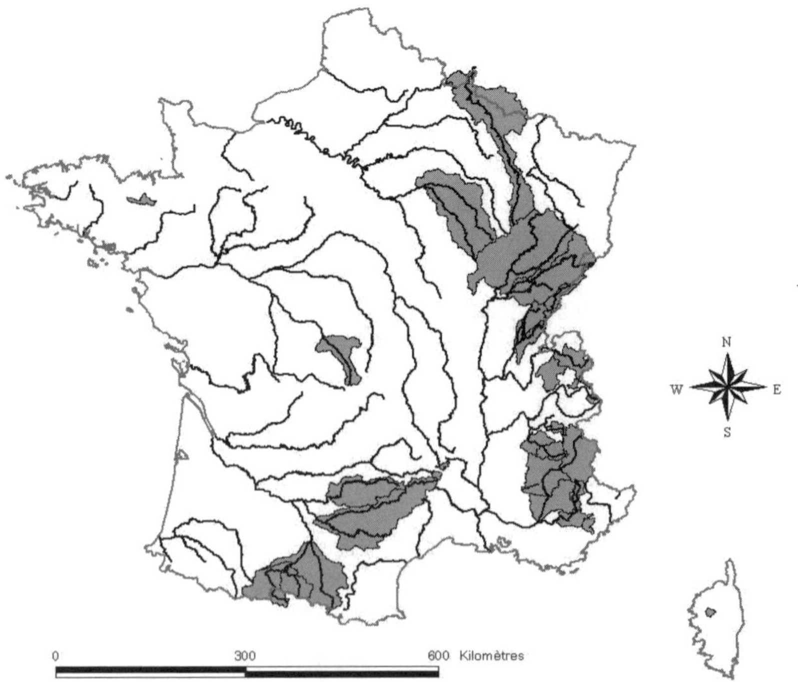

Fig. 1 Watersheds (in grey) where MORDOR models are used for hydrological forecasts.

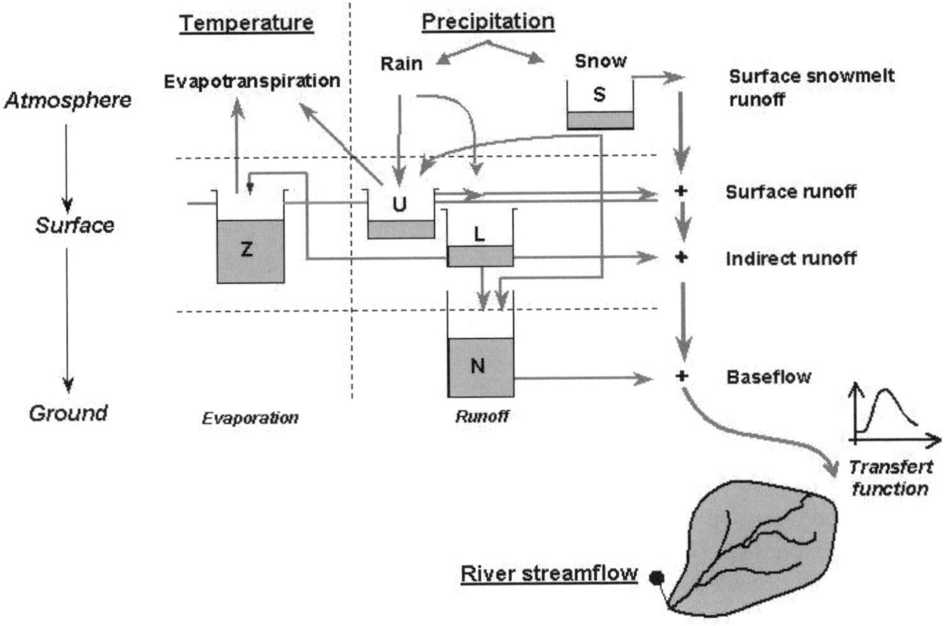

Fig. 2 Schematic representation of the MORDOR model.

Schematic representation of model structure See Fig. 2.

DATA REQUIREMENTS; PARAMETERIZATION

Input data The only inputs are: daily mean air temperature; daily rainfall time series (rainfall is an estimate of areal rainfall, e.g. calculated each day as weighted average on all the available raingauges); the hypsometric curve of the watershed. The model simulates daily flow time series, and daily snow accumulation and snowmelt. Observed daily flow time series are required for model calibration and evaluation.

Overall model parameters The model has up to 23 free parameters: 11 free parameters for the snow module: three for snow accumulation and eight for snow melt; two free parameters for the computation of the potential evapotranspiration; four free parameters for the production function; four free parameters for the routing function; two free parameters for the unit hydrograph (Weibull law).

Sensitivity analysis results The main sensitivity analysis was performed by Mathevet (2005) during research for a PhD. This work was done on a simplified MORDOR model structure, without the snowmelt module and using potential evapotranspiration input, instead of temperature input. This modified structure had ten free parameters. An empirical sensitivity analysis was performed on a sample of 313 watersheds, at the hourly time-step. Within the framework of this sensitivity analysis, the performance of 35 modified structures, with five to ten free parameters, was assessed. Results showed that the structure of the MORDOR model could be simplified and that a six free parameters structure could lead to the same mean model performance over the watershed sample, with a better robustness. This modified structure has not been tested yet in operational conditions.

Calibrated parameters All 23 model parameters are calibrated.

Calibration procedure-algorithm Given the high number of free parameters, the optimization algorithm is based on a global search strategy, using a classical genetic algorithm. This genetic algorithm is based on the generation of parameter vectors by mutations (random change made on a parameter) and crossovers (exchange of corresponding parameter between two vectors). The selection of the best parameter vectors is based on an objective function, such as the Nash-Sutcliffe criterion. This method was compared to the SCE-UA (Duan *et al.*, 1992) and the "step-by-step" (Edijatno *et al.*, 1999) optimization algorithm, on a sample of 313 watersheds at the hourly time-step (Mathevet, 2005). Results showed that this method was able to find parameter sets that have the same level of efficiency, compared to the one found by the two other methods. However, this method requires 3–30 times more objective function evaluations to converge to an optimal parameter set. To calibrate free parameters it is possible to choose among several objective function. Studies have shown that an objective functions based on the Nash-Sutcliffe criterion and the balance error or the streamflow distribution error yield more robust parameter vectors than the use of the Nash-Sutcliffe criterion only.

REFERENCES

Edijatno, Nascimento, N. O., Yang, X., Makhlouf, Z. & Michel, C. (1999) GR3J: a daily watershed model with three free parameters. *Hydrol. Sci. J.* **44**(2), 263–277.

Duan, Q. Y., Sorooshian, S. & Gupta, V. (1992) Effective and efficient global optimization for conceptual rainfall–runoff models. *Water Resour. Res.* **28**(4), 1015–1031.

Garçon, R. (1996) Prévision opérationnelle des apports de la Durance à Serre-Ponçon à l'aide du modèle MORDOR (Operational inflow forecasting of the Durance River at Serre Ponçon using the MORDOR model). *La Houille Blanche* **5**, 71–76.

Garçon, R. (1999) Modèle global pluie-débit pour la prévision et la prédétermination des crues (Overall rain-flow model for flood forecasting and pre-determination). *La Houille Blanche*, **7–8**, 88–95.

Mathevet, T. (2005) Which rainfall–runoff model at the hourly time-step? Empirical development and intercomparison of rainfall–runoff models on a large sample of watersheds. PhD Thesis, ENGREF, Paris, France (in French).

Paquet, E. & Garçon, R. (2002) Caprices du climat et de l'hydrologie en Haute-Durance: nos prévisions d'apports plurimensuels sont-elles encore fiables? (Climate and hydrology vagaries in Haute-Durance: are our several-month supply predictions still reliable?). *La Houille Blanche*, **8**, 62–68.

Paquet, E. (2003) Evolution du modèle hydrologique MORDOR: modélisation du stock nival à différentes altitudes (Improvement of the MORDOR model: toward the modelling of the snowpack distribution in function of the altitude). Société Hydrotechnique de France, section Glaciologie–Nivologie, 12/03/2003, Grenoble, France.

NOAH

GENERAL INFORMATION

Model acronym: Noah LSM
Model full name: Noah Land-Surface Model
Authors first publication: Chen, F., Mitchell, K., Schaake, J., Xue, Y., Pan, H., Koren, V., Duan, Q., Ek, M. & Betts, A. (1996) Modeling of land-surface evaporation by four schemes and comparison with FIFE observations. *J. Geophys. Res.* **101**, 7251–7268.
Original application domain: River and flood forecasting
Type: Lumped
Contact: kenneth.mitchell@noaa.gov

MODEL DESCRIPTION

Brief model description Noah LSM is a land surface model. It has been implemented operationally in the National Center for Environmental Prediction Eta model, a numerical weather prediction model. Noah LSM accounts for energy and water balance. The energy and water balance equations are defined over a single column, with no lateral exchange. Numerical schemes are used to solve the 1-D diffusion type energy and water balance dynamics.

Main hydrological processes Surface runoff when the precipitation exceeds infiltration capacity; baseflow from the bottom of the column; evapotranspiration from canopy, soil columns; energy fluxes; snow processes.

Rainfall–runoff module Richards equation

Transfer function Unit hydrograph

Groundwater/percolation module Darcy

Additional components No erosion and nutrients. Snow is represented by a full energy balance model.

Model applications Numerical weather prediction

Schematic representation of model structure See Fig. 1.

DATA REQUIREMENTS; PARAMETERIZATION

Input data Precipitation, air temperature, specific humidity, surface pressure, incoming short-wave solar radiation, incoming long-wave radiation, wind speed—vertical component, wind speed–horizontal component.

Overall model parameters There are many parameters that are determined using look-up tables based on soil and vegetation classifications.

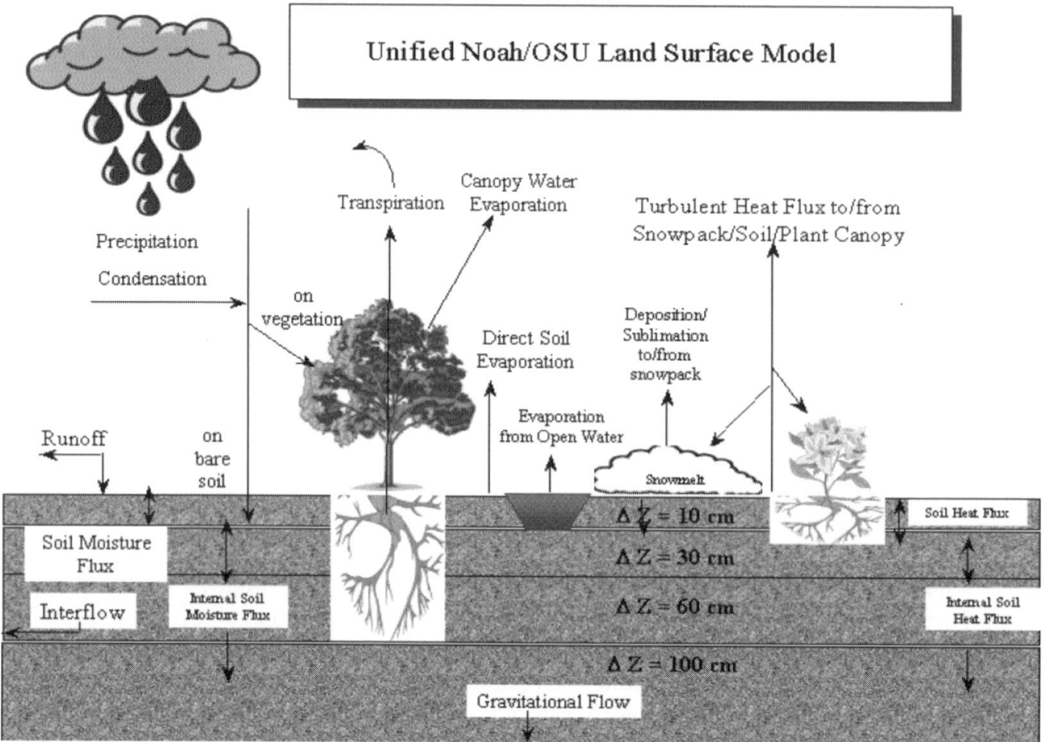

Fig. 1 Schematic representation of NOAH model structure.

Sensitivity analysis results No systematic sensitivity study was performed. Only the parameters considered to be sensitive in terms of runoff generation are considered for optimization.

Calibrated parameters BEXP, Brooke-Corey soil depletion curve exponent; DKSAT, Saturated hydraulic conductivity; DWSAT, Saturated soil matric potential; SMCMAX, Soil porosity; SMCREF, Soil field capacity; SMCWLT, Soil wilting point; KDT, Infiltration parameter–time scale factor.

Calibration procedure-algorithm SCE-UA

RRMT

GENERAL INFORMATION

Model acronym: RRMT
Model full name: Rainfall–Runoff Modeling Toolbox
Authors first publication: Wagener, T., Boyle, D. P., Lees, M. J., Wheater, H. S., Gupta, H. V. & Sorooshian, S. (2001). A framework for development and application of hydrological models. *Hydrol. Earth System Sci.* **5**(1), 13–26.
Original application domaine: Development and application of parsimonious rainfall–runoff models
Type: Lumped
Contact: Thorsten Wagener
Department of Civil and Environmental Engineering, 226B Sackett Building, The Pennsylvania State University, University Park, Pennsylvania 16802, USA
Email: thorsten@engr.psu.edu
Web site: http://www.engr.psu.edu/ce/divisions/hydro/wagener/index.html

MODEL DESCRIPTION

Brief model description A Rainfall–Runoff Modelling Toolbox (RRMT) has been developed within the scope of a model regionalization project to produce parsimonious, lumped model structures with a high level of parameter identifiability. Such identifiability is crucial if relationships between the model parameters representing the system and catchment characteristics (e.g. dominant soil types, land use, etc.) are to be established. RRMT is a modular framework that allows its user to implement different model structures to find a suitable balance between model performance and parameter identifiability. Model structures that can be implemented are lumped, relatively simple (in terms of number of parameters), and of metric (empirical), conceptual or hybrid metric-conceptual type. All structures consist of a moisture accounting and a routing module.

Main hydrological processes Depending on module chosen.

Rainfall–runoff module Empirical and conceptual soil moisture accounting structures. The choice is basically unlimited due to the modular structure of the framework.

Transfer function Both nonlinear and linear transfer functions are available. These include instrumental variable approach (after Peter Young), threshold structures, linear and nonlinear reservoirs, etc.

Groundwater/percolation module Depending on module chosen.

Additional components A snowmelt module is in preparation.

Model applications See References.

Schematic representation of model structure See Fig. 1.

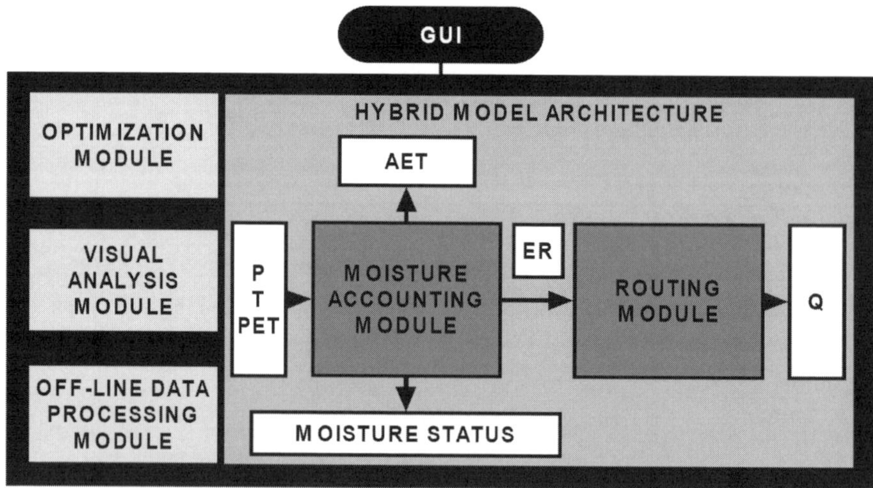

Fig. 1 Schematic representation of RRMT model structure.

DATA REQUIREMENTS; PARAMETERIZATION

Input data Rainfall/precipitation, temperature or potential evapotranspiration (depending on module chosen)

Overall model parameters Depending on module chosen

Sensitivity analysis results Depending on module chosen

Calibrated parameters Depending on module chosen

Calibration procedure-algorithm Currently: Monte Carlo Sampling, Latin Hypercube Sampling, Shuffled Complex Evolution Algorithm, Metropolis Algorithm

REFERENCES

McIntyre, N., Lee, H., Wheater, H. S., Young, A. & Wagener, T. (2005) Ensemble prediction of runoff in ungauged watersheds. *Water Resour. Res.* **41**, W12434, doi:10.1029/2005WR004289.

Wagener, T. & McIntyre, N. (2005) Identification of hydrologic models for operational purposes. *Hydrol. Sci. J.* **50**(5), 735–751.

Wagener, T. & Wheater, H. S. (2005) Parameter estimation and regionalization for continuous rainfall–runoff models including uncertainty. *J. Hydrol.* **320**(1–2), 132–154.

Wagener, T., Boyle, D. P., Lees, M. J., Wheater, H. S., Gupta, H. V. & Sorooshian, S. (2001) A framework for development and application of hydrological models. *Hydrol. Earth System Sci.* **5**(1), 13–26.

Wagener, T., Lees, M. J. & Wheater, H. S. (2002) A toolkit for the development and application of hydrological models. In: *Mathematical Models of Large Watershed Hydrology* (ed. by V. P. Singh & D. K. Frevert), 91–140. Water Resources Publications LLC, USA.

Wagener, T., McIntyre, N., Lees, M. J., Wheater, H. S. & Gupta, H. V. (2003) Towards reduced uncertainty in conceptual rainfall–runoff modelling: Dynamic identifiability analysis. *Hydrol. Processes* **17**(2), 455–476.

Wagener, T., Wheater, H. S. & Gupta, H. V. (2003) Identification and evaluation of watershed models. In *Advances in Calibration of Watershed Models* (ed. by Q. Duan, S. Sorooshian, H. V. Gupta, A. Rousseau & R. Turcotte), 29–47. AGU Monograph.

Wagener, T., Wheater, H. S. & Gupta, H. V. (2004) *Rainfall–Runoff Modelling in Gauged and Ungauged Catchments.* Imperial College Press, London, UK.

Wheater, H. S., McIntyre, N. & Wagener, T. (2006) Calibration, uncertainty and regional analysis of conceptual rainfall–runoff models. In: *Hydrologic Modelling in Arid Regions* (ed. by H. S. Wheater, S. Sorooshian & K. D. Sharma). Cambridge University Press, Cambridge, UK (in press).

SAC-SMA

GENERAL INFORMATION

Model acronym: SAC-SMA
Model full name: Sacramento Soil Moisture Accounting model
Authors-first publication: Burnash, R. J., Ferral, R. L. & McGuire, R. A. (1973) A generalized streamflow simulation system conceptual modeling for digital computers, Technical Report. Joint Federal and State River Forecast Center, US National Weather Service and State of California Department of Water Resources.
Original application domain: River and flood forecasting
Type: Lumped
Contact: Michael Smith
Hydrology Laboratory, NOAA/NWS Office of Hydrologic Development,
1325 East West Highway, Silver Spring, Maryland 20910, USA
Email: michael.smith@noaa.gov

MODEL DESCRIPTION

Brief model description SAC-SMA is a conceptual rainfall–runoff model that simulates runoff response from a watershed, given precipitation and potential evapotranspiration input data. The runoff is consequently converted into streamflow through a unit hydrograph. The model is made up of six water storages, including two in the upper zone and three in the lower zone. The remaining water storage represents the water accumulated over the impervious area from partial saturation. There are 16 model parameters and six state variables. The key function is the percolation equation which determines how water percolates from the upper zone to the lower zone.

Main hydrological processes Direct runoff over impervious area; surface runoff when the upper zone is fully saturated; interflow from upper zone free water storage; primary and supplemental baseflow from the lower zone free water storages; evapotranspiration from all water storages.

Rainfall–runoff module Double storage/reservoir

Transfer function Unit hydrograph

Groundwater/percolation module Storage/reservoir model

Additional components No erosion and nutrients. Snow is represented in a separate model.

Model applications River and flood forecasting, water resources management, climate changes, etc.

Schematic representation of model structure See Fig. 1.

DATA REQUIREMENTS; PARAMETERIZATION

Input data Precipitation and potential evapotranspiration.

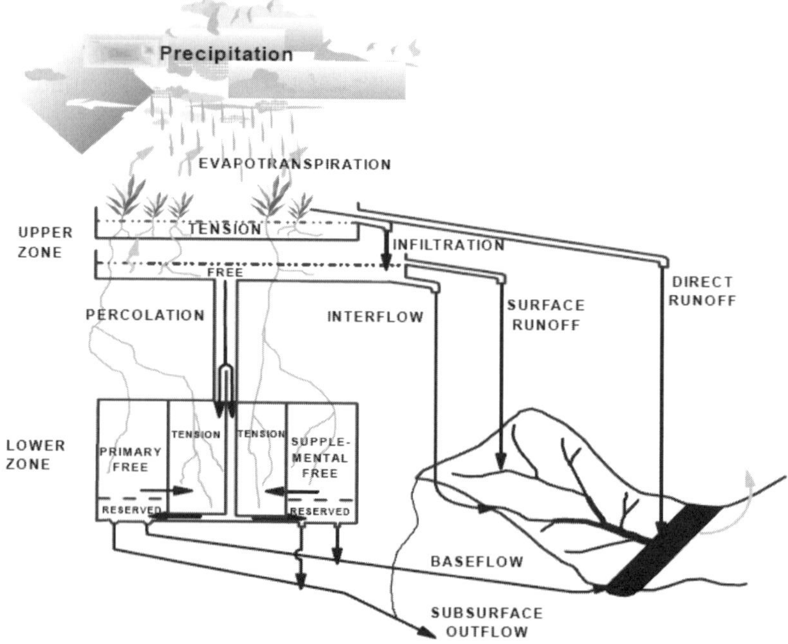

Fig. 1 Schematic representation of SAC-SMA model structure.

Overall model parameters UZTWM, Upper zone tension water storage capacity (mm); UZFWM, Upper zone free water storage capacity (mm); LZTWM, Lower zone tension water storage capacity (mm); LZFPM, Lower zone primary free water storage capacity (mm); LZFSM, Lower zone supplemental free water storage capacity (mm); ADIMP, Additional impervious area (fraction, unitless); UZK, Upper zone free water depletion coefficient (L day^{-1}); LZPK, Lower zone primary free water depletion coefficient (L day^{-1}); LZSK, Lower zone primary free water depletion coefficient (L day^{-1}); PCTIM, Percentage of impervious area (fraction, unitless); ZPERC, Percolation parameter–scaling; REXP, Percolation parameter–exponent; PFREE, Parameter controlling partition of percolated into free water storages and tension water storage; RSERV, Fraction of water not available for runoff or evapotranspiration; RIVA, Parameter controlling riparian water loss; SIDE, Fraction of water lost from river bed.

Sensitivity analysis results The model has 16 parameters, 13 of which are tuneable and included in the calibration.

Calibrated parameters UZTWM, UZFWM, LZTWM, LZFPM, LZFSM, ADIMP, UZK, LZPK, LZSK, PCTIM, ZPERC, REXP, PFREE

Calibration procedure-algorithm SCE-UA

REFERENCES

Burnash, R. J. C., Ferrell, R. L. & McGuire, R. A. (1973) A generalized streamflow simulation system. Jt. Fed.-State River Forecast Center, Sacramento, California, USA.

Burnash, R. J. C. (1995) The NWS River Forecast System—catchment modeling In: *Computer Models of Watershed Hydrology* (ed. by V. P. Singh), 311–366. Water Resources Publications, Highlands Ranch, Colorado, USA.

SMAR

GENERAL INFORMATION

Model acronym: SMAR
Model full name: Soil Moisture Accounting and Routing
Authors first publication: O'Connor, K. M, Goswami, M., Liang, G. C., Kachroo, R. K. & Shamseldin, A. Y. (2001) The development of the Galway Real-Time River Flow Forecasting System (GFFS). In: *Sustainable Use of Land and Water* (Proc. 19th European Regional Conference of the International Commission on Irrigation and Drainage, 4–8 June, Brno and Prague, Czech Republic) (available on CD).
Original application domain: Streamflow simulation and forecasting
Type: Lumped
Contact: Kieran Michael O'Connor
Department of Engineering Hydrology, National University of Ireland, Galway, Galway, Ireland
Telephone: +353 91 492213

MODEL DESCRIPTION

Brief model description The Soil Moisture Accounting and Routing (SMAR) model is a simple lumped conceptual rainfall–evaporation–runoff model.

In the SMAR model, the catchment is visualized as being composed of a set of horizontal soil layers, each of which may contain water up to a maximum depth of 25 mm (for daily data) except for the last (i.e. bottom) layer which may have a maximum depth <25 mm. The total combined water storage depth of these layers is a parameter (Z) of the model.

The evaporation input E to the model, when multiplied by a parameter T (<1), is converted to the estimate of the "potential evaporation demand rate (PE)", expressed as a depth.

Evaporation is considered to occur from the layers only when there is no rainfall or when the rainfall depth R is insufficient to satisfy the PE ($= T \times E$). Any evaporation from the first layer occurs at the full PE rate. On the depletion of the water depth in the first layer, any evaporation from the second layer occurs at the PE rate multiplied by the parameter C (<1). On the depletion of the water depth of the second layer, any subsequent evaporation from the third layer occurs at rate of C^2 and so on. Such evaporation would continue until either the storage of all the layers was depleted or the potential evaporation demand rate (PE) was accounted for.

When rainfall (R) exceeds the PE, runoff is generated. A fraction H' of the excess rainfall X ($= R - PE$) contributes to the generated runoff by producing the direct runoff component r_1. H' is given by $H' = H \times (W_{act}/W_{cap})$ where W_{act} is the total available water depth in the layers, W_{cap} is the maximum combined depth of all layers and H ($0 < H' < H$), which is a parameter of the model, is the constant of proportionality.

Any remainder of the excess rainfall X, after subtraction of r_1, which exceeds the maximum infiltration capacity Y, also contributes to the generated runoff as r_2. The remaining rainfall, after subtraction of r_1 and r_2, replenishes each soil layer in turn beginning from the first (i.e. top) layer downwards until either the rainfall is exhausted or all layers are full. Any still-remaining surplus is further divided into two fractions by a weight parameter G, the first fraction being the groundwater runoff component r_g and the second being the subsurface runoff component r_3. r_3 is added to r_1 and r_2 to produce the total generated surface runoff r_s. This r_s is routed through a two-parameter distribution function. Provision of three 2-parameter distribution function options are available, namely: the classic Negative Binomial distribution, the gamma distribution (Nash-cascade) model, and the Inverse Gaussian distribution, the Nash-cascade involving the shape parameter n and the lag parameter nK being that used in this study.

Main hydrological processes In the structure of the SMAR model, the main hydrological processes are grouped in two distinct complementary components. The first is the nonlinear water balance (soil moisture accounting procedure) component that keeps account of the balance between rainfall, evaporation, runoff and soil storage using a number of empirical and assumed functions which are physically plausible or at least physically inspired. The second is the routing component which simulates the attenuation and the diffusive effects of the catchment by routing through linear time invariant storage systems the different generated runoff components which are the outputs from the water balance part.

Rainfall–runoff module Lumped conceptual

Transfer function Either of three 2-parameter distribution functions, namely, the classic gamma distribution (Nash-cascade) model, its discrete counterpart—the Negative Binomial distribution, and the sharp-peaked Inverse Gaussian distribution.

Groundwater/percolation module Linear reservoir

Additional components None

Model applications Flow simulation, lead-time flow forecasting. The model is versatile, and has been used for flow modelling in catchments all over the world, covering a wide range of climatic types. Recent applications involve catchments from Ireland, France, India, China, Kenya, Bangladesh and Nepal (see References).

Schematic representation of model structure See Fig. 1.

DATA REQUIREMENTS; PARAMETERIZATION

Input data Observed rainfall, evaporation and discharge for flow simulation, Observed rainfall and evaporation over the forecast lead-times for flow forecasting.

Overall model parameters The version of the SMAR model used in this study has nine parameters. Five of these, namely T, H, Y, Z and C, control the overall operation of the water-budget component, one is a weighting parameter G for groundwater routing, and three are routing parameters, namely N, NK and Kg.

Sensitivity analysis results Normally Z is insensitive. The degree of sensitivity of the other eight parameters depends on the catchment system being modelled.

Calibrated parameters Some of the nine parameters of the model may be fixed at appropriately chosen values while the values of the rest are usually estimated

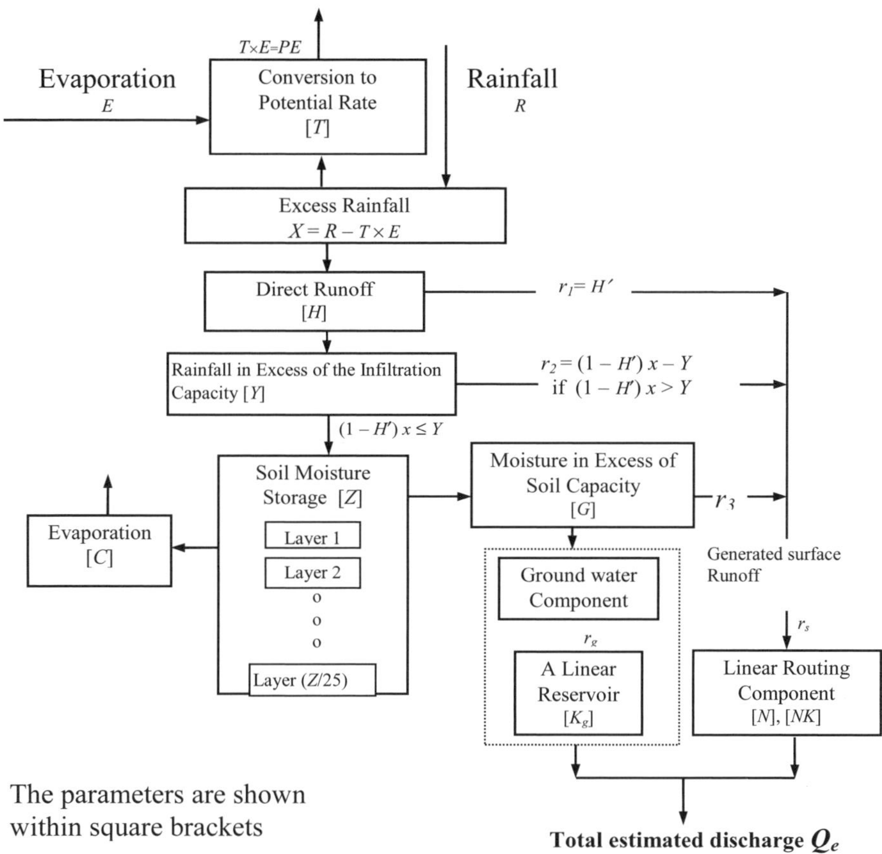

Fig. 1 Schematic representation of SMAR model structure.

empirically by optimization to minimize the selected measure of error between the observed and the model estimated discharges. All nine parameters may also be calibrated simultaneously.

Calibration procedure-algorithm Provision of five automatic optimization procedures/methods has been made. These methods are the Simplex search algorithm, the Rosenbrock direct search method, the Particle Swarm Optimization method, the Simulated Annealing method, and the Shuffled Complex Evolution algorithm. These may be used either individually or sequentially for model calibration.

REFERENCES

Goswami, M, O'Connor, K. M. & Shamseldin, A. Y. (2002) Comparative performance evaluation of rainfall–runoff models, six of black-box type and one of conceptual type from the Galway Flow Forecasting System (GFFS) Package, applied on two Irish catchments. In: *Proc. Conf. XXVII General Assembly of the European Geophysical Society* (Nice, France, 21–26 April, 2002).

Goswami, M, O'Connor, K. M. & Shamseldin, A. Y. (2002) Rainfall–runoff modelling of two Irish catchments (one karstic and one non-karstic). In: *Celtic Water in European Framework* (Proc. Third Inter-Celtic Colloquium on Hydrology and Management of Water Resources, 8–10 July, Galway, Ireland), 151–164.

Goswami, M, O'Connor, K. M. & Shamseldin, A. Y. (2002) Structures and performances of five rainfall–runoff models for continuous river-flow simulation. In: *Proc. first Biennial Meeting of the International Environmental Modelling and Software Society* (24–27 June, Lugano, Switzerland), vol. 1, 476–481.

Goswami, M & O'Connor, K. M. (2005) Real-time flow forecasting in the absence of quantitative precipitation forecasts: a multi-model approach. *J. Hydrol.* (submitted).

Goswami, M. & O'Connor, K. M. (2005) Application of a conceptual rainfall–runoff simulation model to three European catchments characterised by non-conservative system behaviour. In: *Hydrological Perspectives for Sustainable Development* (Proc. Int. Conf., Roorkee, India), vol. 1, 117–130.

Goswami, M., O'Connor, K. M. & Bhattarai, K. P. (2005) Flow simulation in ungauged basins using regionalisation of data and the multi-model approach. *J. Hydrol.* (submitted).

Goswami, M., O'Connor, K. M., Bhattarai, K. P. & Shamseldin, A. Y. (2005) Real-time river flow forecasting for the Brosna catchment in Ireland using eight updating models. *J. Hydrol. Earth System Sci.* **9**(4), 394–411.

O'Connor, K. M, Goswami, M., Liang, G. C., Kachroo, R. K. & Shamseldin, A. Y. (2001) The development of the Galway Real-Time River Flow Forecasting System (GFFS). In: *Sustainable Use of Land and Water* (Proc. 19th European Regional Conference of the International Commission on Irrigation and Drainage, 4–8 June, Brno and Prague, Czech Republic) (available on CD).

O'Connor, K. M., Goswami, M., Bhattarai, K. P. & Shamseldin, A. Y. (2004) A comparison of the lead-time discharge forecasts of the "Perfect" and "Naïve-AR" quantitative precipitation forecast (QPF) input scenarios, to assess the value of having good QPFs. In: *Hydrological Risk; Recent Advances in Peak River Flow Modelling, Prediction and Real-Time Forecasting—Assessment of Impacts of Land Use and Climate Change* (ed by A. Brath, A. Montanari & E. Toth), 187–217. CNR-GNDCI Publ. no. 2858, ISBN 88-7740-378-0, Editoriale Bios, Via A. Rendano, 25-87040 Castrolibero (CS), Italy.

SWAP

GENERAL INFORMATION

Model acronym: SWAP
Model full name: Soil Water–Atmosphere–Plants
Authors first publication: Gusev, Ye. M. & Nasonova, O. N. (1998) The Land Surface Parameterization scheme SWAP: description and partial validation. *Global Planetary Change* **19**(1–4), 63–86.
Original application domain: Simulation of the components of energy and water balance, surface and subsurface state variables, evapotranspiration components.
Type: Land surface model
Contact: Yeugeniy Gusev
Email: gusev@aqua.laser.ru
Olga Nasonova
Email: nasonova@aqua.laser.ru

MODEL DESCRIPTION

The SWAP model is a physically based land-surface model describing heat and water exchange between the land surface and the atmosphere throughout a year at different scales (from local to global) and oriented towards use of atmospheric forcings from the lowest atmospheric layer of GCMs or from any reference height. The main distinctive feature of SWAP is the combination of physically-based treatment of the main processes and the rationality of modelling technique used. The latter is provided by application of analytical methods (contrary to the usual practice of application of numerical ones) to solve the systems of equations and by a relatively small number of model parameters. This allows us to avoid many problems associated with solving the numerical equations (such as instability, great consumption of computer resources and calculational time, and so on) and parameter estimation.

Application of analytical methods has led to the non-traditional structure of SWAP. Thus, in SWAP, a calendar year is divided into the warm and cold seasons. For each season, a separate submodel was developed, then these two submodels were linked into one general model, named SWAP. The cold-season model is used only in the case of the fulfilment of, at least, one of the following conditions: (1) the mean daily air temperature is below 0°C continuously during several days (here, not less than 7 days); (2) the land surface is covered by snow; (3) the soil freezing depth is greater than zero. Otherwise, the warm-season submodel is used.

Main hydrological processes (1) interception of rainfall/snowfall by the canopy; (2) evapotranspiration (including transpiration by plants, soil/snow evaporation, evaporation of intercepted precipitation); (3) formation of snowpack on the ground and on the trees' crowns (including snow accumulation, snow evaporation, snowmelt, water yield of snow cover, refreezing of melt water); (4) formation of surface runoff; (5) formation of drainage; (6) water infiltration into a soil; (7) water exchange between soil layers; (8) interaction between soil water and groundwater.

Rainfall–runoff module Surface runoff in based on Hortonian mechanism (when precipitation rate exceeds infiltration rate).

Transfer function When runoff is simulated, the next problem is to transform it for simulating streamflow at the catchment (or calculational grid cell) outlet. For such a transformation we use an approach based on application of the concept of the 2-D kinematic wave.

Groundwater/percolation module For parameterization of subsurface runoff the concept of saturation excess is used. Modelling the dynamics of the water table.

Additional components Snowpack formation on the ground and on trees' crowns; formation of energy balance; formation of the dynamics of soil freezing and thawing depths.

Model applications See Table below.

Purpose	Location	Climate	Reference
Participation in the international PILPS phase 2a experiment	The Netherlands (Cabauw)	Moderate maritime	Chen *et.al.*, 1997; Gusev & Nasonova, 1998; Qu *et al.*, 1998
Participation in the international PILPS phase 2c experiment	USA (Red-Arkansas River basin)	Subtropical (ranges from arid and semiarid in the west to humid in the east)	Wood *et al.*,1998; Lohmann *et al.*, 1998; Liang *et al.*, 1998; Gusev & Nasonova 2000
Participation in the international PILPS phase 2d experiment	Russia (Valdai, the Usadievskiy catchment)	Moderate continental	Gusev & Nasonova, 2000; Schlosser *et al.*, 2000; Slater *et al.*, 2001; Luo *et al.*, 2003
Participation in the international PILPS phase 2e experiment	Sweden, Finland (Torne-Kalix River basin)	Moderate (transient from oceanic to continental), excessively humid	Bowling *et al.*, 2003; Nijssen *et al.*, 2003
Participation in the international Rhone AGG project	France (Rhone River basin)	Mediterranean, maritime-continental, alpine	Boone *et al.*, 2004
Participation in the international SnowMIP project	France (Col de Porte), Switzerland (Weissfluhjoch)	Moderate highland	Etchevers *et al.*, 2002, 2004
Participation in the international SnowMIP project	Canada (Goose Bay)	Moderate monsoon	Etchevers *et al.*, 2002, 2004
Participation in the international SnowMIP project	USA (Sleepers River)	Moderate, sufficiently wet	Etchevers *et al.*, 2002, 2004
Participation in the international MOPEX project	USA (12 MOPEX river basins)	Climate ranges from subtropical to moderate, from semi-arid to humid	Duan *et al.*, 2006
Participation in the international PILPS-C1 experiment	The Netherlands (Loobos)	Moderate maritime	http://www.pilpsc1.cnrs-gif.fr/
Participation in the international PILPS San Pedro experiment	USA	Subtropical, semi-arid	http://ceefs2.cee.edu/PILPS-SanPedro/PILPS_SanPedro_prelim_v1.pdf
Simulation of soil water content in different ecosystems	Russia (the Kursk region, Petrinka)	Moderate continental	Gusev & Nasonova, 1998
Simulation of runoff from small catchment and river basin	Russia (Valdai, the Tayozhnyi catchment, the Polomet River basin)	Moderate continental	Gusev & Nasonova, 2002; Gusev, Nasonova & Busarova 2003
Simulation of runoff from small catchmets under permafrost and highland conditions	Russia (Kolyma River basin)	Sub arctic continental	Gusev & Nasonova, 2004; Gusev, Nasonova & Dzhogan, 2006
Simulation of the radiation and heat fluxes	USA (Oklahoma)	Subtropical, semi-arid	Gusev, Nasonova & Mohanty, 2004
Participation in the GSWP-2 Project	Globally	All	http://www.iges.org/gswp/

Schematic representation of model structure See opposite. Notation in Figure: representation of water ((a), (c)) and heat ((b), (d)) exchange processes in the natural ecosystem by the SWAP model for summer ((a), (b)) and winter ((c), (d)) seasons.

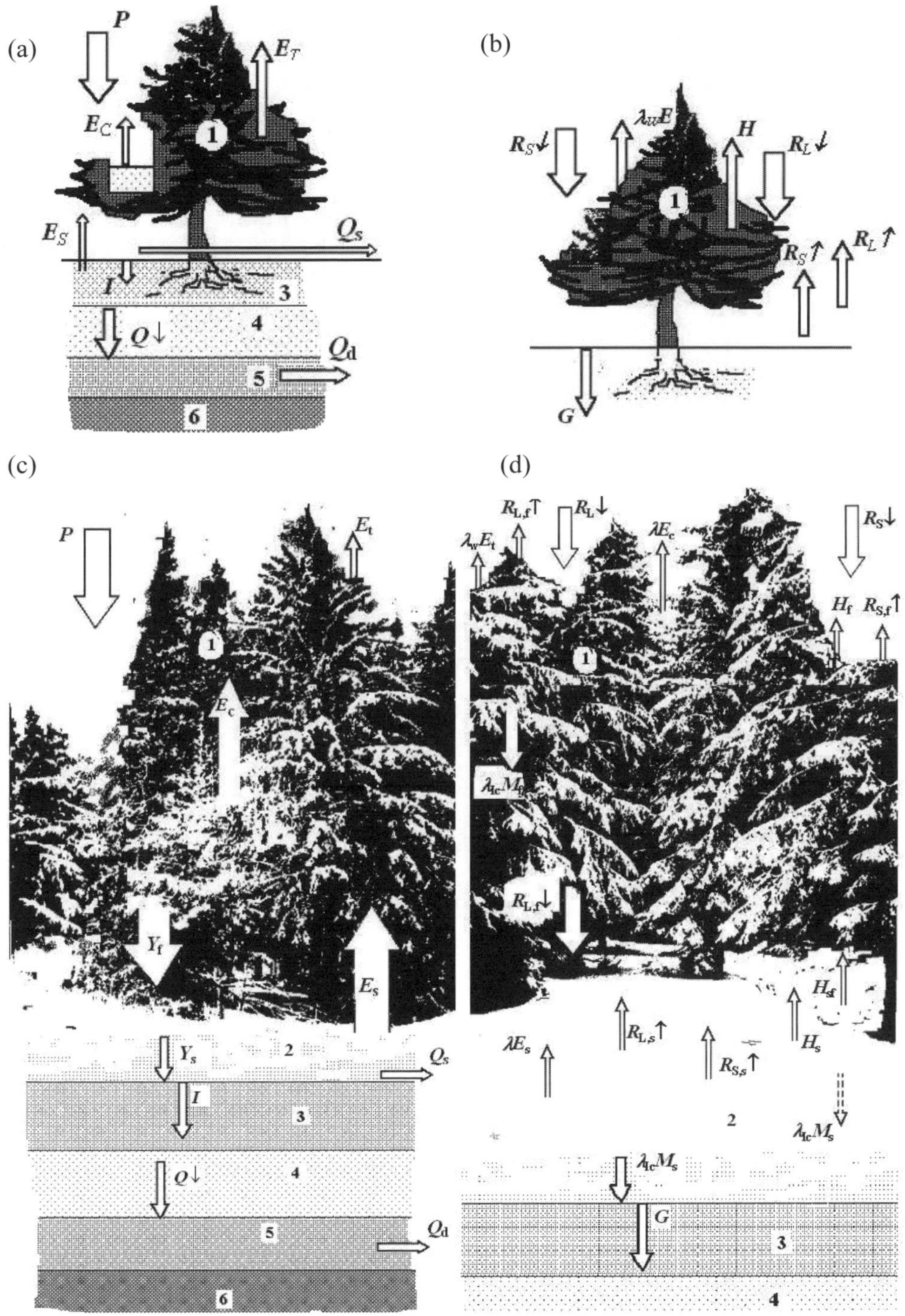

1, vegetation; 2, snow (on the ground and vegetation); 3, the first soil layer (root zone); 4, the second soil layer; 5, groundwater; 6, impermeable layer. Symbols are described in the list following.

List of symbols to schematic representation of model structure:

E	evaporation rate(kg m^{-2} s^{-1})
E_c	rate of evaporation of precipitation intercepted by a canopy (kg m^{-2} s^{-1})
E_s	evaporation rate from surface of soil or snow (kg m^{-2} s^{-1})
E_t	transpiration rate (kg m^{-2} s^{-1})
G	ground heat flux (W m^{-2})
H	sensible heat flux (W m^{-2})
H_f	sensible heat flux from the trees crowns to the atmosphere (W m^{-2})
H_s	sensible heat exchange between the forest floor and the atmosphere (W m^{-2})
H_{sf}	sensible heat exchange between the forest floor and forest crowns (W m^{-2})
I	infiltration rate (kg m^{-2} s^{-1})
M_s	the rate of melting snow on the ground (kg m^{-2} s^{-1})
M_f	the rate of melting snow on the trees' crowns (kg m^{-2} s^{-1})
P	precipitation rate (kg m^{-2} s^{-1})
Q	water exchange between soil and ground water (kg m^{-2} s^{-1})
Q_d	drainage (kg m^{-2} s^{-1})
Q_s	surface runoff (kg m^{-2} s^{-1})
$R_L\downarrow$	downward longwave radiation (W m^{-2})
$R_{L,f}\downarrow$	longwave emission by the trees crowns toward the forest floor (W m^{-2})
$R_{L,f}\uparrow$	upward longwave radiation emitted by trees' crowns (W m^{-2})
$R_{L,s}\uparrow$	upward longwave radiation from the forest floor (W m^{-2})
$R_S\downarrow$	downward shortwave radiation (W m^{-2})
$R_{S,f}\uparrow$	shortwave radiation reflected by trees' crowns (W m^{-2})
$R_{S,s}\uparrow$	shortwave radiation reflected by forest floor (W m^{-2})
Y_f	the rate of water yield of snow cover on trees' crowns (kg m^{-2} s^{-1})
λ =	$\lambda_W + \lambda_{Ic}$ (J kg^{-1})
λ_W	latent heat of vaporization of water (J kg^{-1})
λ_{Ic}	heat of ice fusion (J kg^{-1})

DATA REQUIREMENTS; PARAMETERIZATION

Input data

Forcing data incoming longwave and shortwave radiation, air temperature and air humidity, surface air pressure, wind speed, precipitation.

Overall model parameters See Table opposite

Sensitivity analysis results k_0, saturated hydraulic conductivity, (soil); h_{root}, root zone depth, (vegetation); W_{sat}, porosity, (soil); W_{fc}, water content at the field capacity, (soil); W_{wp}, water content at the plant wilting point, (soil); h_{soil}, the mean depth to the upper impermeable layer, (soil); α_{sn}, "Deep snow" albedo (snow); α_{vg}, Snow-free albedo of vegetation, (vegetation); $\alpha_{vg,sn}$, Albedo of vegetation with intercepted snow on crowns, (vegetation); LAI, SAI, leaf and steam area index (vegetation).

Calibrated parameters Usually we do not calibrate parameters. In the MOPEX experiment (Duan *et al.*, 2006) we have calibrated only one parameter (k_0).

Calibration procedure-algorithm In the MOPEX experiment, manual calibration was used.

Parameters	Symbol
Soil parameters	
Soil water content at the plant wilting point	W_{wp}
Soil water content at the field capacity	W_{fc}
Soil porosity	W_{sat}
Hydraulic conductivity at saturation	k_0
Matric potential of soil water at saturation	ϕ_0
"B" parameter of Clapp and Hornberger	B
Constant soil temperature at the depth where seasonal temperature variations are damped out (°C)	$\tilde{t}$
Snow cover parameters	
"Deep snow" albedo	α_{sn}
Roughness length of snow (m)	z_{0sn}
Vegetation parameters	
Depth of the soil root zone	h_{root}
Snow-free albedo of vegetation	α_{vg}
Albedo of vegetation with intercepted snow on crowns	$\alpha_{vg,sn}$
Effective linear leaf size (m)	l
Interception capacity of vegetation for liquid precipitation (m)	W_{cmax}
Interception capacity of vegetation for solid precipitation (m)	$S_{ch,max}$
Leaf area index	LAI
Stem area index	SAI
Roughness lenght of vegetation (m)	z_{0vg}
Zero plane displacement height (m)	d_0
Parameter β_v for the calculation of transpiration	β_v
Extinction coefficient	η
Other parameters	
The length of the catchment in the direction of mean slope (m)	Δx
The mean slope of watershed ($^0/_{00}$)	i_x
Effective Manning roughness coefficient ($m^{-1/3}$ c)	n
Mean depth from the surface to the impermeable layer (m)	h_{soil}

REFERENCES

Boone, A., Habets, F., Noilhan, J., Clark, D., Dirmeyer, P., Fox, S., Gusev, Y., Haddeland, I., Koster, R., Lohmann, D., Mahanama, S., Mitchell, K., Nasonova, O., Niu, G.-Y., Pitman, A., Polcher, J., Shmakin, A. B., Tanaka, K., van den Hurk, B., Verant, S., Verseghy, D., Viterbo, P. & Yang, Z. -L. (2004) The Rhone-aggregation land surface scheme intercomparison project: An overview. *J. Climate* **17**, 187–208.

Bowling, L. C., Lettenmaier, D. P., Nijssen, B., Graham, L. P., Clark, D. B., Maayar, M. E., Essery, R., Goers, S., Gusev, Ye. M., Habets, F., van den Hurk, B., Jin, J., Kahan, D., Lohmann, D., Ma, X., Mahanama, S., Mocko, D., Nasonova, O., Niu, G. -Y., Samuelsson, P., Shmakin, A. B., Takata, K., Verseghy, D., Viterbo, P., Xia, Y., Xue, Y., & Yang, Z. -L. (2003) Simulation of high latitude hydrological processes in the Torne-Kalix basin: PILPS Phase 2(e). 1: Experiment description and summary intercomparisons. *Global Planetary Change* **38**(1–2), 1–30.

Chen, T. H., Henderson-Sellers, A., Milly, P. C. D., Pitman, A. J., Beljaars, A. C. M., Polcher, J., Abramopoulos, F., Boone, A., Chang, S., Chen, F., Dai, Y., Desborough, C. E., Dickinson, R. E., Dumenil, L., Ek, M., Garratt, J. R., Gedney, N., Gusev, Y. M., Kim, J., Koster, R., Kowalczyk, E. A., Laval, K., Lean, J., Lettenmaier, D., Liang, X., Mahfouf, J.-F., Mengelkamp, H.-T., Mitchell, K., Nasonova, O. N., Noilhan, J., Robock, A., Rosenzweig, C., Schaake, J., Schlosser, C. A., Schulz, J.-P., Shao, Y., Shmakin, A. B., Verseghy, D. L., Wetzel, P., Wood, E. F., Xue, Y., Yang, Z. -L. & Zeng, Q. (1997) Cabauw experimental results from the project for intercomparison of land-surface parameterization schemes. *J. Climate* **10**(6), 1194–1215.

Duan, Q., Schaake, J., Andreassian, V., Franks, S., Goteti, G., Gupta, H. V., Gusev, Y. M., Habets, F., Hall, A., Hay, L., Hogue, T., Huang, M., Leavesley, G., Liang, X., Nasonova, O. N., Noilhan, J., Oudin, L., Sorooshian, S., Wagener, T. & Wood, E. F. (2006) Model parameter estimation experiment (MOPEX): an overview of science strategy and major results from the second and third workshops. *J. Hydrol.* **320**, 3–17.

Etchevers, P., Martin, E., Brown, R., Fierz, C., Lejeune, Y., Bazile, E., Boone, A., Dai, Y.-J., Essery, R., Fernandez, A., Gusev, Y., Jordan, R., Koren, V., Kowalczyk, E., Nasonova, N.O., Pyles, R. D., Schlosser, A., Shmakin, A.B., Smirnova, T.G., Strasser, U., Verseghy, D., Yamazaki, T. & Yang, Z. -L. (2002). SnowMIP, an intercomparison of snow models: first results. In: *Proc. ISSW 2002 meeting* (30 September–4 October, Penticton, Canada) (CD-ROM).

Etchevers, P., Martin, E., Brown, R., Fierz, C., Lejeune, Y., Bazile, E., Boone, A., Dai, Y. -J., Essery, R., Fernandez, A., Gusev, Y., Jordan, R., Koren, V., Kowalczyk, E., Nasonova, N.O., Pyles, R. D., Schlosser, A., Shmakin, A.B., Smirnova, T.G., Strasser, U., Verseghy, D., Yamazaki, T. & Yang Z. -L. (2004) Validation of the energy budget of an alpine snowpack simulated by several snow models (SnowMIP project). *Annals Glaciol.* **38**, 150–158.

Gusev, Ye. M. & Nasonova, O. N. (1998) The land surface parameterization scheme SWAP: description and partial validation. *Global Planetary Change* **19**(1–4), 63–86.

Gusev, Ye. M. & Nasonova, O. N. (1999) The land surface parameterization scheme SWAP: description and validation. In: *Regionalization in Hydrology* (ed. by B. Diekkruger, M. J. Kirkby & U. Shroder), 113–122. IAHS Publ. 254. IAHS Press, Wallingford, UK.

Gusev, Ye. M. & Nasonova, O. N. (2000) An experience of modelling heat and water exchange at the land surface on a large river basin scale. *J. Hydrol.* **233**(1–4), 1–18.

Gusev, Ye. M. & Nasonova, O. N. (2002) The simulation of heat and water exchange at the land-atmosphere interface for the boreal grassland by the land-surface model SWAP. *Hydrol. Processes.* **16**(10), 1893–1919.

Gusev, E. M., Nasonova, O. N. & Busarova, O. E. (2002) Parameterization of heat and moisture exchange on land surface in areas with moderate continental climate. *Water Res.* **29**(1), 99–110.

Gusev, Ye. M. & Nasonova, O. N. (2003) Modelling heat and water exchange in the boreal spruce forest by the land-surface model SWAP. *J. Hydrol.* **280**(1–4), 162–191.

Gusev, E. M., Nasonova, O. N. & Mohanty, B. P. (2004) Estimation of Radiation, Heat, and Water Exchange between Steppe Ecosystems and the Atmosphere in the SWAP Model. *Izvestiya, Atmos. Oceanic Physics* **40**(3), 291–305.

Gusev, E. M. & Nasonova, O. N. (2004) Simulation of Heat and Water Exchange at the Land–Atmosphere Interface on a Local Scale for Permafrost Territories. *Eurasian Soil Sci.* **37**(9), 1077–1092.

Gusev, E. M., Nasonova, O. N. & Dzhogan, L. Ya. (2005) Modeling of heat-, water-, and carbon-exchange processes in a pine forest ecosystem. *Izvestiya, Atmos. Oceanic Physics* **41**(2), 203–216.

Gusev, E. M., Nasonova, O. N. & Dzhogan, L. Ya. (2006) Modelling runoff from small catchments in the permafrost zone using the model SWAP. *Water Res.* **33**(2), 133–145.

Liang, X., Wood, E. F., Lettenmaier, D. P., Lohmann, D., Boone, A., Chang, S., Chen, F., Dai, Y., Desborough, C., Dickinson, R. E., Duan, Q., Ek, M., Gusev, Y. M., Habets, F., Irannejad, P., Koster, R., Mitchell, K. E., Nasonova, O. N., Noilhan, J., Schaake, J., Schlosser, A., Shao, Y., Shmakin, A. B., Verseghy, D., Warrach, K., Wetzel, P., Xue, Y., Yang, Z. -L. & Zeng, Q.-C. (1998) The project for intercomparison of land-surface parameterization schemes (PILPS) phase-2(c) Red-Arkansas River basin experiment: 2. Spatial and temporal analysis of energy fluxes. *Global Planetary Change* **19**(1–4), 137–159.

Lohmann, D., Lettenmaier, D. P., Liang, X., Wood, E. F., Boone, A., Chang, S., Chen, F., Dai, Y., Desborough, C., Dickinson, R. E., Duan, Q., Ek, M., Gusev, Y. M., Habets, F., Irannejad, P., Koster, R., Mitchell, K. E., Nasonova, O. N., Noilhan, J., Schaake, J., Schlosser, A., Shao, Y., Shmakin, A. B., Verseghy, D., Warrach, K., Wetzel, P., Xue, Y., Yang, Z. -L. & Zeng, Q. -C. (1998) The project for intercomparison of land-surface parameterization schemes (PILPS) phase-2(c) Red-Arkansas River basin experiment: 3. Spatial and temporal analysis of water fluxes. *Global Planetary Change* **19**(1–4), 161–179.

Luo, L., Robock, A., Vinnikov, K. Y., Schlosser, A. C., Slater, A. G., Boone, A., Braden, H., Cox, P., de Rosnay, P., Dickinson, R. E., Dai, Y. -J., Duan, Q., Etchevers, P., Henderson-Sellers, A., Gedney, N., Gusev, Ye. M., Habets, F., Kim, J., Kowalczyk, E., Mitchell, K., Nasonova, O. N., Noilhan, J., Pitman, A. J., Schaake, J., Shmakin, A. B., Smirnova, T. G., Wetzel, P., Xue, Y., Yang, Z. -L. & Zeng, Q. -C. (2003) Effects of frozen soil on soil temperature, spring infiltration, and runoff: results from the PILPS 2(d) experiment at Valdai, Russia. *J. Hydrometeorol.* **4**, 334–351.

Nasonova, O. N. & Gusev, Ye. M. (2001) Application of a land surface model for studying the role of boreal spruce forest in land surface–atmosphere interactions. In: *Soil–Vegetation–Atmosphere Transfer Schemes and Large-Scale Hydrological Models* (ed. by A. J. Dolman, A. J. Hall, M. L. Kavvas, T. Oki & J. W. Pomeroy), 65–71. IAHS Publ. 270. IAHS Press, Wallingford, UK.

Nijssen, B., Bowling, L. C., Lettenmaier, D. P., Clark, D. B., Maayar, M. E., Essery, R., Goers, S., Gusev, Ye. M., Habets, F., van den Hurk, B., Jin, J., Kahan, D., Lohmann, D., Ma, X., Mahanama, S., Mocko, D., Nasonova, O., Niu, G. -Y., Samuelsson, P., Shmakin, A. B., Takata, K., Verseghy, D., Viterbo, P., Xia, Y., Xue, Y. & Yang, Z.-L. (2003) Simulation of of high latitude hydrological processes in the Torne-Kalix basin: PILPS Phase 2(e), 2. Comparison of model results with observations. *Global Planetary Change* **38**(1–2), 31–53.

Qu, W., Henderson-Sellers, A., Pitman, A. J., Chen, T. H., Abramopoulos, F., Boone, A., Chang, S., Chen, F., Dai, Y., Diskinson, R. E., Dumenil, L., Ek, M., Gedney, N., Gusev, Y. M., Kim, J., Koster, R., Kowalczyk, E. A., Lean, J., Lettenmaier, D., Liang, X., Mahfouf, J. -F., Mengelkamp, H. -T., Mitchell, K., Nasonova, O. N., Noilhan, J., Robock, A., Rosenzweig, C., Schaake, J., Schlosser, C. A., Schulz, J. -P., Shmakin, A. B., Verseghy, D. L, Wetzel, P., Wood, E. F., Yang, Z. -L. & Zeng Q. (1998) Sensitivity of latent heat flux from PILPS land-surface schemes to perturbations of surface air temperature. *J. Atmos. Sci.* **55**(11), 1909–1927.

Schlosser, C. A., Slater, A. G, Robock, A., Pitman, A. J., Vinnikov, K. Ya., Henderson-Sellers, A., Speranskaya, N. A., Mitchell, K., Boone, A., Braden, H., Chen, F., Cox, P., de Rosnay, P., Desborough, C. E., Dickinson, R. E., Dai, Y. -J., Duan, Q., Entin, J., Etchevers, P., Gedney, N., Gusev, Y. M., Habets, F., Kim, J., Koren, V., Kowalczyk, E., Nasonova, O. N., Noilhan, J., Schaake, J., Shmakin, A. B., Smirnova, T. G., Verseghy, D., Wetzel, P., Xue, Y. & Yang, Z. -L. (2000) Simulations of a boreal grassland hydrology at Valdai, Russia: PILPS Phase 2(d). *Mon. Weath. Rev.* **128**(2), 301–321.

Slater, A. G., Schlosser, C. A., Desborough, C. E., Pitman, A. J., Henderson-Sellers, A., Robock, A., Vinnikov, K. Ya., Mitchell, K., Boone, A., Braden, H., Chen, F., Cox, P. M., de Rosney, P., Dickinson, R. E., Dai, Y. -J., Duan, Q., Entin, J., Etchevers, P., Gedney, N., Gusev, Ye. M., Habets, F., Kim, J., Koren, V., Kowalczyk, E. A., Nasonova, O. N., Noilhan, J., Schaake, S, Shmakin, A. B., Smirnova, T. G., Verseghy, D., Wetzel, P., Xue, Y., Yang, Z. -L. & Zeng, Q. (2001) The representation of snow in land surface schemes: results from PILPS 2(d). *J. Hydrometeorol.* **2**, 7–25.

Wood, E. F., Lettenmaier, D. P., Liang, X., Lohmann, D., Boone, A., Chang, S., Chen, F., Dai, Y., Dickinson, R. E., Duan, Q., Ek, M., Gusev, Y. M., Habets, F., Irannejad, P., Koster, R., Mitchell, K. E., Nasonova, O. N., Noilhan, J., Schaake, J., Schlosser, A., Shao, Y., Shmakin, A. B., Verseghy, D., Warrach, K., Wetzel, P., Xue, Y., Yang, Z. -L. & Zeng, Q. -C. (1998) The project for intercomparison of land-surface parameterization schemes (PILPS) phase-2(c) Red-Arkansas River basin experiment: 1. Experiment description and summary intercomparisons. *Global Planetary Change* **19**(1–4), 115–135.

SWB

GENERAL INFORMATION

Model acronym: SWB
Model full name: Simple Water Balance model
Authors-first publication: Schaake, J. C., Koren, V. I., Duan, Q. Y., Mitchell, K. & Chen, F. (1996) Simple water balance model for estimating runoff at different spatial and temporal scales. *J. Geophys. Res.* **101**(D3), 7461–7475.
Original application domain: River and flood forecasting
Type: Lumped
Contact: Victor Koren
Hydrology Laboratory, NOAA/NWS Office of Hydrologic Development,
1325 East West Highway, Silver Spring, Maryland 20910, USA
Email: victor.koren@noaa.gov

MODEL DESCRIPTION

Brief model description SWB is a conceptual rainfall–runoff model that simulates runoff response from a watershed, given precipitation and potential evapotranspiration input data. The runoff is consequently converted into streamflow through a unit hydrograph. The model has two water storages, with a thin top layer and a thick lower layer. The top layer serves as an interception storage and only evapotranspiration occurs in the upper layer. Both runoff and evapotranspiration occur in the lower zone. There are five model parameters and two state variables.

Main hydrological processes Surface runoff when the precipitation exceeds infiltration capacity; baseflow from the lower water storages; evapotranspiration from all water storages.

Rainfall–runoff module Double storage/reservoir

Transfer function Unit hydrograph

Groundwater/percolation module Not included

Additional components No erosion and nutrients. Snow is represented in a separate model.

Model applications River and flood forecasting, water resources management, climate changes, etc.

DATA REQUIREMENTS; PARAMETERIZATION

Input data Precipitation and potential evapotranspiration

Overall model parameters DMAX, the max soil moisture deficit of bottom layer, in mm; KG, the potential subsurface flow, in mm day^{-1}; ALPSM, the bottom layer part which produces subsurface flow; ALPRT, the upper layer deficit proportion; KDT, the time scale factor.

Sensitivity analysis results The model has five parameters; all are included in the calibration.

Calibrated parameters Five parameters: DMAX, KG, ALPSM, ALPRT, KDT

Calibration procedure-algorithm SCE-UA

REFERENCES

Duan, Q., Schaake, J., Andreassian, V., Franks, S., Gupta, H. V., Gusev, Y. M., Habets, F., Hall, A., Hay, L., Huang, M., Leavesley, G., Liang, X., Nasonova, O. N., Noilhan, J., Oudin, L., Sorooshian, S. & Wood, E. F. (2005) Model parameter estimation experiment: overview and summary of the second and third workshop results. *J. Hydrol.* **320**(1–2), 3–17.

VIC

GENERAL INFORMATION

Model acronym: VIC
Model full name: Variable Infiltration Capacity Macroscale Hydrologic Model
Authors first publication: Liang *et al.* (1994) A Simple hydrologically based model of land surface water and energy fluxes for GSMs. *J. Geophys. Res.* **99**(D7), 14 415–14 428.
Original application domaine: Streamflow simulation and hydrological forecasting
Type: semi-distributed land hydrological model
Contact: Dennis P. Lettenmaier
Surface Water Hydrology Research Group, 202D Wilson Ceramic Laboratory, Box 352700, University of Washington, Seattle, Washington 98195-2700, USA
Email: dennisl@u.washington.edu
Web site: http://www.hydro.washington.edu/Lettenmaier/Models/VIC/VIChome.html

MODEL DESCRIPTION

Brief model description VIC is a macroscale hydrological model that solves full water and energy balances, originally developed by Xu Liang at the University of Washington, USA. VIC is a research model and in its various forms it has been applied to many watersheds including the Columbia River, the Ohio River, the Arkansas-Red Rivers, and the Upper Mississippi Rivers, river basins in China, as well as being applied globally.

Main hydrological processes Soil moisture calculation; surface and subsurface runoff; evaporation; river routing.

Rainfall/runoff module A variable infiltration capacity concept based on soil moisture.

Transfer function The drainage is driven by gravity.

Groundwater/percolation module subsurface runoff follows the Arno model conceptualization.

Additional components Snowmelt module

Model applications See Reference list

Schematic representation of model structure See Fig. 1.

DATA REQUIREMENTS; PARAMETERIZATION

Input data Time series of rainfall, air temperature.

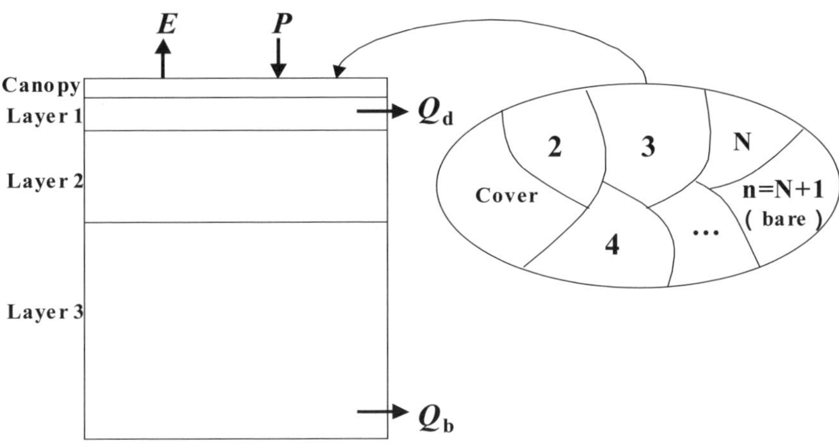

Fig. 1 Schematic representation of VIC model structure.

Overall model parameters

Parameter		Estimating method
Vegetation parameter	Architectural resistance Albedo Minimum stoma resistance Leaf area index Roughness length Zero-plane displacement	Literature
Soil parameter	Porosity Saturated soil potential Saturated hydraulic conductivity Exponent of unsaturated hydraulic conductivity curve Bulk density	Literature
	Exponent of variable infiltration capacity curve Three soil layer thicknesses Maximum velocity of baseflow Fraction of maximum baseflow Fraction of maximum soil moisture content of the third layer	Manual calibration

Sensitivity analysis results

Calibrated parameters Exponent of variable infiltration capacity curve; three soil layer thicknesses; maximum velocity of baseflow; fraction of maximum baseflow; fraction of maximum soil moisture content of the third layer.

Calibration procedure-algorithm During the calibration process, the infiltration parameter (b) and the depths of the three soil layers (d_1, d_2, d_3), which were treated as the primary calibration parameters, were changed to a uniform set of values in a given climate and large river basin zone. Calibrations were performed according to the following procedure: (a) set the estimated values for the depths of the three soil layers (d_1, d_2 and d_3), commonly with deeper depths for arid and semiarid regions and lower depths for humid regions; (b) calibrate the ARNO model parameters (D_m, D_s and W_s) so as to fit the low flow; (c) adjust the infiltration parameter b to match the observed flow peaks, with a higher value to increase the peak and a lower value to decrease the peak; (d) make a fine adjustment on these parameters to get the best simulation results. Consequently, after calibration the texture-based soil hydraulic parameters (k_s and θ_s) varied spatially, while b, d_i (i = 1, 2, 3), D_m, D_s and W_s were constant within each region. Generally, the thickness of the second soil layer was increased to allow for more storage. The calibrated infiltration parameter (b) tended to be smallest in the arid climates, in an effort to reduce runoff production.

REFERENCES

Abdulla, F. A., Lettenmaier, D. P., Wood, E. F. & Smith, J. A. (1996) Application of a macroscale hydrologic model to estimate the water balance of the Arkansas-Red river basin. *J. Geophys. Res.* **101**(D3), 7449–7459.

Bowling, L. C., Pomeroy J. W. & Lettenmaier, D. P. (2004) Parameterization of blowing snow sublimation in a macroscale hydrology model. *J. Hydromet.* **5**(5), 745–762.

Cherkauer, K. A. & Lettenmaier, D. P. (1999) Hydrologic effects of frozen soils in the upper Mississippi River basin. *J. Geophys. Res.* **104**(D16), 19 599–19 610.

Cherkauer, K. A., Bowling, L. C. & Lettenmaier, D. P. (2003) Variable infiltration capacity cold land process model updates. *Global Planetary Change* **38**, 151–159.

Hamlet, A. F. & Lettenmaier, D. P. (1999a) Effects of climate change on hydrology and water resources in the Columbia River basin. *Am. Water Res. Assoc.* **35**(6), 1597–1623.

Hamlet, A. F. & Lettenmaier, D. P. (1999b) Columbia River streamflow forecasting based on ENSO and PDO climate signals. *J. Water Resour. Plan. Manage, ASCE* **125**(6), 333–341.

Hamlet, A. F. & Lettenmaier, D. P. (2000) Long-range climate forecasting and its use for water management in the Pacific Northwest Region of North America. *J. Hydroinformatics* **02.3**, 163–182.

Leung, L. R., Hamlet, A. F. Lettenmaier, D. P. & Kumar, A. (1999) Simulations of the ENSO hydroclimate signals in the Pacific Northwest Columbia River Basin. *Bull. Am. Met. Soc.* **80**(11), 2313–2329.

Liang, X., Lettenmaier, D. P. Wood, E. F. & Burges, S. J. (1994) A Simple hydrologically based model of land surface water and energy fluxes for GSMs. *J. Geophys. Res.* **99**(D7), 14 415–14 428.

Liang, X. (1994) A two-layer variable infiltration capacity land surface representation for general circulation models, *Water Resour. Series*, TR140. University of Washington, Seattle, Washington, USA.

Liang, X., Lettenmaier, D. P. & Wood, E. F. (1996) One-dimensional statistical dynamic representation of subgrid spatial variability of precipitation in the two-layer variable infiltration capacity model. *J. Geophys. Res.* **101**(D16), 21 403–21 422.

Liang, X., Wood, E. F. & Lettenmaier, D. P. (1996) Surface soil moisture parameterization of the VIC-2L model: Evaluation and modifications. *Global Planetary Change* **13**, 195–206.

Liang, X., Wood, E. F. and Lettenmaier, D. P. (1999) Modeling ground heat flux in land surface parameterization schemes. *J. Geophys. Res.* **104**(D8), 9581–9600.

Liang, X. & Xie, Z. (2001) A new surface runoff parameterization with subgrid-scale soil heterogeneity for land surface models. *Adv. Water Resour.* **24**(9–10), 1173–1193.

Liang X. & Xie, Z. (2003) A new parameterization for surface and groundwater interactions and its impact on water budgets with the variable infiltration capacity (VIC) land surface model. *J. Geophys. Res.* **108**(D16), 8613, doi:10.1029/2002-JD003090.

Liang X. & Xie, Z. (2003) Important factors in land-atmosphere interactions: surface runoff generactions and interactions between surface and groundwater. *Global Planetary Change* **38**, 101–114.

Liang, X., Wood, E. F. Lohmann, D. Lettenmaier, D. P. *et al.* (1998) The Project for Intercomparison of Land-surface Parameterization Schemes (PILPS) Phase-2c Red-Arkansas River Basin Experiment: 2. Spatial and temporal analysis of energy fluxes. *Global Planetary Change* **19**, 137–159.

Lohmann, D. Raschke, E. Nijssen, B. & Lettenmaier, D. P. (1998) Regional scale hydrology:II. Application of the VIC-2L model to the Weser river, Germany. *Hydrol. Sci. J.* **43**(1), 143–158.

Mattheussen, B., Kirschbaum, R. L., Goodman, I. A., O'Donnell, G. M. & Lettenmaier, D. P. (2000) Effects of land cover change on streamflow in the interior Columbia river basin (USA and Canada). *Hydrol. Processes* **14**, 867–885.

Maurer, E. P., O'Donnell, G. M., Lettenmaier, D. P. & Roads, J. O. (2001) Evaluation of the land surface water budget in NCEP/NCAR and NCEP/DOE Reanalyses using an off-line hydrologic model. *J. Geophys. Res.* **106**(D16), 17 841–17 862.

Maurer, E. P., O'Donnell, G. M., Lettenmaier, D. P. & Roads, J. O. (2001) Evaluation of NCEP/NCAR reanalysis water and energy budgets using macroscale hydrologic simulations In: *Land Surface Hydrology, Meteorology, and Climate: Observations and Modeling* (ed. by V. Lakshmi, J. Albertson & J. Schaake), 137–158. AGU series in Water Science and Applications.

Nijssen, B. N., Lettenmaier, D. P., Liang, X., Wetzel, S. W. & Wood, E. F. (1997) Streamflow simulation for continental-scale river basins. *Water Resour. Res.* **33**(4), 711–724.

Nijssen, B. N., O'Donnell, G. M., Lettenmaier, D. P. & Wood, E. F. (2001) Predicting the discharge of global rivers. *J. Clim.* **14**, 3307–3323.

Nijssen, B. N., Schnur, R. & Lettenmaier, D. P. (2001) Global retrospective estimation of soil moisture using the VIC land surface model, 1980–1993. *J. Clim.* **14**, 1790–1808.

Su, F. & Xie, Z. (2003) A model for assessing effects of climate change on runoff in China. *Prog. Natural Progress* **13**(9), 701–707.

Xie, Z., Su, F., Liang, X., Zeng, Q. *et al* (2003) Applications of a surface runoff model with Horton and Dunne runoff for VIC. *Adv. Atmos. Sci.* **20**(2), 165–172.

Xie, Z., Liu, Q. & Su, F. (2004) An application of the VIC-3L land surface model with the new surface runoff model in simulating streamflow for the Yellow River basin. In: *GIS and Remote Sensing in Hydrology, Water Resources and Environment* (ed. by Y. Chen, K. Takara, I. D. Cluckie & F. H. De Smedt), 241–248. IAHS Publ. 289. IAHS Press, Wallingford, UK.

Xie, Z., Liu, Q., Yuan, F. & Yang, Hongwei (2004) Macro-scale land hydrological model based on 50 km × 50 km grid system. *J. Hydraul. Engng* **4**, 94–101 (in Chinese).

Xie, Z., Yuan, F., Duan, Q., Zheng, J., Liang, M. & Chen., F. (2006) Regional parameter estimation of the VIC land surface model: methodology and application to river basins in China. *J. Hydrometeorology* (in press).

Yuan, F., Xie, Z., Liu, Q. & Xia, J. (2005) Simulating hydrologic changes with climate change scenarios in the Haihe River Basin. *Pedosphere* **15**(5), 595–600.

Yuan, F., Xie, Z., Liu, Q., Yang, H. & Su, F. *et al* (2004) An application of the VIC-3L land surface model and remote sensing data in simulating streamflow for the Hanjiang River Basin. *Can. J. Remote Sensing* **30**(5), 680–690.

Wood, E. F., Lettenmaier, D., Liang, X., Nijssen B. & Wetzel, S. W. (1997) Hydrological modeling of continental-scale basins. *Ann. Rev. Earth Planet. Sci.* **25**, 279–300.

2 Review papers

Experience from applications of the HBV hydrological model from the perspective of prediction in ungauged basins

STEN BERGSTRÖM

The Swedish Meteorological and Hydrological Institute, SE-601 76 Norrköping, Sweden
sten.bergstrom@smhi.se

Abstract Since its introduction in the 1970s the HBV hydrological model has been applied in more than 50 countries and it is now a standard tool for hydrologists in the Nordic area. It has appeared in many shapes and for a great number of applications. Many of these are in basins with poor data coverage or even ungauged basins. The latter is, for example, the case in the nationwide hydrological mappings carried out in Finland, Norway and Sweden. It has also been used in many river basins where uncalibrated output from a hydrological model is the only realistic option due to shortage of data. These results are produced in a rather pragmatic way, with generalized parameter values, but they are much better than nothing. Ungauged simulations have opened new possibilities in hydrological modelling. In Sweden the total transport of nutrient to the Baltic Sea could be simulated and the step from limited basin studies of impacts of climate change to nationwide assessment was made possible. The paper reviews the history of the HBV model with the focus on discussion of prediction in ungauged basins.

Key words hydrological modelling; prediction in ungauged basins

INTRODUCTION

The first computer based hydrological models appeared in the late 1960s and since then numerous models have been developed for various purposes. Surprisingly good results were obtained by rather simple models. This simplicity was a necessity due to the limited data that were normally available. As the models had some physical background, but subroutines based on a lot of empiricism, they were often referred to as conceptual models. Examples of conceptual hydrological models from the 1970s are the Canadian UBC model (Quick & Pipes, 1976), the Danish NAM model (Nielsen & Hansen, 1973), the Japanese TANK model (Sugawara, 1979), the Swiss-American SRM model (Rango & Martinec, 1979) and two models from the USA: the SSARR model (US Army Corps of Engineers, 1976) and the NWSRFS (Anderson, 1973). Examples of more recent models are the British TOPMODEL (Beven & Kirkby, 1979) the Chinese Xinanjiang model (Zhao, 1992), the Danish MIKE-SHE (Refsgaard & Storm, 1995), the Italian ARNO model (Todini, 1996) and the American VIC model (Liang & Lettenmaier, 1994). An overview of some of the best known hydrological models can be found in the work by Singh (1995).

Most hydrological models need calibration to perform well. But the calibration option is not always there and today at least three situations can be identified with reference to the calibration and application of hydrological models:

(a) applications within the range of model calibration;
(b) applications outside the range of model calibration;
(c) applications without model calibration.

Model applications within the range of calibration are, for example, conventional runoff simulations which may be used to fill in gaps in runoff records or short and long term hydrological forecasting and water balance studies. While most applications within the range of model calibration are quite straightforward and can be done with a high degree of confidence, the use of conceptual models outside the range of calibration is more controversial. This may mean that the model is run on data which are more extreme than those during the period of calibration, e.g. floods, or that input data are from a climate that is not stable. Nevertheless, it is sometimes necessary, when no other reliable methods are at hand. Examples where models regularly are used this way are simulations of design floods for dam safety analysis and studies of the effects of climate change or changing land use on river runoff.

Maybe the most demanding type of application of a hydrological model is to use it without any calibration at all. This is referred to as prediction in ungauged basins, PUB. Although it is often stated that the conventional conceptual model is unsuitable for this role, due to its calibration needs, this is now an everyday type of application. It appears that the need for information about water resources is so strong that model limitations have to be accepted. The information provided by a conceptual hydrological model with a standard set of parameter values is accepted because it is better than no information at all. When the development of the HBV hydrological model started its was intended solely for runoff simulation within or, at least, not far outside its range of calibration. Today it is being used outside its range of calibration and for prediction in ungauged basins as well.

THE HISTORY OF THE HBV MODEL

The history of the HBV hydrological model dates back to 1972 (Bergström & Forsman, 1973). Its development was very much inspired by the strategy suggested by Nash & Sutcliffe (1970). Their work is usually referred to because of the introduction of a criterion of model performance, the R^2-value, but its most important message is a plea for a realistic attitude to model complexity. They realized the risk of overparameterization if the model structure was allowed to grow in complexity, although the term *parameter interaction* was more often used than *overparameterization*. A commonly used tool for analysis of parameter interaction was mapping of the error function, and it is here the R^2-value came into use. If the topography of the error function was not distinct enough the model simply had too many degrees of freedom. This technique, as illustrated in Fig. 1, was an important tool in the early development of the HBV model.

The strategy used for the development of the HBV model can thus be summarized as follows:

(a) The model must be based on a sound physical description but must not be so complex that it has a higher data demand than can be met by our standard climatological and hydrological networks.
(b) The number of free model parameters (used for calibration) shall be kept to a minimum.

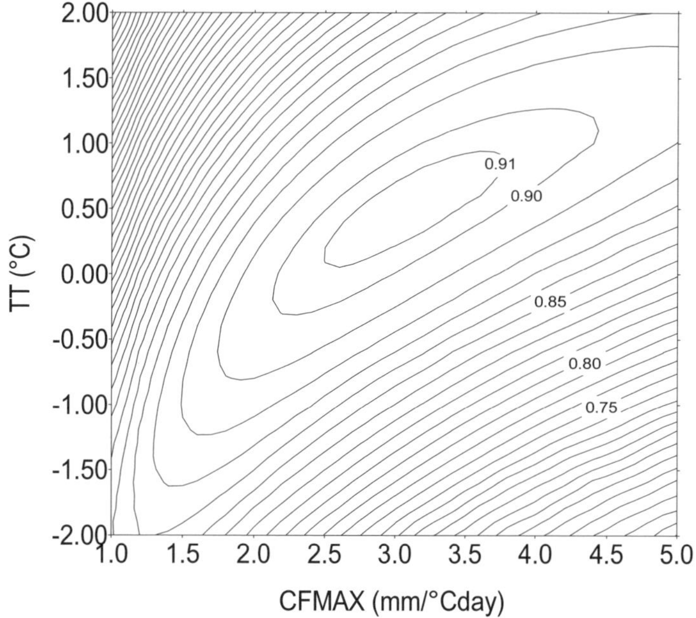

Fig. 1 Example of mapping of the topography of the error function of the two parameters, TT and CFMAX, in the snow routine of the HBV model. In this case the distinct maximum confirms that both parameters contribute to the model performance. The error function is R^2 as suggested by Nash & Sutcliffe (1970).

(c) The performance of the model must be controlled by observations from a period independent of the calibration process.
(d) It must be easy to understand and to use the model.

In the 1970s and 1980s a number of well controlled model intercomparisons were carried out. Of great importance for the development of the HBV model was the intercomparison organized by the World Meteorological Organization (WMO, 1986). From the conclusions of that project the following is worth citation: "*On the basis of available information, it was not possible to rank tested models or classes of models in order of performance. The complexity of the structure of the models could not be related to the quality of the simulation results.*" (WMO, 1986, page 35). These results meant a lot as they confirmed the experience gained during the early development of the HBV model. When trying to improve results by introducing more complexity, we soon reached a point where this increased complexity simply could not be justified by improved performance. When introducing too many parameters the calibration of the model became like manoeuvring a car with individuals steering each of its four wheels.

The first successful run with the HBV hydrological model was carried out in 1972 (Fig. 2, Bergström & Forsman, 1973,) and a routine for snow accumulation and melt was introduced in 1975 (Bergström, 1975). So far the model was lumped, even though area-elevation zoning was introduced in its snow routine. Since then it has been improved in small steps during its more than 30 years of existence, but its initial basic structure remains the same. The most robust part is probably its soil moisture accounting, which has proved very versatile in applications all over the world. Today

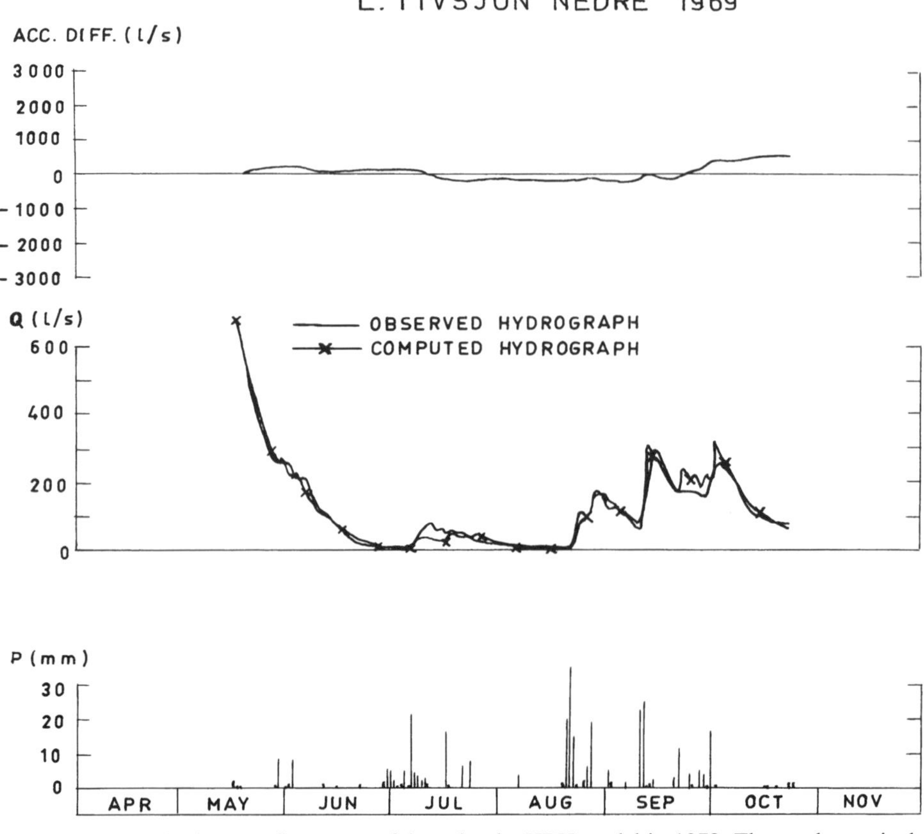

Fig. 2 The very first successful run by the HBV model in 1972. The catchment is the Lilla Tivsjön research basin in northern Sweden, with an area of 13 km^2 (from Bergström & Forsman, 1973).

the HBV model has developed from being a forecasting model to become a more general tool in many types of applications, whenever there is a need to transform meteorological observations into runoff. Even though the model has not undergone any dramatic changes, its scope of application has widened substantially over the years.

The recent HBV model is best characterized as a semi-distributed conceptual model (Lindström *et al.*, 1997). Like most hydrological models it consists of three main components: subroutines for snow accumulation and melt, subroutines for soil moisture accounting and response and river routing subroutines. It uses sub-basins as primary hydrological units and an area-elevation distribution and a crude classification of land use. The sub-basin option is used in geographically or climatologically heterogeneous basins or in the presence of large lakes. Depending on the choice of the modeller and model version the number of parameters to calibrate are normally 2–4 in the snow routine, 3 in the soil moisture routine and 4–5 in the response function.

Input variables to the HBV model are normally 24-hourly values of precipitation and air temperature and some estimate of potential evapotranspiration, which can either be daily or of lower resolution in time. A version of the model with hourly resolution in time is also available. The model has been applied to a wide range of scales, from small research basins less than 1 km^2 in size up to the continental scale representing of the whole land area of 1 729 000 km^2 draining into the Baltic Sea (Graham, 2004).

Applications of the HBV model

Over time the HBV model became very popular and it is now a standard tool for runoff simulations in Sweden, Norway and Finland. The number of applications outside the Nordic region is also growing and it has been applied in more than 50 countries all over the world. Some of these applications are made by modified model versions of the model developed in, for example, Norway, Finland, Germany and Switzerland.

The HBV model is nowadays applied within the range of model calibration, outside this range and even without calibration, in a PUB mode. This development has to a great degree been driven by demands. In Scandinavia hydrological models came into operational use for flood warning and forecasts of the inflow to the reservoirs of the hydropower systems in the mid 1970s. In 1990 new guidelines on design flood determination were adopted, which meant a new role for the HBV model, as it became part of the design procedure (Flödeskommittén, 1990; Bergström *et al.*, 1992; Norstedt *et al.*, 1992). This also meant that model results extrapolated outside the range of model calibration were trusted as the basis for design of Sweden's high hazard dams.

Starting from the 1990s the growing concern about the risks of climate change stimulated the use of hydrological models as tools for analysis of impacts on water resources. The HBV soon came into use for this purpose in the Nordic countries (Vehviläinen & Lohvansuu, 1991; Saelthun *et al.*, 1998; Bergström *et al.*, 2001). These are other examples of use of the HBV model outside its range of calibration. Still others were when the HBV model was used for analyses of the effects of deforestation based on empirical experience of its parameter values (Brandt *et al.*, 1988) and a study of the effects of drainage of wetlands, based on an improved version of the HBV model with a more realistic description of runoff generation, was carried out by Johansson & Seuna (1994).

From the 1980s and onwards the HBV model has also been developed to meet the needs of the environmental sector. Initially acidification was in focus (Bergström *et al.*, 1985b), but later nonpoint source pollution and transport of nutrients from land to sea became a major field of application (Arheimer & Brandt, 1998; Andersson *et al.*, 2003). This development was driven by a growing concern about eutrophication of the Baltic Sea and a need for analysis tools for counter measures. Later on the needs of the environmental sector were to become a major motivation for starting to use HBV without calibration.

Ungauged basins

Prediction in ungauged basins (PUB) is the ultimate goal of many hydrological modellers. But it has long been argued that a simple conceptual model like HBV would lack potential in this respect, due to problems in the identification of its empirical coefficients (parameters). Nevertheless, this was a type of application that started already in the early 1990s in Sweden (Johansson, 1992). The reason was obvious. There was a need for assessments of river flow at an arbitrary point and it is not realistic to believe that we can meet that demand by means of measurements. There was also a need for retrospective simulations in cases where water chemistry had been observed, but simultaneous runoff records were lacking. The development towards prediction in ungauged basins was thus mainly driven by the environmental sector, which needed runoff data to estimate the transport of nutrients based on observations of their concentrations.

It was realized that there were basically three possible ways of estimating runoff in rivers without data. The first was to simply use information from neighbouring rivers through statistical methods. The second option was to get so much experience with a conceptual model that we can map the optimum values of its parameters, or relate them to catchment characteristics. The third was to use a model that is so physically correct that it does not need calibration at all. The first option had been used for a long time at the Swedish Meteorological and Hydrological Institute and the use of a conceptual hydrological model was considered to be a realistic supplement to that method. It was, however, not considered realistic to introduce the physically correct modelling option, simply because there was no such model available. And anyhow, such a model would not be feasible due to the lack of proper and unbiased input data.

A first nationwide mapping of applications of the HBV model did not show a discouraging variability of its parameter values (Bergström, 1990) and subsequent evaluations of blind tests also looked promising (Johansson, 1992). One additional finding, which helped stabilize the variations in parameter values, was the break-out of major lakes in the model structure. This was shown to give more stable recession coefficients (Bergström *et al.*, 1985a). The introduction of a method for regional model calibration was also a great help (see, for example, Lindström *et al.*, 2005). This generated regional standard sets of parameter values, which could easily be used for basins in between the gauged ones. The key to success in ungauged prediction with standard parameter values is to keep the number of such parameters as low as possible. The new release of the HBV-96 model helped a bit further along that route as it reduced the complexity of the response function of the model by one empirical parameter (Lindström *et al.*, 1997). Finally it was found that the success in ungauged prediction is strongly dependent on a consistent database of meteorological input to the hydrological model. Such a database was developed for Sweden by Johansson (2002).

Attempts to relate model parameters to catchment characteristics were also carried out. The idea was that clear relationships would help the assessing of parameter values from the physiographic setting in the basin. The studies did, however, show that most optimum model parameters of this conceptual model were relatively insensitive to catchments characteristics. The most significant signal came from lakes and vegetation cover (Johansson, 1994). The difficulty of relating the empirical parameter values of the HBV model to physiographic conditions was later confirmed by Arheimer (2005) in a European evaluation of the model's ability to predict both water quality and quantity in ungauged basins.

So in the end the pragmatic solution to the problem of PUB turned out to be to rely upon empirical experience from the vast number of HBV applications and to supplement these with regional model calibrations. This made it possible to carry out nationwide application with complete areal coverage and thus a new era in hydrological modelling could start. HBV model-based national hydrological maps are now available for Sweden. They are based on model simulations in some 1000 basins, after calibration of the model to runoff records available at almost 400 sites (Figs 3 and 4). Similar work has been carried out in Norway and Finland. The database behind these maps has opened the possibility to assess the total load of nutrients from Sweden to the Baltic Sea and to carry out nationwide vulnerability studies of the impacts of climate change on water resources among many other applications (Andréasson *et al.*, 2004).

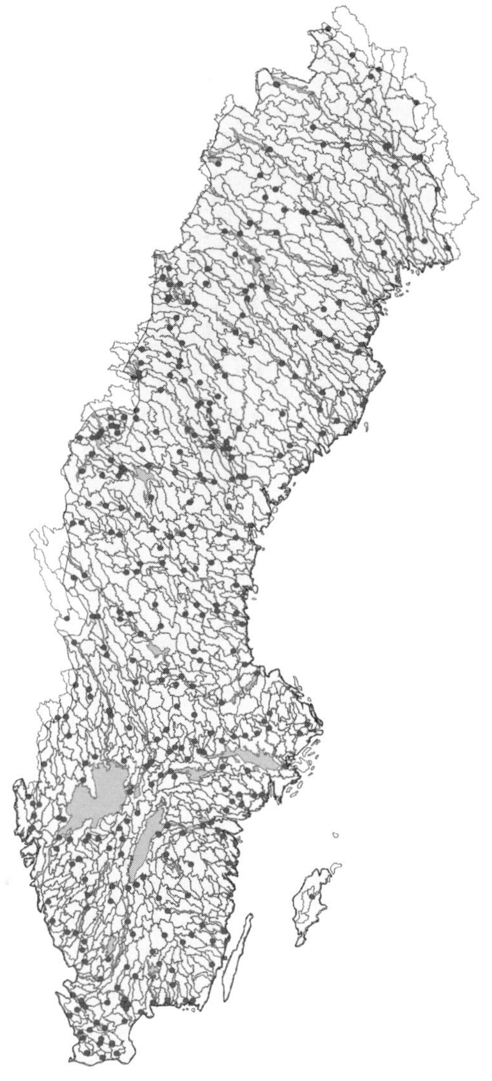

Fig. 3 The runoff stations used (dots) for calibration of the HBV model and generalization of model parameters, and basins used for the ungauged simulations for production of nationwide runoff maps. The area of Sweden is around 450 000 km^2.

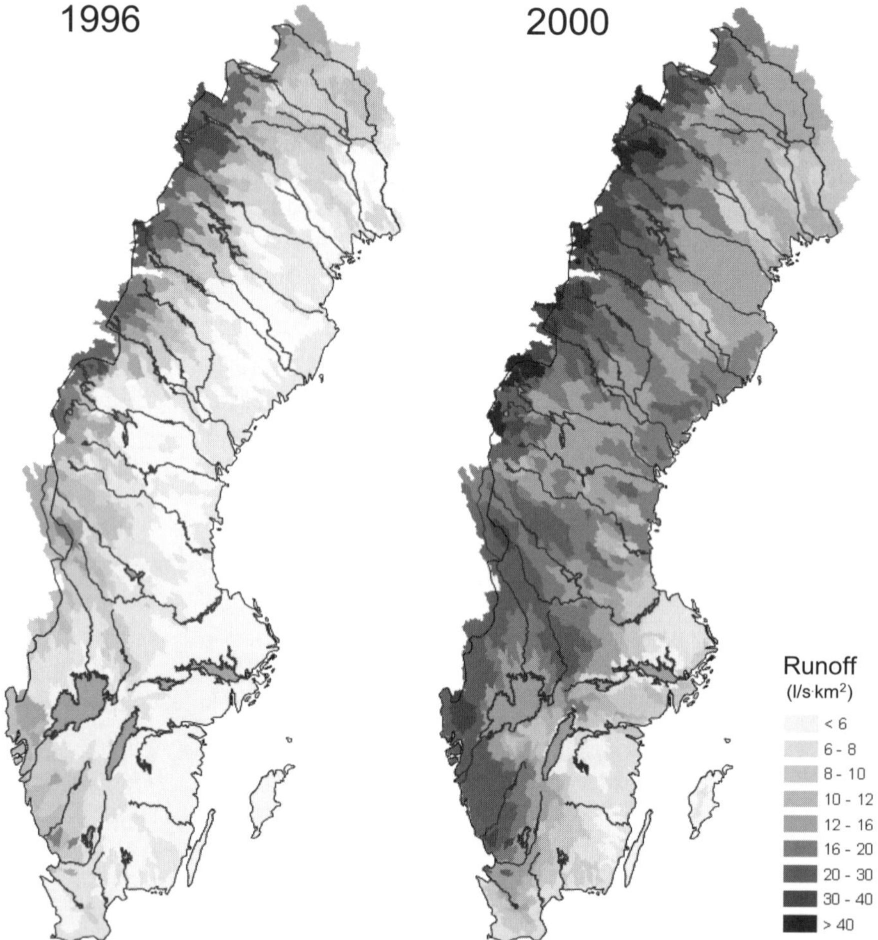

Fig. 4 These runoff maps of Sweden for a dry (1996) and a wet (2000) year are examples of products that can easily be generated by a database consisting of uncalibrated model runs with generalized parameter values in some 1000 basins.

DISCUSSION

Conceptual hydrological models have proved to be very cost-efficient tools for many water resources applications. Their best merits are that they are cheap compared to measurements and that they help us combine hydrometeorological information in a clever way. Thus we can make optimal use of regionally scattered runoff data and climatological records. Sometimes models are the only realistic way of obtaining runoff data. The main limitations are related to the need for model calibration and stationarity of the model in a non-stationary world.

The need for runoff data may sometimes force us to run the hydrological models without a complete calibration of their parameters. This can be done if a robust model structure with few empirical parameters is used. Values for these parameters can be set by calibration of the model to rivers in the neighbourhood or to rivers of similar character elsewhere. But we have to be aware that this is a compromise which yields data of lower quality than ordinary measurements. One important limitation is that the models need input data. Therefore the quality of the ungauged prediction has to be judged in the light of the quality of the data used for running the hydrological model.

The large number of applications of the HBV model in the world have proved that this relatively simple conceptual model is surprisingly general. The key processes seem to be described in a reasonable way. The model has the advantage of limited demand on data coverage and computer facilities, which makes it very versatile. However, generalization of optimum values of parameter values of a model like HBV is not trivial. Differences in climatological conditions in a basin are sometimes accounted for by the set up of the model structure in sub-models. These sub-models may have different parameter values. There also exist different versions of the HBV model and the different groups who apply it may have developed their own calibration practise. Further, the parameter values depend on the type of input data used, computational details in the model version and, to some degree, on the calibration skill of the user. All these factors have to be considered when trying to chose parameter values for the ungauged simulation. Thus the safest thing to do is to carry out a new regional calibration of a number of basins in the surrounding area of the point of interest.

CONCLUSIONS

The HBV model has found a multitude of applications that were hardly anticipated when the model was developed in 1972. Prediction of runoff in ungauged basins is probably one of the most demanding of these. Experience shows that a conceptual model of this type does have a potential for predictions under ungauged conditions, even though it is conceptual and requires calibration to obtain simulations of the highest quality. With *a priori* knowledge from generalized standard values of the parameter values it is possible to obtain added value by blind simulations. The proof is the widespread use of the HBV model under ungauged conditions today. However, it must not be forgotten that model simulations in the normal case are inferior to observations and that output from calibrated models is almost always more reliable than results from uncalibrated models. We also have to realize that the use of hydrological models without calibration means that we have to rely almost entirely upon information from input data. This requires a critical assessment of the climate database.

The application of the HBV model under ungauged conditions would have been improved further with a robust relationship between the physiographic conditions in the basin and the optimum values of the model's parameters. This is not easily obtained. The values of the parameters are affected by many other factors, like compensation for systematic errors in observations and compensating errors that most likely mask the possible signals from catchment characteristics.

The use of the HBV model with generalized model parameters has grown unexpectedly fast. In this process a database of model parameters developed from regional model calibration has been very important. Equally important is the access to a consistent database of meteorological data used as input to the model. These two factors have opened new possibilities for nationwide hydrological mapping and for many other aspects of large scale hydrological modelling and climate impact studies.

Acknowledgements The author is grateful to all friends and colleagues, in Sweden and abroad, who have participated in the long process of model development and application. Thanks are also due to all those who have funded our research, development and applications.

REFERENCES

Anderson, E. A. (1973) National Weather Service River Forecast System—Snow accumulation and ablation model. NOAA Technical Memorandum NWS HYDRO-17, US Dept of Commerce, Silver Spring, Maryland, USA.

Andersson, L., Arheimer, B., Larsson, M., Olsson, J., Pers, B.C., Rosberg, J., Tonderski, K., Ulén, B. (2003) HBV-P: a catchment model for phosphorus transport. In: *Proc. Quantifying the Agricultural Contribution to Eutrophication* (COST 832 Final Meeting, 31 July–2 August, Cambridge, UK), 59–60.

Andréasson, J., Bergström, S., Carlsson, B., Graham, L. P. & Lindström, G. (2004) Hydrological change—climate change impact simulations for Sweden. *Ambio* **33**(4–5), 228–234.

Arheimer, B. (2005) Evaluation of water quantity and quality modelling in ungauged European basins. In: *Predictions in Ungauged Basins: Promises and Progress* (ed. by M. Sivapalan, T. Wagener, S. Uhlenbrook, E. Zehe, V. Lakshmi, X. Liang, Y. Tachikawa & P. Kumar) (Proc. Symp. S7 held during the Seventh IAHS Scientific Assembly at Foz do Iguaçu, Brazil, April 2005), 99–107. IAHS Publ. 303. IAHS Press, Wallingford, UK.

Arheimer, B. & Brandt, M. (1998) Modelling nitrogen transport and retention in the catchments of southern Sweden. *Ambio* **27**(6), 471–480.

Bergström, S. (1975) The development of a snow routine for the HBV-2 model. *Nordic Hydrol.* **6**, 73–92.

Bergström, S. (1990) Parametervärden för HBV-modellen i Sverige. Erfarenheter från modellkalibreringar under perioden 1975–1989 (Parameter values for the HBV model in Sweden. Experience from calibrations during the period 1975–1989). SMHI, Rapport Hydrologi nr 28 (in Swedish).

Bergström, S. & Forsman, A. (1973) Development of a conceptual deterministic rainfall-runoff model. *Nordic Hydrol.* **4**, 147–170.

Bergström, S., Brandt, M. & Carlsson, B. (1985a) Hydrologisk och hydrokemisk modellberäkning i sjörika skogsområden (Hydrological and hydrochemical simulation in basins dominated by forests and lakes). *Vatten* **41**(3), 164–171 (in Swedish).

Bergström, S., Carlsson, B., Sandberg, G. & Maxe, L. (1985b) Integrated modelling of runoff, alkalinity and pH on a daily basis. *Nordic Hydrol.* **16**, 89–104.

Bergström, S., Harlin, J. & Lindström, G. (1992) Spillway design floods in Sweden. I. New guidelines. *Hydrol. Sci. J.* **37**(5), 10/1992, 505–519.

Bergström, S., Carlsson, B., Gardelin, M., Lindström, G., Pettersson, A. & Rummukainen, M. (2001) Climate change impacts on runoff in Sweden—assessments by global climate models, dynamical downscaling and hydrological modelling. *Climate Res.* **16**, 101–112.

Beven, K. J. & Kirkby, M. J. (1979) A physically-based variable contributing area model of basin hydrology. *Hydrol. Sci. Bull.* **24**(1), 43–69.

Brandt, M., Bergström, S. & Gardelin, M. (1988) Modelling the effects of clearcutting on runoff—examples from central Sweden. *Ambio* **17**(5), 307–313.

Flödeskommittén (1990) Slutrapport från Flödeskommittén. (Final report from the Swedish committee on spillway design). Swedish State Power Board, Swedish Power Association and SMHI, Norrköping (in Swedish).

Graham, L. P. (2004) Climate change effects on river flow to the Baltic Sea. *Ambio* **33**(4–5), 235–241.

Johansson, B. (1992) Vattenföringsberäkningar i recipientkontrollpunkter—en utvärdering av PULS-modellen (Runoff calculations in ungauged catchments—An evaluation of the Pulse model). *Vatten* **48**(2), 111–116.

Johansson, B. (1994) The relationship between catchment characteristics and the parameters of a conceptual runoff model. A study in the south of Sweden. In: *FRIEND: Flow Regimes from International Experimental and Network Data* (ed. by P. Seuna, A. Gustard, N. W. Arnell & G. A. Cole), 475–482. (Second Int. Conf., Braunschweig 11–15 October, 1993). IAHS Publ. 221. IAHS Press, Wallingford, UK.

Johansson, B. (2002) Estimation of areal precipitation for hydrological modelling in Sweden. PhD Thesis, Earth Sciences Centre, Göteborg University, Göteborg, Germany. Report no. A76.

Johansson, B. & Seuna, P. (1994) Modelling the effects of wetland drainage on high flows. *Aqua Fennica* **24**(1), 59–68.

Liang, X. & Lettenmaier, D.P. (1994) A simple hydrologically based model of land surface water and energy fluxes for general circulation models. *J. Geophy. Res.* **99**(D7), 14 415–14 428.

Lindström, G., Johansson, B., Persson, M., Gardelin, M. & Bergström, S. (1997) Development and test of the distributed HBV-96 model. *J. Hydrol.* **201**, 272–288.

Lindström, G., Rosberg, J. & Arheimer, B. (2005) Parameter precision in the HBV-NP model and impacts on nitrogen scenario simulations in the Rönneå River, southern Sweden. *Ambio* **34**(7), 533–537.

Nash, J. E. & Sutcliffe, J. V. (1970) River flow forecasting through conceptual models. Part I. A discussion of principles. *J. Hydrol.* **10**, 282–290.

Nielsen, S. A. & Hansen, E. (1973) Numerical simulation of the rainfall-runoff process on a daily basis. *Nordic Hydrol.* **4**, 171–190.

Norstedt, U., Brandesten, C.-O., Bergstrom, S., Harlin, J. & Lindströrn, G. (1992) Re-evaluation of hydrological dam safety in Sweden. *International Water Power and Dam Construction*, June 1992.

Quick, M. C. & Pipes, A. (1976) A combined snowmelt and rainfall runoff model. *Can. J. Civil Engng* **3**(3), 449–460.

Rango, A. & Martinec, J. (1979) Application of a snowmelt-runoff model using Landsat data. *Nordic Hydrol.* **10**, 225–238.

Refsgaard, J. C. & Storm, B. (1995) MIKE-SHE. In: *Computer Models of Watershed Hydrology* (ed. by V. P. Singh), 809–846. Water Resources Publications, Highland Ranch, Colorado, USA.

Saelthun, N. R., Aittoniemi, P., Bergström, S., Einarsson, K., Jóhannesson, T., Lindström, G., Ohlsson, P. -E. Thomsen, T., Vehviläinen, B. & Aamodt, K. O. (1998) Climate change impacts on runoff and hydropower in the Nordic countries. Final report from the project "Climate Change and Energy Production". *Tema Nord* **1998**, 552, Oslo, Norway.

Singh, V. J. (1995) (ed.) *Computer Models of Watershed Hydrology*. Department of Civil and Environmental Engineering, Louisiana State University, Water Resources Publications, Highlands Ranch, Colorado, USA.

Sugawara, M. (1979) Automatic calibration of the TANK-model. *Hydrol. Sci. Bull.* **24**(3), 375–388.

Todini, E. (1996) The ARNO rainfall-runoff model. *J. Hydrol.* **175**, 339–382.

US Army Corps of Engineers (1976) Development and application of the SSARR model. *Summaries of technical reports, North Pacific Division, Portland, Oregon, USA.*

Vehviläinen, B. & Lohvansuu, J. (1991) The effects of climate change on discharges and snow cover in Finland. *Hydrol. Sci. J.* **36**, 2, 4, 109–121.

WMO (1986) Intercomparison of models of snowmelt runoff. *Operational Hydrology Report no. 23, WMO-No. 646.* WMO, Geneva, Switzerland.

Zhao, R. (1992) The Xinanjiang model applied in China. *J. Hydrol.* **135**, 371–381.

Has basin-scale modelling advanced beyond empiricism?

C. MICHEL[1], C. PERRIN[1], V. ANDRÉASSIAN[1], L. OUDIN[2] & T. MATHEVET[1]

1 *Cemagref, Hydrosystems and Bioprocesses Research Unit, Parc de Tourvoie, BP 44, 92163 Antony cedex, France*
claude.michel@cemagref.fr

2 *Université Pierre-et-Marie Curie, UMR SISYPHE, 4 place Jussieu, 75252 Paris cedex 05, France*

Abstract Today, due to the increasing availability of spatial representations of our environment, there seems to be a growing feeling that hydrological modelling can eventually produce efficient distributed physically-based rainfall–runoff (RR) models. However, the improvement brought by such models is still to be demonstrated. Several authors have already sounded the alarm bell on the development and application of these models, apparently to no avail. In this paper, we argue that hydrological modelling has not gone far beyond an empirical view of the way basins transform rainfall into streamflow at their outlet. We address simple questions relative to: (a) the unit physical object that should be represented by RR models; (b) the actual limits of lumped RR models; (c) the impact of the time step on the structure of RR models; (d) the necessity of *a priori* conceptualization in the design of RR model structures; and last (e) what defines a "good" model. We are convinced that, up to now, the empirical modelling approach has not been studied with the rigour that would help to discriminate between the numerous mathematical tools at the hydrologist's disposal to build model structures. Thus, the purpose of this paper is to recall that before considering complex models, a lot remains to be done in the area of simple lumped models.

Key words catchments; empiricism; lumped modelling; rainfall–runoff modelling; time-step

INTRODUCTION

In this paper the only study variable that will be considered is the water flow observable through a given section of a stream. The prediction of streamflow requires modelling the transformation of precipitation that has fallen over the basin area upstream of the point of interest. Two opposed approaches are available to build such a model: the upward approach (also called bottom-up, reductionist or mechanistic) and the downward approach (also called top-down, empirical or non-reductionist). This distinction, which reflects a well-known debate of scientific methodology (see for example von Bertalanffy, 1968), was discussed at length by Klemeš (1983), who defines the downward approach as the route that "*starts with trying to find a distinct conceptual node directly at the level of interest (or higher) and then looks for the steps that could have led to it from a lower level*". Conversely, in the upward approach that dominates modern science, basin properties are considered to be a summation of the hillslope and stream channel properties, at all scales.

To assist hydrologists in their model-building enterprise, today there is an unprecedented amount of data and information on the catchment, supplied by various measuring devices ranging from ground networks to satellites, at various time and space scales, arranged into powerful databases and displayed by user-friendly geographical information systems (see e.g. Babovic, 2005). Although most of the hydrologically-relevant information remains unknown to us (especially subsurface basin characteristics), this wealth of information appeals like a mirage to those of the reductionist modellers who consider that every cube of earth can be physically described and modelled, at any resolution in time and space. This may be possible in the idealized conditions of laboratories where parameter values and boundaries conditions can be measured and recorded to allow the application of the well-established concepts of hydrodynamics (see Beven, 2001, for examples).

However, such conditions do not exist in the real world and these concepts are inapplicable to solving the problem of rainfall–runoff transformation at the catchment scale. Indeed, in spite of the apparent plethora of existing data, these requirements cannot be satisfied over natural catchments because of the impossibility of monitoring the huge and intricate three-dimensional (3-D) landscape involved in the rainfall–flow transformation. To cope with our ignorance of the key basin characteristics, reductionist modellers are forced to introduce hypotheses and subjective parameters that can be far from reality: their physically-based models then fall back to the level of simple conceptual or empirical models.

Despite the alarm bells sounded by authors such as Beven (1989), Bergström (1991), Jakeman & Hornberger (1993) and Young (1998), many hydrologists believe that reductionist modelling is the best way, i.e. the way that will yield significant progress in rainfall–runoff (RR) modelling (see e.g. Loague & VanderKwaak, 2004). Our view is that, even though some progress has been made over the past decades in the understanding of the hillslope processes, little advance has been made in catchment scale RR modelling (see e.g. Sivapalan, 2003). And, we think that the best way to advance our hydrological modelling, is through an *empirical lumped approach* (following a downward approach) very similar to the data-based modelling approach advocated by Young (2003).

In this paper, we discuss several of the generally accepted ideas in hydrological modelling and some of the fundamental questions that today seem insufficiently or inadequately dealt with by the hydrological community. Our paper is organized around five successive questions: we first discuss the unit physical object on which RR models should be based. Second, we examine the actual limitations of lumped models, in comparison to the distributed models. Third, we discuss issues of the time step in modelling. Fourth, we contest the widely accepted belief that conceptualization is a prerequisite to model building, and advocate data-based approaches. Fifth, we expose the desirable statistics on which we can base our judgement to consider a model as "good".

WHAT UNIT PHYSICAL OBJECT SHOULD RAINFALL–RUNOFF MODELS BE BASED ON?

Unfalsifiable distributed models

The object of the study of hydrology is the catchment: an extraordinarily complex, 3-D system (a catchment is even 4-D as it is not always stationary), subjected to a similarly complex climatic forcing, and producing visible (streamflow) and invisible (atmospheric and underground water fluxes) outputs. The constituent unit of a hydrological model depends on the modelling approach: the elementary unit will be a grid cell in a distributed model, a basin in a lumped model.

The problem we see with the use of grid cells as elementary units, is that the modelling of these components is necessarily hypothetical. Making each constituent unit work properly is a necessary *but not sufficient* condition to make the whole model mimic the actual basin functioning. The construction of the full model is not a simple matter of straightforward aggregation of all the constituent units (see e.g. Sivapalan, 2003). The laws that govern the interactions between these elementary units remain to be discovered. Indeed it is necessary to figure out how fluxes originating from one element are integrated by the adjoining elements and how the state variables of one block interact with those of the neighbouring blocks. Thus, there are numerous mingled hypotheses, those made about the constituent blocks and those made about the interactions between them. Only when the whole structure has been devised does it become possible to confront the single model output to reality, i.e. the observed discharge at the basin outlet.

The problem with physically-based models is that all constituent hypotheses *taken separately* are reasonable. Thus, there cannot be any strong refutation of the whole construction, and if a few tuning parameters are left to the model user, it will be easy to adapt the whole construction to achieve a good simulation of flows. A scientific approach would require that one questions the model, but the numerous and varied hypotheses cannot be challenged individually. We have thus obtained a model that yields reasonable results, but that is not questionable. Clearly, such a model is not falsifiable (i.e. it cannot be corroborated or refuted in the sense of hypothesis testing) whereas it should be (see e.g. Oreskes *et al.*, 1994; Refsgaard & Henriksen, 2004). It is all the more unfalsifiable that it is generally applied on a single basin with no consideration for possible falsification on other basins.

Model complexity

In addition to these problems, the aggregation of grid cells systematically raises the issue of model complexity. As noted by Sivapalan (2003), the complexity of the resulting model can be higher than the complexity of individual units, and this will result in problems of over parameterization. However, in the bottom-up approach, today there is no guideline to rationally reduce this complexity during the aggregation process. This is probably partly due to the "frustration" that such simplifications may cause, as noticed by Bergström (1991); "*going from complex to simpler model structures requires an open mind, because it is frustrating to have to abandon seemingly elegant concepts and theories*".

This upward approach for model construction therefore seems unable to design parsimonious models. However, many results indicate that a limited number of free parameters (less than six) are sufficient to simulate the RR relationship (see e.g. Jakeman & Hornberger 1993; Perrin *et al.*, 2001; Young, 2003). Simple models are often as good as complex models in simulating the RR relationship. As mentioned by

Martin (1996), "*the prediction obtained with a complex model often points to a simpler model which could have been used in the first place*".

Going back to the basin

Our overriding feeling is that there is a need to go back to the basics and to study as a whole the object responsible for the observed flow, i.e. the basin. During the past 20 years, hydrologists have considered that the necessary basis for any modelling effort was to look at the basin as a sum of grid cells (or sub-basins). We believe they have overlooked the actual challenges presented to our science. Before partitioning the basin into smaller areas, it seems advisable to look at the entire object itself and to try and get a general idea of its functioning. To put it another way, the first action to be taken is to improve lumped models. Such models are simple, limit the problems related to undue complexity and can be easily tested and falsified.

CAN WE ACHIEVE FURTHER PROGRESS WITH LUMPED MODELS?

Why do we need lumped models?

Lumped modelling searches for the mechanisms relevant *at the basin scale* in order to describe the *overall hydrological behaviour* of a catchment. As an individual concerned with his health will first consult a physician who will handle him as a person (and not as a sum of cells), we believe that any hydrological study should start with an investigation of the basin as a single functional entity. The empirical lumped approach recognizes the impossibility of monitoring the great complexity of the 3-D geographical domain involved in the rainfall–flow transformation. Thus, it chooses to give a central part to the data in the design of a model structure, which will reveal itself by successive hypothesis testing, avoiding constraints in the model development by employing (often wrong) preconceived ideas.

In a lumped context, it seems obvious that the model cannot be other than an abstract representation of reality. The links between the parameters of this abstract structure with the physical reality cannot be set explicitly and *a priori*: they remain implicit, to be discovered *a posteriori*. The reason is that phenomena observable at the plot scale are no longer relevant when applied at the basin scale in a lumped mode, in the same way that the mechanisms of neuronal functioning are not directly relevant to explain human behaviour at the individual's scale. One has to think of entirely new tools suitable for basin-scale modelling, raising basic questions such as:

(1) Is a subtraction of rainfall input (infiltration model) more relevant than a multiplication by a number less than one (runoff coefficient)?
(2) Is it necessary to split the outflows into more than two components?
(3) Is the bucket concept with an upper bound to its effective inputs an efficient way to model the RR transformation at the basin scale?

Naturally, these simple questions may seem disturbing, as distributed modelling has pre-designed answers to them, but lumped modelling has none. We believe that this is the reason why lumped modelling is all the more suitable for scientific investigation: it raises questions, leaves room for doubting, and allows easy tests of hypotheses.

Lumped models as scientific investigation tools

To judge the suitability of lumped models, it can be interesting to look at them from an uncertainty analysis point of view, which is indeed an important aspect of hydrological modelling. Up-to-date techniques such as Kalman filtering are currently used to track model states and parameters. Which models lend themselves to these uncertainty analyses? Simple lumped models! For instance, in their demonstration of the use of sequential data assimilation, Moradkhani *et al.* (2005) resort to the HyMOD model, a simple five-parameter model composed of the probability distributed moisture model (Moore, 1985) and of linear reservoir routing. Instead of disappearing from the scientific literature, these simple models are employed more often than before (see e.g. Sivapalan *et al.*, 2003). The reason could be that lumped conceptual models are subject to more effective scientific scrutiny than distributed models whose scientific foundation is taken for granted. However, a lot of work remains to be done to discriminate among the numerous models published in the hydrological literature. Lumped modelling is definitely not a thing of the past and we think it can be the cornerstone on which future hydrology will rise.

DOES THE TIME STEP OF A MODEL MATTER?

Surprisingly, the time step of model inputs and outputs is often considered as incidental information. Very few hydrologists (see e.g. Jothityangkoon *et al.*, 2001; Eder *et al.*, 2003; Mouelhi, 2003) have questioned the influence of the time step on model structure. Modellers generally present their models as if the time step of its functioning was just a matter of data availability, or an exclusively computational question. In fact, the influence of the time step on model structure is far more important than the influence of spatial discretization. If a model works satisfactorily in a lumped mode, it will similarly do so when used in a semi-distributed mode. In contrast, a model successful at a monthly time step may yield poor performances when it is run at a daily time step. Each time scale should have its proper RR modelling characteristics. Generally, the longer the time step, the less complex the model structure and the smaller the number of free parameters. Therefore, it is far simpler to develop an annual model than a daily model. However, the link between model structures at different time steps is not straightforward (Jothityangkoon *et al.*, 2001; Mouelhi, 2003).

IS PROCESS DESCRIPTION THE GOAL?

"Processes" at the basin scale

If hydrologists tend to relinquish the strict physically-based approach for model development, they generally stick to the necessity of having a clear view of the actual processes taking place in a basin (a predefined "perceptual" model). They fear that if a model does not explicitly take into account the well-known small-scale processes, this model will lose its credibility, and if it still happens to work, it will be so for the "wrong reasons". This is quite a weird idea, and this argumentation does look circular. Generally, modelling hypotheses are justified by the fact that the model based on them achieves better results than another model that rejects them. Processes are very complicated, intertwined and highly variable over the basin (see e.g. Joerin *et al.*, 2005). To the basin-scale modeller, the important thing is the end result of this small-scale complexity, i.e. the discharge measured at the gauging station. The problem to be solved is to find the combination of mathematical equations that is the most successful in predicting the discharges.

Flows and other hydrological variables

Once a model has been developed based on a conceptualization approach, the whole structure is often taken for granted, i.e. every part of it is believed to have its counterpart in the real basin (see e.g. Günter *et al.*, 1999). As far as lumped empirical modelling is concerned, such a consideration has no sound foundation because the only concern of lumped modelling is to produce discharges that are close to the observed discharges. Given lumped models generally have a single target variable (the discharge), there is no point at all in trying to find *a posteriori* links between model states and other variables observed on the basin. As pointed out by Beven (2001), it is misguided to give more consideration to searching for "explanatory depth" (i.e. internal physical realism) in the model than to searching for improved "predictive power" (i.e. the capacity to satisfactorily simulate the target variable).

As an afterthought, some modellers have tried to include secondary modelling objectives, i.e. secondary outputs (e.g. soil moisture, piezometric level, etc.), in order to constrain those parameters that appeared too poorly identified (see e.g. Lamb *et al.*, 1998). We believe that if two or more outputs are to be predicted, the whole model development has to be started again. This implies the definition of new objective functions and addressing difficult issues such as the guidelines for dealing with Pareto optima. It is an altogether new problem that requires a lot more data than are generally available to hydrologists.

The reason for distrusting the internal functioning of a lumped model is that aggregation of processes is unavoidable. In a parsimonious model developed following the Okham's razor rule (i.e. discarding model components not leading to a clear advantage in terms of performance), processes elicited by model development can overlap many elementary physical processes singled out in field studies. A lumped model should not and cannot be a compendium of all the processes that can be observed in the real world. It is rather an abstract structure that is only justified by its ability to simulate satisfactorily streamflow at the outlet of any basin, provided that its parameters are adjusted to suitable values. For these reasons, we consider that the prediction of discharges at the basin outlet is the primary objective of RR models and, unfortunately, there is still a lot of work to efficiently simulate these discharges.

"GOOD" OR "BETTER" MODELS?

Limits in testing a single model

It is often said that when the Nash & Sutcliffe (1970) criterion is above 0.8, the model fit can be considered as good. It is easy to show the limitations of such a statement. That a model yields a Nash-Sutcliffe criterion equal to 0.8 on a given basin, for a given period, is in no way a validation: it is rather a judgement on the basin and on the period. If it happens that this period has been particularly wet, there is a high probability of obtaining poorer results in a drier period. If the same model were to be applied to other basins, even within the same region, there is an even higher probability of obtaining a range of Nash-Sutcliffe criteria spanning from low to high values. Therefore, model performance on a single basin and a single period provides little information on the actual model value.

Also note that, if the Nash-Sutcliffe criterion is widely used as a derivative of the root mean square error (RMSE), it nevertheless does not constitute an absolute reference. The model user may prefer other criteria such as the mean absolute error calculated on discharges or on the logarithms of discharges and so on.

Sometimes, a model is considered as "good" when the RMSE obtained in simulation ("validation") on a period different from the calibration period is not clearly larger than the RMSE obtained in calibration. But actually both RMSEs are generally quite close. In that sense, should all models be considered as satisfactory? Or is it better to acknowledge that there cannot be any absolute measure of the quality of a model and, consequently, there is no possibility to consider a model as "good".

Model comparison as objective assessment

Many years ago, one of the founding fathers of hydrological modelling, Ray Linsley (1982), pointed out that "*almost any model with sufficient free parameters can yield good results when applied to a short sample from a single basin, [so that] effective testing requires that models be tried on many basins of widely differing characteristics, and that each trial cover a period of many years*". We concur with this statement: first, we think that a model can be properly assessed only if it is applied on a large sample of basins representing a wide range of hydro-climatic conditions (see Mathevet *et al.*, 2006). This gives clear ideas on its robustness and generality. Second, we think that a model cannot be considered as *good* but only as *better* than another model. Therefore, the model should be assessed comparatively to another model, as also advocated by Seibert (2001). If $NS_A(k)$ is the Nash-Sutcliffe criterion obtained by model A on basin k and if $NS_B(k)$ is the Nash-Sutcliffe criterion obtained by model B on the same basin, then a statistically founded opinion can be reached if a statistical test can demonstrate that there is a high probability that the mean of $NS_A(k)$, $k = 1, 2, \ldots, N$ be different from the mean of $NS_B(k)$, $k = 1, 2, \ldots, N$ over the whole sample of N basins. Only such a comparative framework on large samples of basins will give reliable conclusions on the actual merits of RR models.

DISCUSSION: WHAT IS THE STATE OF THE ART OF LUMPED RAINFALL–RUNOFF MODELLING?

Since a lumped model is unable to reproduce all processes taking place in a basin, its simulations are bound to be in error (actually, this statement applies to all kinds of models). It is rare to obtain a Nash-Sutcliffe criterion larger than 0.9. This means that the RSME is commonly greater than 0.32 times the standard deviation of observed discharges. Thus, there is plenty of room for progress.

It is unlikely that a unique structure can emerge as the best model, just because there is no physical reason for that. For the time being, there are a lot of competing structures that are approximately equivalent, although some are definitely inferior (e.g. Perrin *et al.*, 2001). However, the hydrological community has not yet recognized which models are in the upper class and which ones are to be forsaken. The reason is that, in spite of several efforts to compare model structures (e.g. WMO, 1975; Smith *et al.*, 2004), a lot of questions have yet to be answered. In particular, there is no indication as to which tools (unit hydrographs, linear and nonlinear stores, etc.) are the best suited to the loss and transfer sub-models. In order to represent the hydrological behaviour of a basin, it seems as if the choice between several unit hydrographs, different sorts of routing reservoirs, etc., is a matter of sheer convenience. It is probably not.

CONCLUSION

In this paper, we have tried to answer some simple questions on the present state of RR modelling. We have the overriding impression that lumped empirical modelling has so far not received the attention it deserves and requires. At the basin scale, it seems necessary to grow out of the analysis of small-scale processes that cannot be accounted for in a tractable way, and to explore the emergent properties of the catchment behaviour: this quest must necessarily start with simple lumped models. Although most of the constituent tools needed to represent the emerging properties of a drainage basin are already known, hydrologists have not yet identified which are the most promising ones and understood their roles in the whole model structure. We definitely think that at this stage of hydrological modelling, empirical research remains the most promising approach. Large databases and model comparisons should be key tools to support future progress.

REFERENCES

Babovic, V. (2005) Data mining in hydrology. *Hydrol. Processes* **19**(7), 1511–1515.

Bergström, S. (1991) Principles and confidence in hydrological modelling. *Nordic Hydrology*, **22**, 123–136.

Beven, K. J. (1989) Changing ideas in hydrology—the case of physically-based models. *J. Hydrol.* **105**, 157–172.

Beven, K. (2001) On explanatory depth and predictive power. *Hydrol. Processes* **15**(15), 3069–3072.

Eder, G., Sivapalan, M. & Nachtnebel, H. P. (2003) Modelling water balances in an Alpine catchment through exploitation of emergent properties over changing time scales. *Hydrol. Processes* **17**(11), 2125–2149.

Güntner, A., Uhlenbrook, S., Seibert, J. & Leibundgut, C. (1999) Multi-criterial validation of TOPMODEL in a mountainous catchment. *Hydrol. Processes* **13**(11), 1603–1620.

Jakeman, A. J. & Hornberger, G. M. (1993) How much complexity is warranted in a rainfall–runoff model? *Water Resour. Res.* **29**(8), 2637–2649.

Joerin, C., Beven, K. J., Musy, A. & Talamba, D. (2005) Study of hydrological processes by the combination of environmental tracing and hill slope measurements: application on the Haute-Mentue catchment. *Hydrol. Processes* **19**(16), 3127–3145, doi:10.1002/hyp.5836.

Jothityangkoon, C., Sivapalan, M. & Farmer, D. L. (2001) Process controls of water balance variability in a large semi-arid catchment: downward approach to hydrological model development. *J. Hydrol.* **254**(1–4), 174–198.

Klemeš, V. (1983) Conceptualization and scale in Hydrology. *J. Hydrol.* **65**, 1–23.

Lamb, R., Beven, K. and Myrabø, S. (1998) Use of spatially distributed water table observations to constrain uncertainty in a rainfall-runoff model. *Adv. Water Resour.* **22**(4), 305–317.

Linsley, R. K. (1982) Rainfall–runoff models—an overview. In: *Proc. Int. Symp. Rainfall–Runoff Modelling* (ed. by V. P. Singh), 3–22. Water Resources Publications, Littleton, Colorado, USA.

Loague, K. & VanderKwaak, J. E. (2004) Physics-based hydrologic response simulation: platinium bridge, 1958 Edsel, or useful tool. *Hydrol. Processes* **18**(15), 2949–2956.

Martin, P. H. (1996) Physics of stamp-collecting? Thoughts on ecosystem model design. *Sci. Total Environ.* **183**, 7–15.

Moore, R. J. (1985) The probability-distributed principle and runoff production at point and basin scales. *Hydrol. Sci. J.* **30**(2), 273–297.

Moradkhani, H., Hsu, K-L., Gupta, H. & Sorooshian, S. (2005) Uncertainty assessment of hydrological model states and parameters: Sequential data assimilation using the particle filter. *Water Resour. Res.* **41**, W05012, doi:10.1029/2004WR003604.

Mouelhi, S. (2003) Vers une chaîne cohérente de modèles pluie-débit conceptuels globaux aux pas de temps pluriannuel, annuel, mensuel et journalier. PhD Thesis, ENGREF, Cemagref Antony, France (in French).

Nash, J. E. & Sutcliffe, J. V. (1970) River flow forecasting through conceptual models. Part I—A discussion of principles. *J. Hydrol.* **27**(3), 282–290.

Oreskes, N., Shrader-Frechette, K. & Belitz, K. (1994) Verification, validation and confirmation of numerical models in the Earth Sciences. *Science* **263**, 644–646.

Perrin, C., Michel, C. & Andréassian, V. (2001) Does a large number of parameters enhance model performances? Comparative assessment of common catchment model structures on 429 catchments. *J. Hydrol.* **242**, 275–301.

Refsgaard, J. C. & Henriksen, H. J. (2004) Modelling guidelines—terminology and guiding principles. *Adv. Water Resour.* **27**(1), 71–82.

Seibert, J. (2001) On the need for benchmarks in hydrological modelling. *Hydrol. Processes* **15**(6), 1063–1064.

Sivapalan, M. (2003) Process complexity at hillslope scale, process simplicity at the watershed scale: is there a connection. *Hydrol. Processes* **17**(3), 1037–1041.

Sivapalan, M., Zhang, L., Vertessy, R. & Blöschl, G. (eds) (2003) Downward approach to hydrological prediction. *Hydrol. Processes* **17**(11), 2099–2326.

Smith, M. B., Seo, D. J., Koren, V. I., Reed, S. M., Zhang, Z., Duan, Q., Moreda, F. & Cong, S. (2004) The distributed model intercomparison project (DMIP): motivation and experiment design. *J. Hydrol.* **298**(1–4), 4–26.

Young, P. C. (1998) Data-based mechanistic modelling of environmental, ecological, economic and engineering systems. *Environ. Modelling Software* **13**, 105–102.

Young, P. (2003) Top-down and data-based mechanistic modelling of rainfall–flow dynamics at the catchment scale. *Hydrol. Processes* **17**(11), 2195–2217.

von Bertalanffy, L. (1968) *General Systems Theory*. Georges Braziller, New York, USA.

Regionalization methods in rainfall–runoff modelling using large catchment samples

RALF MERZ, GÜNTER BLÖSCHL & JURAJ PARAJKA

Institute for Hydraulic and Water Resources Engineering, Vienna University of Technology, Vienna, Austria

merz@hydro.tuwien.ac.at

Abstract In ungauged catchments, model parameters are usually transposed from gauged catchments as no runoff data are available for calibrating them. Parameters can be either transposed from nearby catchments or, alternatively, functional relationships between model parameters and available catchment attributes can be derived from gauged catchments that may be further away. This article summarizes the most important methods and recent findings from the literature with a focus on the relative performance of the regionalization methods. As performance is strongly influenced by the local conditions, general findings on the suitability of regionalization methods for a given climate region can only be obtained by analysing a large number of catchments at the same time. Even with large catchment samples, the differences between the regionalization methods tend to be small. Suggestions for improving current regionalization methods are given.

Key words rainfall–runoff modelling; regionalization; similarity; ungauged catchments

INTRODUCTION

Modelling the rainfall–runoff behaviour of ungauged catchments is of interest both for understanding systems behaviour and as a basis of sustainable water resources management. The main challenge with rainfall–runoff modelling in ungauged catchments is the lack of local runoff data to calibrate the model parameters. Parameter calibration is important because most model equations are empirical in nature, model output depends on the initial and boundary conditions that are poorly known and, probably most importantly, because most of the flow processes take place in the subsurface where media characteristics are heterogeneous and unknown. Soil properties can change dramatically in space but change very little in time, so parameter calibration can significantly enhance model performance.

In ungauged catchments, model parameters have to be estimated from other sources of information. An appealing way to estimate model parameters in ungauged catchments is to glean the model parameters from hydrologically similar catchments. The concept of hydrological similarity assumes that the runoff response to a given rainfall input in two different catchments will be similar if similar rainfall–runoff processes occur.

These processes are not known in full detail and thus different similarity concepts have been proposed in the literature. The process of transferring parameters from hydrologically similar catchments to a catchment of interest is generally referred to as regionalization (Blöschl & Sivapalan, 1995). The two most widely used concepts for regionalizing model parameters are hydrological similarity as a function of spatial proximity and similarity as a function of catchment attributes.

Regionalization based on spatial proximity

The rationale of this group of methods is that catchments that are close to each other will have a similar response as the climate and catchment conditions will only vary smoothly in space. The notion of spatial proximity is by no means trivial as it can be defined in a number of ways and the choice of method in any particular case is usually not obvious. An example of this group is the delineation of spatially contiguous regions with approximately homogeneous model parameters. The regions are found from an analysis of a number of gauged catchments and available hydrological information using statistical tools such as cluster analysis, principal component analysis and multiple regression (Nathan & McMahon, 1990). Hydrological information to assist in delineating homogeneous regions may consist of hydrogeological maps, climate maps, soil and vegetation maps and process indicators such as the seasonality of hydrological processes (Merz *et al.*, 1999). Yu & Yang (2000) propose to establish regional flow duration curves, which comprise a family of regression equations relating the flow of various exceedence percentages with catchment area. The regional flow duration curves can then be used to calibrate model parameters in ungauged catchments.

An alternative to homogeneous regions are geostatistical methods, such as kriging. The main strength of kriging is that it is a best linear unbiased estimator (BLUE); best meaning that the mean squared error is a minimum, linear meaning that the estimate is a weighted mean of the data in the area, and unbiased meaning that the mean expected error is zero (Merz & Blöschl, 2004).

Regionalization based on catchment attributes

The analysis of observed hydrological behaviour often reveals small scale variability but catchments that are far apart may still be hydrologically similar (e.g. Pilgrim, 1983), so alternatives to the spatial proximity concepts have been proposed. These concepts are often based on similarity of catchment attributes that are available in both gauged and ungauged catchments. Runoff is not considered as a catchment attribute that will make this group of methods applicable to the ungauged catchment case. Catchment attributes include catchment size, information on topography, land use, geology, elevation, soil characteristics, as well as climate variables such as mean annual precipitation, and are thought of as surrogates of the hydrological processes within a catchment. The rationale of this approach is that catchments with similar attributes may also behave hydrologically similarly (Acreman & Sinclair, 1986). The catchment attributes can be used in regionalization methods of various structures.

The first type of methods use a distance measure of hydrological similarity which is a function of the differences in catchment attributes of two catchments. The distance measure is zero if the catchment attributes are identical and increases as the attributes get more dissimilar. The distance measure can be used in statistical methods such as cluster analysis, principal component analysis, and classification trees (Breiman *et al.*, 1984; Nathan & McMahon, 1990; Bates, 1994) to group the catchments. Once the groups are identified, the model parameters can be transferred from an analogue gauged catchment within the same group to the ungauged catchment of interest. A particular variant is the Region of Influence approach (Burn & Boorman, 1993), where for each catchment of interest a separate pooling group is formed.

The second way of using catchment attributes are regression analyses between model parameters and catchment attributes. Regression relationships are black-box models, although some degree of process reasoning can come in. Due to the availability of catchment attributes in geographic information systems, correlations between model parameters and catchment attributes are widely used in regionalization (e.g. Sefton & Howarth, 1998, Seibert, 1999; Kokkonen *et al.*, 2003; Merz & Blöschl, 2004). In multiple regressions, one may encounter the problem of multicollinearity, i.e. when at least one of the attributes is highly correlated with another attribute or with some linear combination of them. If multicollinearity is present, the regression coefficient can be highly unstable and unreliable (Hirsch *et al.*, 1992). One therefore limits the number of catchment attributes used in the regression, sometimes combining a number of attributes into an index, which is assumed to be representative of one aspect of the rainfall–runoff relationship (such as the base flow index, IH, 1999). Sequential regression may assist in identifying robust parameter estimates (Lamb & Kay, 2004). Sometimes the delineation of homogeneous regions and regression analyses are combined (e.g. Burn & Boorman, 1993). A formal way of combining regression analyses and the delineation of homogeneous regions are Classification And Regression Tree (CART) models (Breiman *et al.*, 1984; Laaha & Blöschl, 2006).

ANALYSES USING LARGE SAMPLE SIZES

Testing the performance of various parameter regionalization methods is usually done by withholding the runoff data in a catchment of interest, transferring the model parameters from neighbouring gauged catchments, simulating runoff in the catchment of interest and, finally, comparing the simulation performance in the catchment of interest against the observed runoff data in that catchment. This type of test is useful as it fully emulates the ungauged catchment case and one can analyse the relationship between model parameters and catchment attributes in a subjective way. However, the performance of the regionalization methods is strongly influenced by the local conditions and it is hence very difficult to arrive at general conclusions. Only if a large number of catchments is analysed at the same time can more general conclusions be inferred on the suitability of regionalization methods in a given climate region. It is therefore very important to use a large number of samples in comparative regionalization analyses. Although numerous regionalization studies are reported in the literature, only a few of them use large catchment samples. Clearly, it is difficult to obtain reliable data sets containing many catchments that are not affected by human intervention. A number of studies of this type are summarized below.

Sefton & Howarth (1998) compared calibrated parameters of the IHACRES model with catchment attributes of 60 catchments in England and Wales. The best correlations they obtained were $R^2 = 0.59$ between a routing parameter and percentage of aquifers, and $R^2 = 0.69$ between an evaporation parameter and mean annual precipitation. For the storage parameters, no significant correlations were obtained. Peel *et al.* (2000) related, separately, the calibrated parameters of the SYMHID model to four indices that were to reflect climate, terrain and soil characteristics in 331 Australian catchments. They used the ratio of mean annual rainfall to the mean annual areal potential evaporation as an index for climate and the difference of the 90th percentile and the 10th percentile elevation in a catchment as an index of topographical relief. Soil depth and plant available water holding capacity were used as soil indices to reflect the relationship of model parameters and soil characteristics. The highest coefficient of determination, $R^2 = 0.2$, was found for the correlation of one parameter to the climate index. The correlations of model parameters to other indices were smaller and not statistically significant at the 0.05 level. As climate was suggested to be the main driving factor, the data set was subdivided into three climate regions but this did not increase the correlations much.

Young (2006) regionalized the model parameter of the PDM water balance model based on a data set of 260 catchments in the UK. The model parameters of 179 catchments were calibrated against 10 years of observed daily runoff and 81 catchments were retained as an independent evaluation catchment data set in a split sample test. Two regionalization methods, based on the similarity of catchment attributes, were tested. First, in a regression analysis, each of the six model parameters was related individually to the available catchment attributes in a way to maximize the explained variance. In the second regionalization method, termed nearest neighbour approach, the complete set of model parameters was transposed from a number of donor catchments that were most similar in terms of catchment attributes. Runoff time series were simulated using the transposed parameters and the final time series was taken as the arithmetic mean of the individual series. Thus, the nearest neighbour approach retains the structure of the donor catchment parameters. For both methods, the similarity between the catchments was mainly based on catchment attributes derived from a hydrological response classification of soils (Hydrology of Soil types, HOST, Boorman *et al.*, 1995). The median Nash-Sutcliffe model efficiency using regression-based model parameters was ME = 0.70 as compared to ME = 0.71 for locally calibrated simulations. For the nearest neighbour method the efficiencies were slightly lower.

Beldring *et al.* (2002) used 141 catchments in Norway for calibrating a version of the HBV model. They then treated 43 additional catchments as ungauged and regionalized the model parameters as a function of land-use classes. For both sets of catchments they found median Nash-Sutcliffe efficiencies of 0.68 and concluded that the regionalization method represented the main features of the landscape well. However, for 20% of the second set of stations the efficiencies were less than 0.3.

Vandewiele & Elias (1995) derived the parameters of a monthly water balance model for 75 catchments in Belgium from neighbouring catchments. For a case where they regionalized parameters using kriging, their model performed well for 72% of the catchments while it was only 44% when transferring parameters from the nearest catchment.

Hundecha & Bárdossy (2004) applied the HBV model to 95 catchments in the Rhine basin to analyse the effect of land-use change on runoff. Instead of regionalizing the model parameters directly, the regionalization was carried out by initially assuming a functional form of the relationship of model parameters and catchment attributes and then calibrating the parameters of the functional form for many catchments simultaneously. The Nash-Sutcliffe model efficiency was between 0.79 and 0.9 for 30 calibration catchments and between 0.76 and 0.92 for the validation catchments. The established functional form of model parameters and catchment attributes could be used to estimated model parameters in ungauged catchments.

Parajka *et al.* (2005a), which is an extension of Merz & Blöschl (2004), examined the relative performance of a range of parameter regionalization methods. They simulated the daily water balance dynamics of 320 Austrian catchments using a semi-distributed conceptual catchment model, following the structure of HBV. They evaluated the predictive accuracy of the regionalization methods by jack-knife cross-validation against observed daily runoff and snow cover data. They compared nearest neighbours, inverse distance weighting and kriging as variants of spatial proximity-based methods and multiple regressions and the transposition of a set of model parameters from a donor catchment as variants of catchment attribute-based methods (Fig. 1). Two methods performed best. The first was a kriging approach where the model parameters were regionalized independently from each other. The second was a similarity approach where the complete set of model parameters was transposed from the donor catchment that was most similar in terms of a number of catchment attributes. For both methods, the median Nash-Sutcliffe model efficiency of daily runoff for the 11 year calibration period was ME = 0.67 as compared to ME = 0.72 for locally calibrated simulations. For the verification period, the corresponding efficiencies were 0.62 and 0.66. The regionalization of model parameters based on multiple regression performed poorer with the corresponding efficiencies of 0.65 and 0.60 for the calibration and verification periods, respectively.

An overview of studies on regionalization of rainfall–runoff model parameters using large catchment samples is given in Table 1.

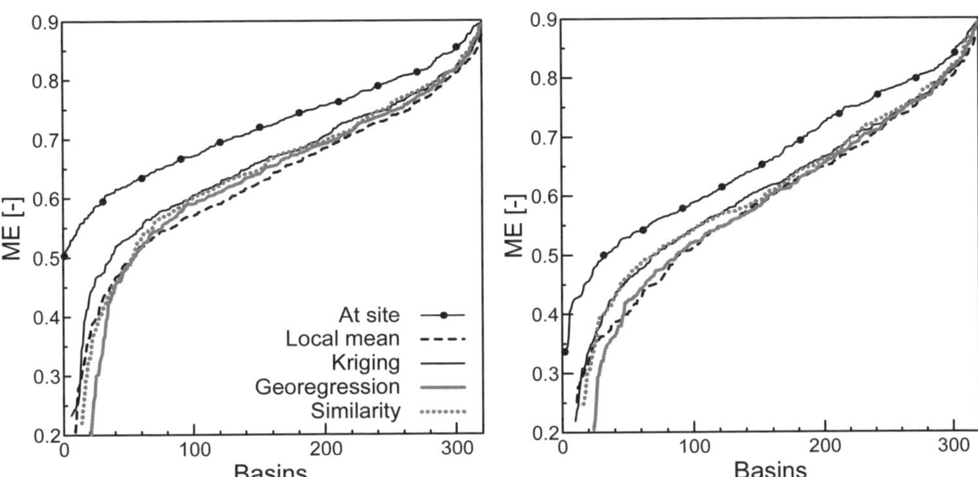

Fig. 1 Cumulative distribution functions of Nash-Sutcliffe runoff model efficiencies (ME) using different regionalization methods in 320 Austrian catchments. Calibration (left) and verification (right) periods. From Parajka *et al.* (2005a).

Table 1 Overview of rainfall–runoff regionalization studies using large catchment samples.

Reference	Country	Model	Number of catchments used	Time step	Methods
Sefton & Howarth (1998)	UK	IHACRES	60	Daily	Regression with catchment attributes
Peel *et al.* (2000)	Australia	SIMHYD	331	Monthly	Regression with catchment attributes
Beldring *et al.* (2002)	Norway	HBV	186	Daily	Regression with catchment attributes
Parajka *et al.* (2005a)	Austria	HBV (semi lumped)	320	Daily	Regression with catchment attribute Similarity concepts Spatial proximity methods (Kriging, nearest neighbour)
Merz & Blöschl (2004)	Austria	HBV (lumped)	308	Daily	Regression with catchment attributes Spatial proximity methods (Kriging, nearest neighbour)
Young (2006)	UK	PDM	260	Daily	Regression with catchment attributes Nearest neighbour approach based on similarity in catchment attributes
Hundecha & Bárdossy (2004)	Rhine basin	HBV	95	Daily	Regional calibration of parameter of regression function of model parameters and catchment attributes
Vandewiele & Elias (1995)	Belgium	3 parameter water balance model	60	Monthly	Spatial proximity methods (Kriging, nearest neighbour)

DISCUSSION AND CONCLUSIONS

All the studies examined here reported relatively low correlations between model parameters and catchment attributes. One explanation for the low correlations is that the measurable catchment attributes may not be relevant for the dominant processes. Most of the rainfall–runoff processes take place in the subsurface while most of the catchment attributes (such as topographic characteristics and vegetation) are available at the land surface only and contain very little information about the subsurface. Soil type or geological characteristics as available in regional studies, often, are not representative of subsurface flow processes that can be dominated by peds and cracks. Also, pieces of information such as geology and soils are static indicators of hydrological response as they do not vary much over hydrological time scales. What would be more valuable are indicators that better capture the dynamics of hydrological processes. An attempt at obtaining more hydrologically justifiable catchment attributes is the classification of soil types in the UK (Boorman *et al.*, 1995). Certainly, the successful application of regression analyses hinges on a sensible parameterization of subsurface processes. In the Austrian case study (Parajka *et al.*, 2005a), for example, where information on soil hydrology was unavailable, the regression performed poorer than in the UK study where such information was available (Young, 2006). It would also seem possible to choose catchment attributes by hydrological reasoning based on the dominant hydrological processes. The process typology of regional floods of Merz & Blöschl (2003), for example, could be used to guide selection of catchment attributes to be used in statistical regionalization methods.

If no hydrologically justifiable catchment attributes are available, spatial proximity may be a better measure of catchment similarity (e.g. Parajka *et al.*, 2005a). The delineation of spatially contiguous regions according to climate is often a first step in a regionalization approach or can be combined with regression analyses between model parameters and catchment attributes (Peel *et al.*, 2000). However, it may be difficult to assign ungauged catchments to one of the delineated homogeneous regions, particularly for ungauged catchments close to the region boundaries. An alternative to the homogeneous region approach are geostatistical methods. Although these methods tend to exhibit little bias, random errors can be large (Parajka *et al.*, 2005a).

One of the problems with the use of geostatistical methods in hydrology is that they have evolved in the mining industry, where one is interested in cubic blocks. Unlike mining blocks, catchments are nested and water follows a stream network. It is therefore clear that upstream and downstream catchments would have to be treated differently from neighbouring catchments that do not share a subcatchment. Based on ideas of Sauquet *et al.* (2000), Skøien *et al.* (2005) proposed the TOP-KRIGING approach of geostatistical regionalization that takes into account the organization of the landscape into nested catchments. They showed that the proposed method outperformed the traditional Euclidian framework when regionalizing flood frequencies to ungauged catchments and they were able to assign uncertainty bounds to the regionalized values. More work along the lines of hydrologically justified spatial statistics seems to be warranted to contribute to improved parameter regionalization.

Another problem in regionalizing model parameters is the significant uncertainty of the calibrated parameters in the gauged catchments *per se*. There are methods of accounting for parameter uncertainty in the regionalization of model parameters (see, e.g. Campell & Bates, 2001). To better constrain model parameters, additional data on state variables can be used in the calibration procedure (Mroczkowski *et al.*, 1997; Gupta *et al.*, 1998). Parajka *et al.* (2005a) used observed snow depths to constrain model parameters and Parajka *et al.* (2005b) used soil moisture information from scatterometer satellite data. An alternative to the use of observed state variables is regional calibration in which the model parameters of a number of catchments are calibrated simultaneously (e.g. Fernandez *et al.*, 2000; Szolgay *et al.*, 2003). The idea of regional calibration is to obtain more robust parameters which may reduce regionalization uncertainty but the potential of this approach has not yet been fully evaluated.

In terms of the overall runoff simulation performance, the comparisons of the different regionalization studies based on many catchments suggest that there is no single best regionalization method. Results of some of the studies are seemingly conflicting. In Young (2006), the best regionalization method was a regression of individual model parameters to catchment attributes, while the transposition of the complete data set using a nearest neighbour approach based on catchment similarity performed slightly poorer. In contrast, the results of Parajka *et al.* (2005a) indicated that the transposition of a complete set of model parameters performed better than the regression method. It should be noted, however, that the differences between the methods were small and may be related to the particularities of the data sets used. The crucial point, perhaps, is not the statistical method to be used but the measures of similarity. Traditional catchment attributes such as catchment size, land use, geology, topographic elevation, soil characteristics, as well as climate variables such as mean annual precipitation, are perhaps less representative of the driving hydrological processes than is usually thought. It may, therefore, be worthwhile to focus future work on hydrologically meaningful catchment attributes (such as Boorman *et al.*, 1995) and hydrological distance measures (Skøien *et al.*, 2005). Analyses such as Merz & Blöschl (2003) may assist in understanding processes at the regional scale and help derive such attributes. Although scientific regionalization studies are usually directed towards formalized regionalization methods, as discussed above, it is important to remember that the quantity and quality of expert judgement included in any regionalization approach, be it through delineating regions, selecting catchment attributes or other avenues, play a very significant role in maximizing regionalization performance.

Acknowledgements We would like to thank the Austrian Academy of Sciences (project no. HÖ31 and APART [Austrian Programme for Advanced Research and Technology] – fellowship) for financial support.

REFERENCES

Acreman, M. C. & Sinclair, C. D. (1986) Classification of drainage basins according to their physical characteristics; an application for flood frequency analysis in Scotland. *J. Hydrol.* **84**, 365–380.

Bates, B. C. (1994) Regionalization of hydrologic data: a review. Cooperative Research Centre for Catchment Hydrology, Monash University, Victoria, Australia.

Beldring, S., Roald, L. A. & Voksø, A. (2002) Avrenningskart for Norge (Runoff map for Norway, in Norwegian), Norwegian Water and Energy Directorate, Report No. 2, Oslo, Norway.

Blöschl, G. & Sivapalan, M. (1995) Scale issues in hydrological modelling—a review. *Hydrol. Processes* **9**, 251–290.

Boorman, D. B., Hollis, J. M. & Lilly, A. (1995) Hydrology of Soil Types: A Hydrologically Based Classification of the Soils of the United Kingdom. *IH Report No. 126.* Institute of Hydrology, Wallingford, UK.

Breiman, L., Friedman, J. H., Olshen, R. & Stone, C. J. (1984) *Classification and Regression Trees*. Wadsworth International Group, Belmont, California, USA.

Burn, D. H. & Boorman, D. B. (1993) Estimation of hydrological parameters at ungauged catchments. *J. Hydrol.* **143**, 429–454.

Campbell, E. P. & Bates, B. C. (2001) Regionalization of rainfall–runoff model parameters using Markov Chain Monte Carlo samples. *Water Resour. Res.* **37**(3), 731–739.

Fernandez, W., Vogel, R. M. & Sankarasubramanian, A. (2000) Regional calibration of a watershed model. *Hydro. Sci. J.* **45**(5), 689–707.

Gupta, H. V., Sorooshian, S. & Yapo, P. O. (1998) Toward improved calibration of hydrologic models: multiple and noncommensurable measures of information. *Water Resour. Res.* **34**(4), 751–763.

Hirsch, R. M., Helsel, D. R., Cohn, T. A. & Gilroy, E. J. (1992) Statistical analysis of hydrological data. In: *Handbook of Hydrology* (ed. by R. Maidment), 17.1–17.55. McGraw-Hill, New York, USA.

Hundecha, Y. & Bárdossy, A. (2004) Modeling of the effect of land use changes on the runoff generation of a river basin through parameter regionalization of a watershed model. *J. Hydrol.* **292**, 281–295.

Institute of Hydrology (IH) (1999) *Flood Estimation Handbook*. Institute of Hydrology, Wallingford, UK.

Kokkonen, T. S., Jakeman, A. J., Young, P. C. & Koivusalo, H. J. (2003) Predicting daily flows in ungauged catchments: model regionalization from catchment descriptors at the Coweeta Hydrologic Laboratory, North Carolina. *Hydrol. Processes* **17**, 2219–2238.

Laaha, G. & Blöschl G. (2006) A comparison of low flow regionalization methods—catchment grouping. *J. Hydrol.* (in press).

Lamb, R. & Kay, A. L. (2004) Confidence intervals for a spatially generalized, continuous simulation flood frequency model for Great Britain. *Water Resour. Res.* **40**, W07501, doi:10.1029/2003WR002428.

Merz, R. & Blöschl, G. (2003) A process typology of regional floods. *Water Resour. Res.* **39**(12), 1340, doi:10.1029/2002WR001952.

Merz, R. & Blöschl, G. (2004) Regionalization of catchment model parameters. *J. Hydrol.* **287**, 95–123.

Merz, R., Piock-Ellena, U., Blöschl, G. & Gutknecht, D. (1999) Seasonality of flood processes in Austria. In: *Hydrological Extremes: Understanding, Predicting, Mitigating* (ed. by L. Gottschalk, J. C. Olivry, D. Reed & D. Rosbjerg) (Proc. Birmingham Symp., July 1990), 273–278. IAHS Publ. 255. IAHS Press, Wallingford, UK.

Merz, R. & G. Blöschl (2005) Flood frequency regionalization—spatial proximity vs. catchment attributes. *J. Hydrol.* **302**(1–4), 283–306.

Mroczkowski, M., Raper, G. P. & Kuczera, G. (1997) The quest for more powerful validation of conceptual catchment models. *Water Resour. Res.* **33**(10), 2325–2335.

Nathan, R. J. & McMahon, T. A. (1990) Identification of homogeneous regions for the purpose of regionalization. *J. Hydrol.* **121**, 217–238.

Parajka, J., Merz, R. & Blöschl, G. (2005a) A comparison of regionalization methods for catchment model parameters, *Hydrol. Earth System Sci.* (HESS) **9**, 157–171.

Parajka, J., Naeimi, V., Blöschl, G., Wagner, W., Merz, R. & Scipal, K. (2005b) Assimilating scatterometer soil moisture data into conceptual hydrologic models at the regional scale. *Hydrol. Earth System Sci.* (HESS) Discussion, SRef-ID: 1812-2116/hessd/2005-2-2739.

Peel, M. C., Chiew, F. H. S., Western, A. W. & McMahon, T. A. (2000) Extension of unimpaired monthly streamflow data and regionalization of parameter values to estimate streamflow in ungauged catchments. Report prepared for the National Land and Water Resources Audit, in: Australian Natural Resources Atlas website: http://audit.ea.gov.au/anra/water/docs/national/Streamflow/Streamflow.pdf.

Pilgrim, D. H. (1983) Some problems in transferring hydrological relationships between small and large drainage basins and between regions. *J. Hydrol.* **65**, 49–72.

Sauquet, E., Gottschalk, L. & Leblois, E. (2000) Mapping average annual runoff: a hierarchical approach applying a stochastic interpolation scheme. *Hydrol. Sci. J.* **45**, 799–815.

Sefton, C. E. M. & Howarth, S. M. (1998) Relationships between dynamic response characteristics and physical descriptors of catchments in England and Wales. *J. Hydrol.* **211**, 1–16.

Seibert J. (1999) Regionalization of parameters for a conceptual rainfall–runoff model. *Agric. Forest Meteorol.* **98–99**, 279–293.

Skøien, J., Merz, R. & Blöschl, G. (2005) Top-Kriging—geostatistics on stream networks. *Hydrol. Earth System Sci.* Discussions **2**, 2253–2286. SRef-ID: 1812-2116/hessd/2005-2-2253.

Szolgay, J., Hlavčová, K., Kohnová, S. & Danihlik, R. (2003) Estimating regional parameters of a monthly balance model. *J. Hydrol. Hydromech.* **51**(4), 256–273.

Vandewiele, G. L. & Elias, A. (1995) Monthly water balance of ungauged catchments obtained by geographical regionalization. *J. Hydrol.* **170**, 277–291.

Young, A. (2006) Stream flow simulation within UK ungauged catchments using a daily rainfall–runoff model. *J. Hydrol.* (in press).

Yu, P. S. & Yang, T. C. (2000) Using synthetic flow duration curves for rainfall–runoff model calibration at ungauged sites. *Hydrol. Processes* **14**(1), 117–133.

A review of Australian model parameterization studies using large basin samples

WALTER BOUGHTON
Griffith University, 11 Preston Place, Brookfield 4069, Queensland, Australia
wboughto@bigpond.net.au

Abstract The five major studies in Australia, using 168, 195, 221, 221 and 175 data sets, are described. The basins range in size from 1 to 8400 km^2 with an average of a few hundred km^2. The paper describes the models, data sets, methods used and results. The most promising approach shifts the problem of estimating model parameter values to that of estimating average annual runoff. Given an estimate of average annual runoff and daily rainfall and evaporation data, the AWBM rainfall–runoff model self calibrates and calculates daily runoff for the period of input data. Relationships of average annual runoff to average annual rainfall and evaporation have been established for the whole of Australia and results from these relationships with the AWBM are presented. The main problem in Australia is the estimation of areal rainfall for input to the models.

Key words AWBM; hydrological modelling; rainfall–runoff models; runoff; ungauged catchments

INTRODUCTION

Rainfall–runoff modelling in Australia has been reviewed by Boughton (2005). Because of the large area of the country (7 682 000 km^2) and small population (20 million) the network of streamgauging stations is barely adequate in the main agricultural regions, and sparse in the sparsely populated regions. This has prompted many studies directed towards estimating runoff from ungauged basins. There were some early studies in the 1970s with different models, but all with a small number of data sets and all without any useful results. More recently, there have been five studies with large numbers of data sets, and two with smaller numbers. Those studies are the subject of this paper. The models used in the major studies (SFB, MOSAZ, AWBM and SIMHYD) are described in more detail in Boughton (2005). All of these models are run on daily data.

MAJOR STUDIES

The SFB model has three parameters, with one mainly determining the amount of runoff and two determining the division of runoff between surface runoff and baseflow. Nathan & McMahon (1990a,b) calibrated the model on 168 basins of 1 to 250 km^2 in area, with median annual rainfall 600 to 1200 mm, in New South Wales and Victoria. The areal rainfall data were derived by Thiessen weights. Potential evapotranspiration (PET) was based on Morton's wet environment areal evapotranspiration (Morton,

1983). Parameters were optimized using a Simplex search algorithm with several sets of starting values. Some 37% of the calibrations had either $r^2 < 0.60$ or differences between observed and calculated total flow >±10%. These poor calibrations were usually in basins in which the water balance problems indicated that the rainfall and PET data were not representative of the basin.

Scatter plots between calibrated model parameter values and basin characteristics were used to look for visual evidence of correlation, without success. Multiple linear regression equations were also developed between parameter values and basin characteristics (Nathan & McMahon, 1991). At the time of this study, the SFB model was in widespread use in Australia, but has since been replaced, mainly by the AWBM, and is little used at the time of writing.

The MOSAZ (Modified Semi-arid Zone) model is very simple with two parameters, one for moisture storage capacity and one for baseflow discharge. Nathan *et al.* (1996) calibrated the model on 195 basins of 4 to 8400 km^2 in area, with average annual rainfall 450 to 2300 mm, in Victoria. The areal rainfall data were derived by Thiessen weights. Mean monthly PETs were estimated by Morton's complementary procedure for regional evapotranspiration. Parameters were calibrated by selecting several sets of starting values, and then optimizing using a Simplex search algorithm to minimize the square root of difference between observed and calculated monthly flows.

Linear regressions were used to look for correlations between parameter values and nine measurable characteristics. Ten data sets were not used in determining the regressions, but retained for independent testing. The coefficients of efficiency between estimated and actual runoff on the test data sets varied from 0.015 to 0.83. The authors expressed concern about the "weak physical significance" of the equations. This was the only major study in which the MOSAZ model was used. It is not in wide use.

The AWBM model has three parameters with one determining the amount of runoff, one determining the division between surface runoff and baseflow, and one determining the rate of baseflow discharge. Boughton & Chiew (2003) calibrated the model on 221 basins of 50 to 2000 km^2 in area, with annual rainfall 300 to 2800 mm, spread over much of the main agricultural regions of Australia. The self-calibrating version of the model that automatically calibrates to a set of daily data without action by the user was used (Boughton 2003). They made no correlations between model parameters and basin characteristics; instead they tabulated the calibrated values for all 221 basins. They recommended selecting the calibrated parameter values for several of the basins nearest to the ungauged basin of interest, using tabulated basin characteristics as a guide. Variation in results from the different calibrations in a region gives some indication of potential error in estimating runoff. Taylor (2004) extended the work for the State of Tasmania by calibrations of the model on more basins than were used by Boughton & Chiew (2003) in that State.

At the time of writing, this is the only published procedure that has potential for use over most of Australia. It is too soon since its introduction to assess how much it will be used in practical applications, but the follow-on study by Taylor (2004) shows there is interest in the approach. The AWBM is in common use for rainfall–runoff modelling, and the publication of calibrated parameter values for 221 basins offers a simple extension of existing technology.

The AWBM model was used in another study by Boughton & Chiew (2006). The single parameter determining the amount of runoff can be calibrated to an estimate of average annual runoff, and the model can then be used with daily rainfall and evaporation data to estimate daily runoff with the same average annual runoff. The authors established linear equations to estimate average annual runoff from average annual rainfall and evaporation anywhere in Australia. The model has two baseflow parameters that affect the timing of runoff (but not the amount). The tabulated values of these parameters in Boughton & Chiew (2003), with the regression equations for determining the amount of runoff, provide a simple method for estimating daily runoff anywhere in Australia.

Six catchment characteristics (two topographic, two soil characteristics and two relating to vegetation cover) were tested in turn with the regression equations for estimating average annual runoff, but none produced any improvement. The lack of any significant correlation between the characteristics and runoff is anomalous. Many other studies reported in the literature show such correlations, e.g. the difference in runoff from forested and grassed catchments is extensively reported, and so the lack of correlation between runoff and the percent of woody vegetation is a significant anomaly. The basin characteristics were estimated from satellite observations and from broad scale soil mappings, not from field measurements, so there is a possible explanation in the methods used to estimate the characteristics.

The SIMHYD model (Chiew *et al.*, 2002) was used by Chiew (2003) to generalize the model's seven parameters on 175 basins in southeast Queensland. The basins were divided into nine "hydrologic regions" and one set of parameter values was determined for each region to give good overall agreement between modelled and recorded monthly flows for basins in that region. The results were intended for use only within the region of study and not elsewhere.

STUDIES WITH SMALL SAMPLES

Two studies with small numbers of data sets provide additional information on model parameterization in Australia. Ibrahim & Cordery (1995) calibrated a monthly rainfall–runoff model to data from 18 basins in New South Wales for the purpose of estimating runoff and recharge volumes on ungauged basins. Post & Jakeman (1996) calibrated the daily IHACRES model on 16 small basins in Victoria with an objective of regionalizing the model parameters. Neither study produced any practical application.

MOST PROMISING APPROACH

The AWBM has only one parameter (average surface storage capacity) determining the amount of runoff. This should simplify the establishment of relationships between the parameter and basin characteristics; however, small errors in rainfall data cause very big changes in the calibrated value, sufficient to confuse any relationships. Boughton (1996) scaled rainfall data to simulate errors and showed that change of ±20% in rainfall produced changes of +98% and –68%, respectively, in the calibrated value of average surface storage capacity. Other models with a single parameter determining the amount of runoff (such as the Curve Number) show similar sensitivity. Errors of ±20% in areal rainfall data are commonplace (Boughton, 2006). This makes the calibrated values of parameters suspect for relating to basin characteristics.

However, the AWBM average surface storage capacity can be accurately calibrated to an estimate of average annual runoff in order to calculate that amount of runoff. The average capacity is simply increased and decreased by trial and error until the calculated runoff matches the estimate of runoff to any required degree of accuracy. The AWBM model, used with estimates of average annual runoff and tabulated values of baseflow parameters, offers the simplest and most robust method for estimating daily runoff from ungauged basins in Australia. This approach transfers the problem of estimating values of model parameters to estimating average annual runoff.

Table 1 shows the regression equations developed by Boughton & Chiew (2006) for estimating average annual runoff in six of the major Drainage Divisions of Australia, and for all mainland data lumped together. The coefficients of determination (r^2) and the F statistics indicate the relative accuracy of the runoff estimates on ungauged basins. The regression equations are based on groups of average annual rainfall, and the r^2 values are generally highest in the higher rainfall ranges and lowest in the lower ranges, as expected.

The AWBM has two baseflow parameters in addition to the average surface storage parameter—the baseflow index (BFI) that determines the division of runoff between surface runoff and baseflow, and the daily baseflow recession constant (Kbase) that determines the rate of baseflow discharge from storage. Boughton & Chiew (2003) documented calibrated values of these parameters on 221 basins in Australia, and Table 2 summarizes those values as median, 10 percentile and 90 percentile values for the six Drainage Divisions in Table 1. These values are used with the average surface storage capacity to estimate daily runoff from ungauged basins.

Boughton & Chiew (2006) tested the procedure using 23 years of rainfall and PET data from the 108 km^2 Boggy Creek basin in Drainage Division IV. Average annual values of rainfall, PET and runoff for the 23-year period were 1185, 1085 and 324 mm year^{-1}, respectively. Figure 1 shows a comparison of the estimated monthly and yearly runoff

Table 1 Regression equations for estimating average annual runoff.

Region	Rainfall range mm year^{-1}	N	Regression	r^2	F	df	Prob.
Div I	700–1730	12	Q = 0.544P-350	0.959	234	10	2.9E-8
Div II	>1000	41	Q = 0.641P-0.0717E-361	0.903	193	38	1.3E-20
	700–1000	40	Q = 0.619P-0.157E-206	0.620	31	37	1.1E-8
	<700	7	Use mainland				
Div III	546–2062	11	Q = 0.773P-0.902E+401	0.983	235	8	7.8E-8
Div IV	>1000	22	Q = 0.861P-0.0395E-661	0.932	129	19	8.6E-12
	700–1000	47	Q = 0.502P-0.259E+4	0.601	33	44	6.4E-7
	<700	15	Q = 0.276P-0.139E+47	0.811	23	12	7.5E-5
Div V	490–850	8	Q = 0.351P-0.171E+27	0.976	101	5	9.1E-5
Div VI	850–1050	5	Q = 0.684P-497	0.673	6	3	0.089
	390–850	5	Q = 0.124P-37	0.759	9	3	0.054
Mainland	>1000	71	Q = 0.659P-0.073E-382	0.902	333	68	7.3E-36
	700–1000	100	Q = 0.571P-0.119E-212	0.566	63	97	3.1E-18
	<700	31	Q = 0.211P-0.078E+11	0.550	19	28	7.2E-6

N, number of catchments in sample; F, F statistic; df, degrees of freedom; Prob, probability of chance result; Q: average annual runoff mm year^{-1}; P, average annual rainfall mm year^{-1}; E, average annual areal PET mm year^{-1}.

Table 2 Values of baseflow parameters by Drainage Division.

Division	BFI			Kbase		
	90%	Median	10%	90%	Median	10%
I	0.11	0.17	0.45	0.813	0.950	0.987
II	0.21	0.33	0.60	0.915	0.980	0.993
III	0.23	0.32	0.57	0.920	0.966	0.991
IV	0.18	0.41	0.63	0.910	0.976	0.989
V	0.20	0.29	0.45	0.950	0.958	0.985
VI	0.30	0.56	0.63	0.900	0.956	0.981

BFI, baseflow index; Kbase, daily baseflow recession constant.

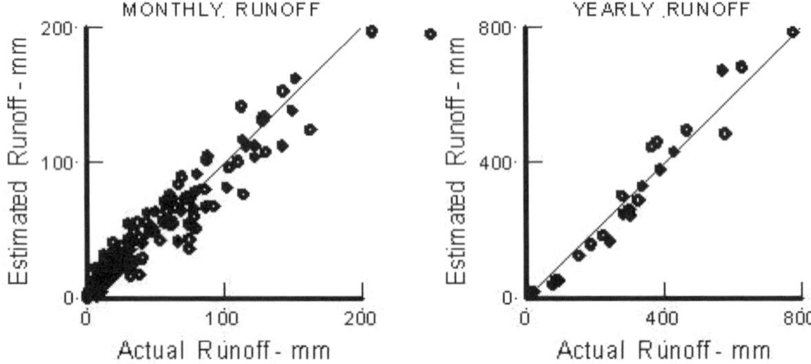

Fig. 1 Estimated and actual monthly and yearly runoff on Boggy Creek basin.

with actual values. The monthly and yearly coefficients of efficiency are 0.918 and 0.908, respectively. These Boggy Creek results are typical of the better quality results in the higher rainfall ranges, and are neither the best results nor outstanding in any way.

The statistical characteristics of the regression equations for estimating runoff from rainfall and evaporation give a measure of the possible error in the estimate of total runoff. The accuracy of daily runoff depends mainly on the accuracy of the values of the baseflow index (BFI) and the daily baseflow recession constant (Kbase). While the calibrated values in Table 2 give a guide for ungauged basins, there are substantial ranges between the 10 and 90 percentile values. At present, no relationships have been found between these parameters and basin characteristics.

FUTURE DIRECTION

The accuracy of the method using estimates of average annual runoff with the AWBM depends mainly on the accuracy of the estimate of average annual runoff. The linear regressions relating runoff to rainfall and PET in Table 1 were based on 221 data sets covering much of Australia. There are more than 10 times that number of streamgauging stations in Australia, so there is considerable potential for improving the regressions by use of more data. Other studies such as Gan *et al.* (1990) provide other approaches to the estimation of average annual runoff. The substantial variability of the Australian climate makes the estimation of runoff on ungauged catchments more difficult than in many temperate zone countries. In the study with the SFB model and the two studies with the AWBM model, there were problems of input data quality in one-third of the data sets available for use. The problems were mainly associated with the rainfall data not being representative of the areal rainfall that produced the runoff. Typical density of raingauges in Australian basins is in the order of one gauge in 10 to 100 km^2, giving a sample of one part in 300 million to 3 billion of the basin rainfall. In addition to the sampling error, there are periods of missing data, and errors and mistakes in the records. Some techniques are available for checking the consistency of rainfall and runoff data (Boughton, 1996, 2006) but these have not been used much to date. There is considerable scope for improving the studies directed towards ungauged basins by giving more attention to the quality of input data.

More than 30 years of worldwide effort directed to establishing relationships between model parameter values and basin characteristics has produced little of practical use. The substantial effect of errors in input data on the calibrated values of model parameters is one of the major problems. The transfer of the problem of estimating values for model parameters to that of estimating average annual runoff is a major change of approach that has potential for simplifying the modelling of runoff from ungauged basins.

REFERENCES

Boughton, W. C. (1996) Detecting errors in rainfall–runoff data sets. *Tech. Rept. 96/2, CRC for Catchment Hydrology, Monash Univ. 01-23.*

Boughton, W. (2003) The Australian water balance model. *Environ. Modell. & Software* **19**, 943–956.

Boughton, W. C. (2005) Catchment water balance modelling in Australia 1960–2004. *Agric. Water Manage.* **71**, 91–116.

Boughton, W. C. (2006) Calibrations of a daily rainfall–runoff model with poor quality data. Accepted for publication *Environ. Modell. & Software* **21**, 1114–1128.

Boughton, W. C. & Chiew, F. (2003) Calibrations of the AWBM for use on ungauged catchments. *Tech. Rept. 03/15, CRC for Catchment Hydrology, Monash Univ. 01-37.*

Boughton, W. & Chiew, F. (2006) Estimating runoff in ungauged catchments from rainfall, PET and the AWBM model. Accepted for publication *Environ. Modell. & Software* (in press).

Chiew, F. H. (2003) Generalisation of rainfall–runoff model parameters for modelling applications. In: *Proc. 28th International Hydrology & Water Resources Symposium* **2**, 203–210. Inst. Engrs Aust., Canberra, Australia.

Chiew, F. H., Peel, M. C. & Western, A. W. (2002) Application and testing of the simple rainfall–runoff model SIMHYD. In: *Mathematical Models of Small Watershed Hydrology and Application* (ed. by V. P. Singh & D. K. Frevert), 335–367. Water Resources Publication, Littleton, Colorado, USA.

Gan, K. C., McMahon, T. A. & O'Neill, I. C. (1990) Errors in estimated streamflow parameters and storages for ungauged catchments. *Water Resour. Bull.* **26**(3), 443–450.

Ibrahim, A. G. & Cordery, I. (1995) Estimation of recharge and runoff volumes from ungauged catchments in eastern Australia. *Hydrol. Sci. J.* **40**(4), 499–515.

Morton, F. I. (1983) Operational estimates of actual evapotranspiration and their significance to the science and practice of hydrology. *J. Hydrol.* **66**, 1–76.

Nathan, R. J., Austin, K., Crawford, D. & Jayasuriya, N. (1996) The estimation of monthly yield in ungauged catchments using a lumped conceptual model. *Aust. J. Water Resour.* **1**(2), 65–75.

Nathan, R. J. & McMahon, T. A. (1990a) The SFB model Part I—validation of fixed model parameters. *Civ. Eng. Trans., Inst. Engrs Australia* **CE32**(3), 157–161.

Nathan, R. J. & McMahon, T. A. (1990b) The SFB model Part II—operational considerations. *Civ. Eng. Trans., Inst. Engrs Australia,* **CE32**(3), 162–166.

Nathan, R. J. & McMahon, T. A. (1991) Estimating low flow characteristics in ungauged catchments. Centre for Environ. Applied Hydrology, Dept. Civ. & Agric. Eng., University of Melbourne, Melbourne, Australia.

Post, D. A. & Jakeman, A. J. (1999) Predicting the daily streamflow of ungauged catchments in S. E. Australia by regionalizing the parameters of a lumped conceptual model. *Ecol. Modelling* **123**, 91–104.

Taylor, H. (2004) AWBM Calibrated parameters for Tasmanian catchments. SEN713 Project Report, School of Engineering & Technology, Deakin University, Melbourne, Australia.

3 Model parameterization experiments using the MOPEX database

Distributed hydrological modelling with lumped inputs

VINCENT FORTIN[1], NANEE CHAHINIAN[2], ALBERTO MONTANARI[3], GRETA MORETTI[3] & ROGER MOUSSA[4]

1 *Numerical Prediction Research, Meteorological Research Division, Environment Canada, 2121 North Service Road, Trans-Canada Highway, Dorval, Quebec H9P 1J3, Canada*
vincent.fortin@ec.gc.ca

2 *Institut de Recherche pour le Développement(IRD), UMR HydroSciences–Université Montpellier II, Case Courrier MSE, Place Eugène Bataillon, F-34095 Montpellier cedex 5, France*

3 *Università di Bologna, Viale del Risorgimento 2, 40136 Bologna, Italy*

4 *Laboratoire d'étude des Interactions entre Sol, Agrosystème et Hydrosystème (LISAH), INRA, 2 Place Pierre Viala, F-34060 Montpellier cedex 1, France*

Abstract This paper presents three different spatially distributed hydrological models and discusses the possibility of using them for flow simulation when only lumped information is available for the meteorological input. The three models, namely, AFFDEF, HYDROTEL and ModSpa, are applied on three French catchments from the MOPEX database. The purpose of the study is to provide examples to support the use of spatially-distributed approaches for real world applications even when fully distributed information is not available. The use of distributed models is advisable when at least a piece of distributed information is available such as a digital elevation model, a soil map or land-use data as for the three catchments and the models considered in this study. Overall, the applications presented here show that spatially distributed models can be successfully applied by using mixed lumped/distributed information. The performance of the models is comparable with the results that are usually obtained in real world applications of lumped models in similar situations.

Key words distributed models; lumped models; parameterization; ungauged catchments

INTRODUCTION

Rainfall–runoff hydrological models are largely used in applied hydrology for various applications such as river flow simulation, flood prediction, drought mitigation, catchment management, sediment and solute transport, and the design of hydraulic structures. Flow simulation on ungauged catchments is also a typical application of distributed models (Refsgaard, 1997). However, there are relatively few studies regarding this issue in the literature.

Hydrological models can be classified into two major types, lumped and distributed. Lumped models were developed since the 1960s (e.g. the Stanford catchment model, Crawford & Lindsey, 1966). They consider the catchment as an undivided entity and use lumped values of input variables and parameters. For the most part (for a review, see Singh, 1995), they have a conceptual structure based on the interaction between storage elements representing the different processes with mathematical functions to describe the fluxes between the storage (e.g. HSPF, Donigian *et al.*, 1995;

GR, Perrin *et al.*, 2003). In the last two decades, lumped models were challenged by distributed models whose spatial structure allows the taking account of the spatial variability of processes within catchments and consequently the prediction of local hydrological responses for points within the catchment. Some distributed models are physically-based (e.g. SHE, Abbott *et al.*, 1986) while others maintain a distributed description of catchment responses but in a much simpler way (e.g. TOPMODEL, Beven & Kirkby, 1978; HYDROTEL, Fortin *et al.*, 2001). In distributed models, parameters need to be defined for every spatial element and for each process representing equation. In principle, parameter adjustment should not be necessary for this type of model because parameters should be related to the physical characteristics of the surface, soil and land use. However, in practical applications, calibration procedures are required for both lumped and distributed models; consequently the models require effective or equivalent values for some parameters.

Despite these difficulties, there has been a strong surge in the use of distributed modelling for applied hydrology over the last decade. However, in most practical applications, little geographical and spatial information is available.

This paper presents three spatially distributed models that can be applied to a wide range of real world applications and questions the use of distributed models for flow simulation taking into account the amount of information necessary to run them. We address this issue by applying the models (AFFDEF, Moretti & Montanari, 2006; HYDROTEL, Fortin *et al.*, 2001; ModSpa, Moussa, 1991) on three catchments, both for gauged catchments on which the models can be calibrated, and for ungauged catchments for which no, or little, flow information is available. This paper is structured in three sections. The first presents the three catchments used in the applications, the second presents the structure and the parameters of the three models, and the third discusses the parameterization strategies and the results of the simulations.

THE STUDY SITES

Three catchments were used in the applications (Fig. 1): Le Guillec at Trézilidé located in Brittany, western France; Le Toulourenc at Malaucène, a tributary of the Rhone River, located in the Vaucluse; and Le Loup at Villeneuve Loubet located in the Alps Mountains in Alpes Maritimes, southern France. Hydrological data are presented in Chahinian *et al.* (The MOPEX 2004 database, this volume) and Table 1 shows the main characteristics of the three catchments. Catchment areas range between 45 and 279 km^2, and the outlet altitude ranges between 35 m (Le Guillec) and 2000 m (Le Loup). For all three catchments, the mean annual rainfall is of the same order and ranges between 1000 and 1200 mm, while the mean annual evapotranspiration varies between 700 and 1100 mm.

The three catchments have various hydrometeorological regimes. The Guillec catchment is located in western France, with an oceanic humid climate where baseflow is the major component of the hydrograph. The two other catchments have a Mediterranean climatic regime, characterized by a succession of drought and high intensity rainfall periods and by the high spatio-temporal variability of precipitation distribution

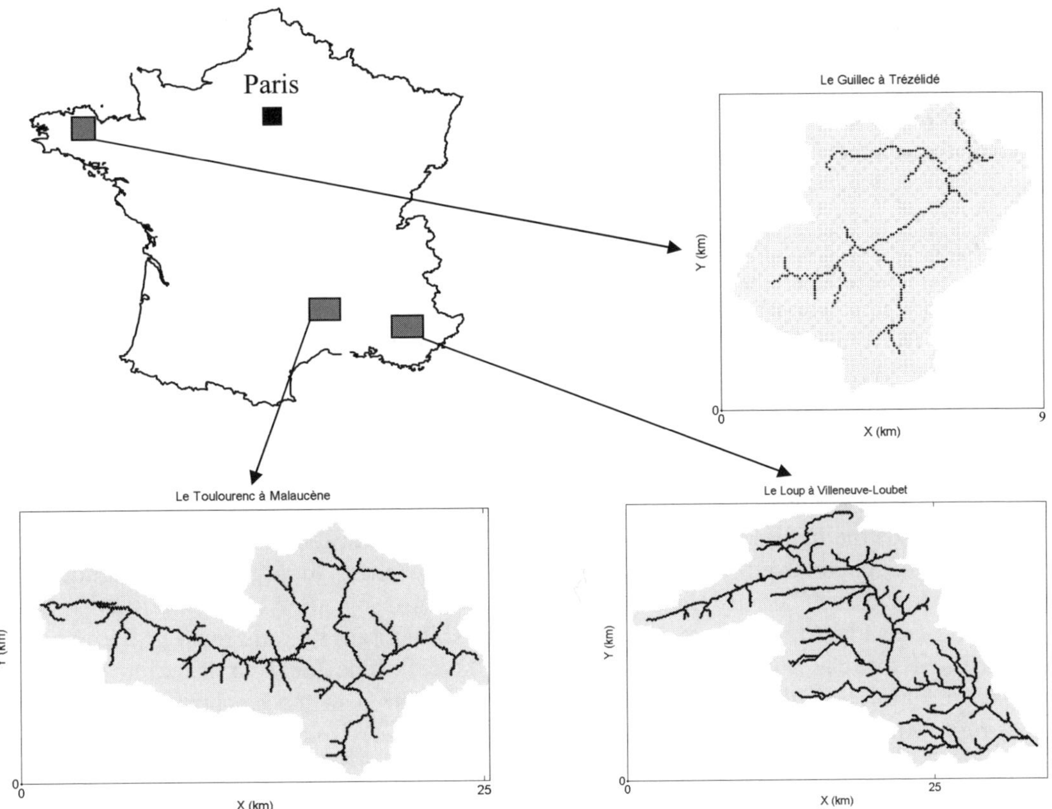

Fig. 1 Basin locations.

Table 1 Catchment characteristics.

Catchment	Area (km^2)	Outlet altitude (m NGF)	Mean annual rainfall 1995–2002 (mm $year^{-1}$)	Median of specific annual discharge ($m^3 s^{-1}$)	Mean annual evapotranspiration 1995–2002 (mm $year^{-1}$)
Le Guillec à Trézilidé (J3024010)	43	35	1015	0.67	686
Le Toulourenc à Malaucène (V6035010)	150	311	1059	1.32	1081
Le Loup à Villeneuve-Loubet (Y5615030)	279	2000	1159	4.47	1120

within and between years. The rainfall spatial variability is accentuated with altitude in the mountainous Loup catchment. During flood periods, overland flow is the main hydrological process, while during drought periods the main hydrological processes are evapotranspiration and baseflow. Note that while all models were used to simulate flows on all catchments in an ungauged mode (without calibration), HYDROTEL was only calibrated on the Guillec and Toulourenc catchments.

MODEL DESCRIPTIONS

This section presents the main structure, the hydrological processes, and the parameters of the three spatially distributed hydrological models (AFFDEF, HYDROTEL and ModSpa) used in the applications.

AFFDEF

AFFDEF is a spatially-distributed, continuous (in time) rainfall–runoff simulation model. The main characteristic of the model is that long simulation runs can be performed in limited computational times.

AFFDEF is raster-based. It takes as input the digital elevation model (DEM) of the catchment in raster form, as a rectangular matrix covering the catchment. The cells of the DEM can be of any size. It also needs as input rainfall and temperature data collected at an arbitrary number of raingauges and thermometers .

Many of the hydrological processes involved in the rainfall–runoff transformation have been schematized by using conceptual approaches. In particular, the model computes the local contribution at the surface runoff by applying a modified CN method (see Fig. 2(a)). In order to compute the soil storativity, one must provide the matrix of the CN numbers for any given DEM cell. The local contribution to the surface runoff and the groundwater flows are transferred to the catchment outlet by using a Muskingum-Cunge model with variable parameters, which are determined on the basis of the "matched diffusivity" concept (Orlandini & Rosso, 1996). The model has ten parameters (see Moretti & Montanari, 2006): the channel width/height ratio for the hillslope and channel network, the Strickler coefficients for the hillslope and the channel network, the critical source area, the saturated hydraulic conductivity, the width of the rectangular cross section of the subsurface water flow, the bottom discharge parameter for the infiltration reservoir capacity, the multiplying parameter for the infiltration reservoir capacity and the multiplying parameter for the interception reservoir capacity. Some of the model parameters have a well defined physical meaning and can be estimated on the basis of *in situ* surveys; the remaining parameters have to be optimized by calibration on the basis of some historical hydrometeorological records.

Although it can be used for any kind of catchment, it should be noted that AFFDEF simplifies the modelling of groundwater flows. Therefore it is best suited for basins where the runoff production is mainly due to infiltration excess.

The model may be freely downloaded at the web site http://www.costruzioni-idrauliche.ing.unibo.it/people/alberto, along with a user guide. A routine for performing automatic calibration that makes use of the SCE-UA genetic algorithm is included in the code.

HYDROTEL

HYDROTEL is a distributed hydrological model compatible with remote sensing and GIS (Fortin *et al.*, 2001). It is used operationally for flood forecasting in Québec

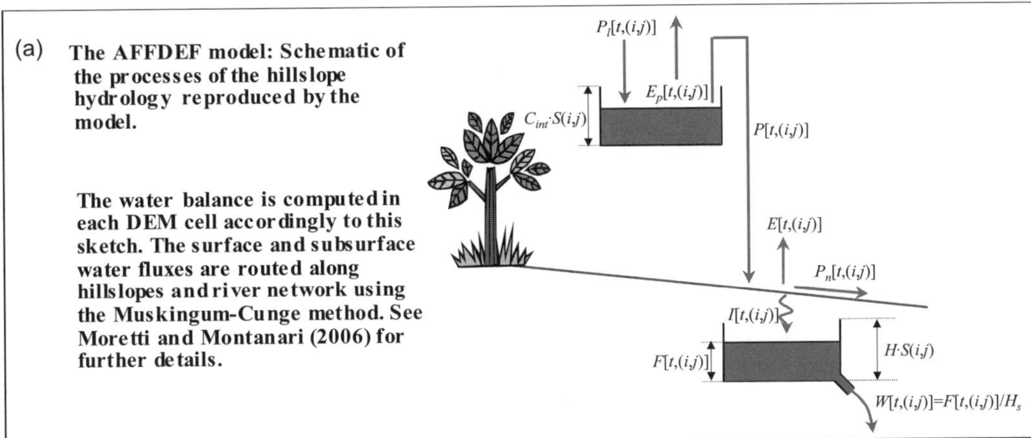

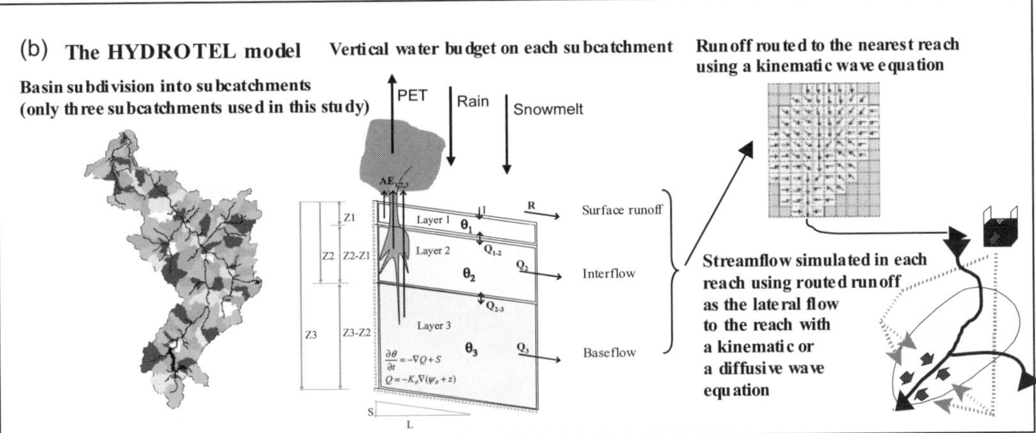

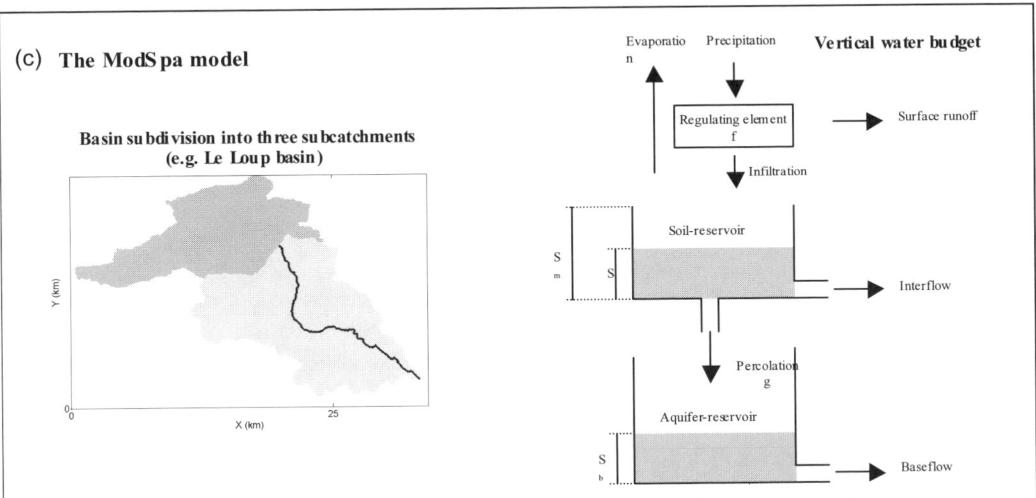

Fig. 2 The structure of the three spatially distributed mdodels AFFDEF (Moretti & Montanari, 2005), HYDROTEL (Fortin *et al.*, 2001) and ModSpa (Moussa, 1991).

(Turcotte *et al.*, 2004). Algorithms are derived as much as possible from physical processes, together with more conceptual or empirical algorithms. Natural units were chosen for the simulations: small subcatchments for the vertical water budget and flows towards the outlet of the unit, and river reaches for channel flow (Fig. 2(b)). Each subcatchment is characterized by a mean soil class and by the percentage of each land-use class (e.g. forest, agricultural zone, urban area, lakes, etc.).

The water budget procedure was developed so as to approximately represent the macro-processes related to infiltration and vertical redistribution of water in the "soil column" corresponding to the simulation unit. The soil column is divided into three layers (Fig. 2(b)). The surface layer is relatively shallow (5–20 cm), so as to represent the soil layer affected by evaporation over bare ground. While the first layer controls infiltration, the second layer can be associated with interflow and the third layer with baseflow. The Richards-1D equation is used to simulate flux exchange between the three layers. At each time step, the state variables calculated are the water content (θ_1, θ_2 and θ_3) of the three layers. HYDROTEL also simulates accumulation and melt of the snowpack using a mixed degree-day/energy-budget approach, and potential evapotranspiration using various methods such as Thornthwaite, Linacre, Penman-Monteith, Priestly-Taylor and the Hydro-Québec method. The kinematic wave model is used to simulate surface and subsurface flow on each subcatchment. Two simulation options are available for channel routing: the diffusive wave and the kinematic wave models. For this study, the Hydro-Québec method (Fortin, 1999) was used to compute potential evapotranspiration from daily minimum and maximum temperature, and the kinematic wave model was used for channel routing. Hence, the only meteorological inputs used in this study were temperature and precipitation. Note that the Hydro-Québec method for computing potential evapotranspiration includes a multiplicative parameter, c_{PET}, that is subject to estimation.

Over each subcatchment, the model parameters are the depth of each layer, the soil hydrodynamic properties of each soil layer (hydraulic conductivity at natural saturation, water content at natural saturation, residual water content, relationships between the hydraulic conductivity, the hydraulic head and the water content), the leaf area index, rooting depth of each land-use class and the geometric characteristics (area, slope, etc.). Each reach is characterized by the celerity of the flow. All of the hydrodynamic, land cover and geometric characteristics are estimated using a GIS and are typically not subject to estimation.

ModSpa

The main structure of the ModSpa (Modèle Spatialisé) spatially distributed hydrological was described by Moussa (1991). Digital elevation models are used to subdivide the catchment into right-banks, left-banks or source/head subcatchments and to extract the channel network. Over each subcatchment, the vertical water budget is computed using a two-layer model (Fig. 2(c)). The first layer, noted "soil reservoir", represents the upper soil layer where surface runoff, infiltration, interflow, percolation and evapotranspiration occurs. Infiltration is modelled using a reservoir Diskin-Nazimov model function of the precipitation, the total storage of the soil reservoir (S_m), the hydraulic conductivity at natural saturation of the soil (K_s), the maximum value of the hydraulic conductivity ($K_{max} = \alpha.K_s$ with α a parameter) and the soil humidity (S/S_m where S is the reservoir level). Evapotranspiration is calculated as a function of the above-listed parameters and the leaf area index (LAI) and the interflow is calculated as a function of the soil characteristics and the subcatchment surface slope. The second layer, noted "aquifer reservoir", represents the aquifer from which the baseflow is calculated as a function of the level (S_b) of this reservoir and a constant k representing the recession curve of this reservoir. Three state variables are calculated as a function of time: the regulating function (f) which separates rainfall into surface runoff and infiltration, the level (S) in the soil-reservoir and the level (S_b) in the aquifer-reservoir. Then, a transfer function, based on the diffusive wave equation, is used to route flows on each subcatchment (surface runoff, interflow and baseflow) and then through the channel network. This equation depends on the celerity (C) and diffusivity on the subcatchments and on each reach. In the applications, we do not consider the spatial variability of the parameters and only six parameters were calibrated, five for the vertical water budget (LAI, Ks, S_m, α and k) and one for the transfer function (C).

MODEL APPLICATIONS

This section presents applications of HYDROTEL and ModSpa referring to both the ungauged mode, where the model parameters are estimated without using hydrometric measurements for optimization, and the gauged mode, where the hydrological model is calibrated by using a split-sample procedure. The model was calibrated by using the data observed in the 1 August 1995–31 July 1999 time period and validated by referring to the 1 August 1999–31 July 2002 period. The objective functions used are the Nash and Sutcliffe efficiency criteria, the bias and the Root Mean Square Error. The detailed equations are presented in Chahinian *et al.* "Compliation of the MOPEX 2004 results" (this volume).

AFFDEF was applied in a mixed gauged/ungauged mode. The application of AFFDEF was developed by following the same procedure described above for the gauged mode, but the calibration was performed using a limited data set. Only one flood event for each catchment was used for calibrating the model, by using a trial and error procedure (Table 2(a)). A limited calibration data set was used in order to test the model capability of providing a robust simulation even when a limited data set is available. Therefore, the modality of the AFFDEF application differs with respect to the other two models. The ungauged application of AFFDEF was not performed as no preliminary knowledge was available about the distribution of the AFFDEF parameters with reference to the hydrological behaviour of the three catchments. In fact, the previous applications of AFFDEF referred to catchments located in Italy and Germany. Therefore no information was available for French catchments or for catchments that can be considered similar to those considered here.

A priori parameter estimation (ungauged mode)

HYDROTEL is very sensitive to some parameters, but as most of them have a physical meaning, the *a priori* values of these parameters was based on expert experience in

Table 2 Comparison between the estimated parameters (ungauged mode) and the calibrated parameters (gauged mode) for the AFFDEF, HYDROTEL and ModSpa models.

(a) AFFDEF Model parameters	Ungauged mode	Gauged mode Guillec (J3024010)	 Toulourenc (V6035010)	 Loup (Y5615030)
Channel width/height ratio for the hillslope (-)	---	400	100	600
Strickler roughness for the hillslope ($m^{1/3}\ s^{-1}$)	---	0.5	2.0	0.1
Channel width/height ratio for the channel network (-)	---	20	20	30
Strickler roughness for the channel network ($m^{1/3}\ s^{-1}$)	---	15	25	8
Constant critical source area (km^2)	---	0.5	0.5	0.5
Saturated hydraulic conductivity ($m\ s^{-1}$)	---	0.01	0.01	0.005
Width of the rectangular cross section of the sub-surface water flow (m)	---	0.5	0.5	0.5
Bottom discharge parameter for the infiltration reservoir capacity (s)	---	1000000	50000	79095
Multiplying parameter for the infiltration reservoir capacity (-)	---	1.10	1.00	0.20
Multiplying parameter for the interception reservoir capacity (-)	---	0.10	0.60	0.75

(b) HYDROTEL Model parameters subject to calibration	Ungauged mode Guillec (J3024010)	 Toulourenc (V6035010)	Gauged mode Guillec (J3024010)	 Toulourenc (V6035010)
c_{PET}	1.5	1.34	1.52	1.61
z_2	40 cm	37.5 cm	1.9 m	44.5 cm
z_3	80 cm	75 cm	2 m	89 cm

(c) ModSpa Model parameters	Ungauged mode	Gauged mode Guillec (J3024010)	 Toulourenc (V6035010)	 Loup (Y5615030)
K_s ($\times 10^{-7}$ m s^{-1})	7.00	0.69	5.56	4.40
S_m (m)	0.50	0.42	0.30	0.51
α	7.0	7.3	6.0	4.8
k ($\times 10^{-7}$ 1 s^{-1})	1.50	1.51	1.48	1.80
C (m s^{-1})	0.20	0.01	0.20	0.56
LAI	1.00	0.85	2.45	1.00

hydrological modelling in France using Rawls & Brakensiek (1989) pedotransfer functions. Also, since potential evapotranspiration was provided as an input in the meteorological database for the MOPEX experiment, the coefficient c_{PET} was set to a value such that the average potential evapotranspiration (PET) computed by HYDROTEL matched the average PET in the meteorological database. The parameters that were subjectively estimated by experts were the channel width, the vegetation rooting depth, and the depth of the third layer of soil (z_3), The depth of the bottom of the second layer (z_2) was set to half of z_3, and the depth of the first layer was kept at 5 cm.

ModSpa was applied by subdividing the catchment into three subcatchments: one source subcatchment, one right-bank subcatchment and one left-bank subcatchment as shown in Fig. 2(c). Parameters were considered constant on the three subcatchments and considered equal to those calibrated by Moussa (1991) on French catchments.

Table 2 shows the estimated parameters for HYDROTEL and ModSpa. The results show that the Nash and Sutcliffe efficiency of the two models ranges between 46% and 64% for the Toulourenc and the Loup catchments while negative values were obtained for Le Guillec catchment. The results can be considered satisfactory, especially if one considers that the hydrological observations available on these catchments were not used in any way to calibrate the models.

Parameter estimation through calibration (gauged mode)

AFFDEF was calibrated by following a trial and error procedure, as mentioned above. The values of the calibrated parameters are shown in Table 2(a) and Fig. 3 shows a comparison between measured and calculated discharge. HYDROTEL was calibrated by minimizing by trial and error the sum of the absolute relative bias on the daily flows and the absolute bias on the coefficient of variation of the daily flows (Fig. 4). Hence, individual streamflow observations were not used directly for model calibration: only the first two moments of daily flows (mean and variance) are used when computing the objective function. It is hoped that reasonable values for these two statistics can be estimated through statistical regionalization techniques, thus allowing one to use this calibration technique even for ungauged catchments. Manual calibration of HYDROTEL is quite tedious. Hence, only the parameters c_{PET}, z_2 and z_3 were calibrated, being amongst the most sensitive parameters in the model (Table 2(b)). An automated calibration algorithm is now available for HYDROTEL (see Turcotte *et al.*, 2003), but this technique was not available when the current study was performed. For this reason, manual calibration was only performed for the Guillec and Toulourenc catchments. The ModSpa calibration procedure was performed on six parameters *Ks*, *Sm*, α, *k*, *C* and LAI using a manual trial and error method minimizing the Nash and Sutcliffe criteria on the daily flows for the first five year period. Table 2(c) shows the calibrated parameters of ModSpa and Fig. 5 shows the simulation results.

Overall, the performances of AFFDEF, HYDROTEL and ModSpa are satisfactory when compared with similar applications presented in the literature for both lumped and distributed models: The Nash-Sutcliffe criteria ranges between 55% and 82% for the three models on the three catchments. Evaluation of the performances of AFFDEF should take into consideration the fact that this model was calibrated using a very limited data record.

AFFDEF

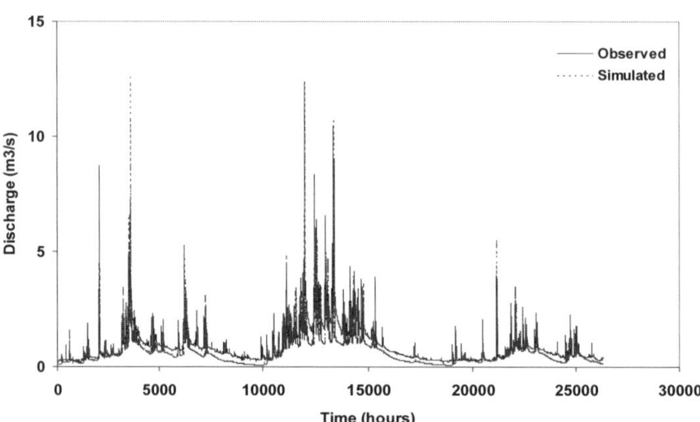

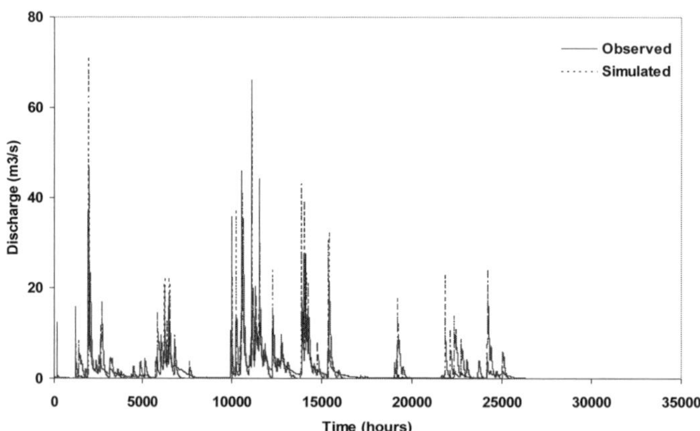

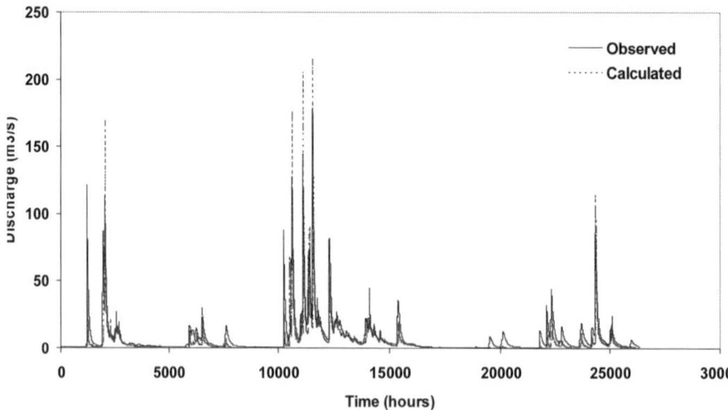

Fig. 3 Comparison between observed and calculated discharge for the AFFDEF model for the gauged mode.

HYDROTEL

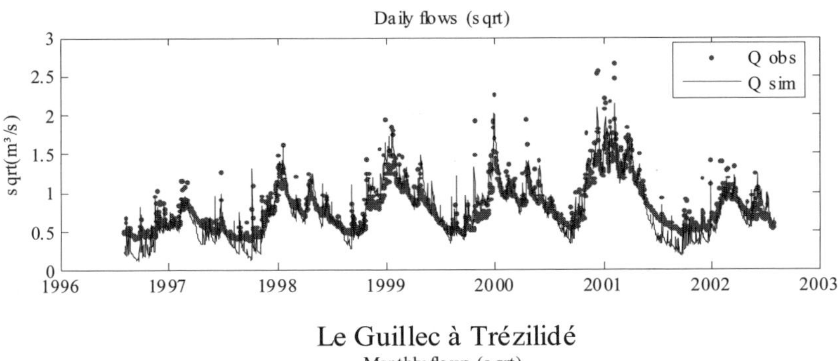

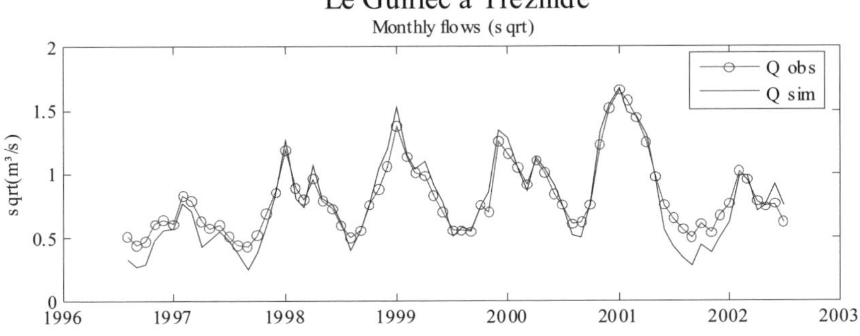

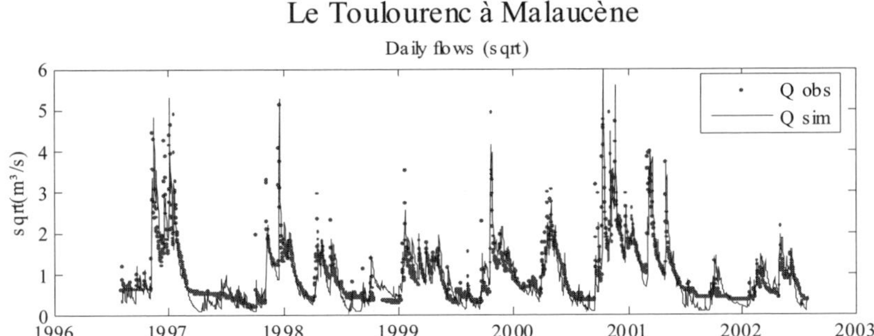

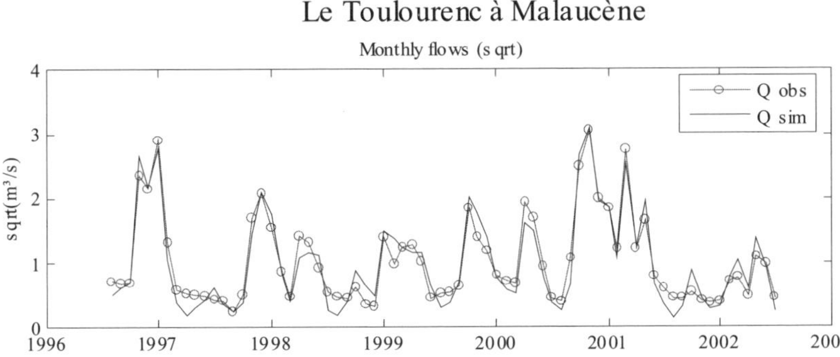

Fig. 4 Comparison between observed and calculated discharge for the HYDROTEL model for the gauged mode.

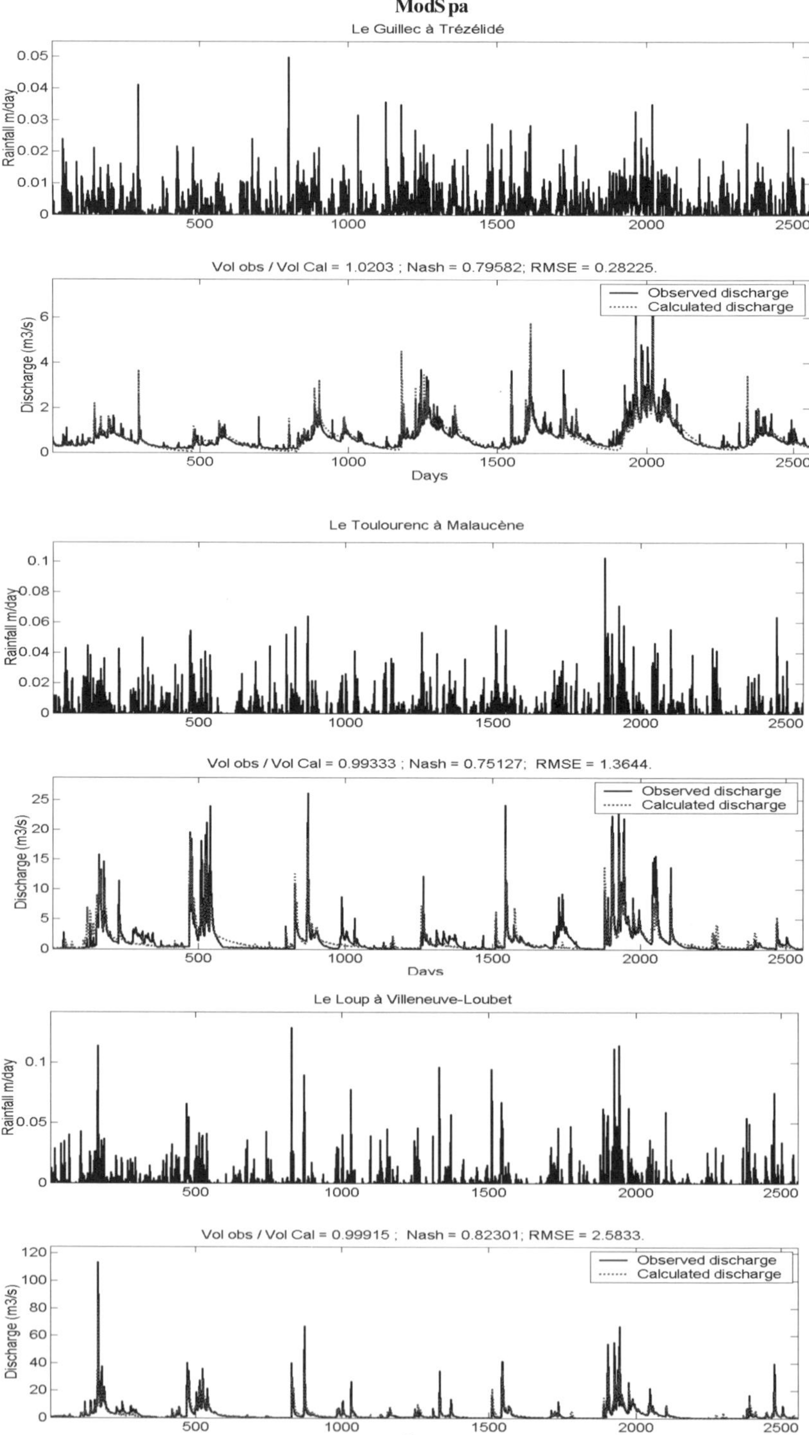

Fig. 5 Comparison between observed and calculated discharge for the ModSpa model for the gauged mode.

DISCUSSION AND CONCLUSIONS

The purpose of this study is to present and apply the three spatially distributed models described above by using only lumped values of input data such as precipitation and potential evapotranspiration. The aim is to discuss the usefulness of the application of these models for operational hydrology.

When only lumped information is available, or when applying hydrological models on ungauged catchments, the drawbacks in using a spatially distributed model are the impossibility of fully using the distributed information, the difficulty of obtaining values for distributed parameters, the model set-up time and the computation time. However, the reasons for using distributed models when the inputs are lumped are the possibility of downscaling some input variables such as precipitation and temperature as a function of the altitude available from the DEM (e.g. as in HYDROTEL) and the possibility of getting an insight to the hydrological response of the catchment from the DEM (e.g. as in ModSpa and AFFDEF). The use of spatially distributed models also allows one to identify and deal with data inconsistencies when flow measurements are available on internal subcatchments.

Overall, the applications presented here show that spatially distributed models can be successfully applied by using mixed lumped/distributed information. In fact, the performances obtained by the models are satisfactory and comparable with the results that are usually obtained in real world applications of lumped models in similar situations. The fact that HYDROTEL was calibrated only using the first two moments of daily flows, is also interesting, as this means that the data requirements for calibration are fairly low for this model. The same consideration applies to AFFDEF, which was calibrated using data observed during only one flood event and therefore equivalent to a nearly ungauged situation. The model produced satisfactory results for all catchments.

The results obtained in these applications confirm that the use of distributed models is advisable when at least a piece of distributed information is available. For instance, in the case considered here, a DEM was available for the three catchments and the models considered in this study benefited from the valuable information that the DEM itself can provide.

REFERENCES

Abbott, M. B., Bathurst, J. C., Cunge, J. A., O'Connell, P. E. & Rasmussen J. (1986) An introduction to the European Hydrological System–Système Hydrologique Européen, SHE. 2: Structure of a physically-based, distributed modelling system. *J. Hydrol.* **87**(1), 61–77.

Beven, K. J. & Kirkby, M. J. (1979) A physically based variable contributing area model of basin hydrology. *Hydrol. Sci. Bull.* **24**(1), 44–69.

Crawford, N. H. & Lindsey, R. K. (1966) Digital simulation in hydrology: Stanford watershed model IV. *Tech. Rep. no. 39,* Stanford, Univ., Stanford, California, USA.

Donigian, A. S. D., Bicknell, B. R. & Imhoff, J. C. (1995) Hydrological simulation program—Fortran (HSPF). In: *Computer Models of Watershed Hydrology* (ed. by V. P. Singh), 395–442. Water Resource Publications, Highlands Ranch, Colorado, USA.

Fortin, J. P., Turcotte, R., Massicotte, S., Moussa, R. & Fitzback, J. (2001) Distributed Watershed model compatible with remote sensing and GIS data. I : Description of the model. *J. Hydrol. Engng, ASCE* **6**(2), 91–99.

Fortin, V. (1999) Le modèle météo-apport HSAMI: historique, théorie et application. Rapport de recherche, révision 1,5. Institut de recherche d'Hydro-Québec (IREQ), Varennes, Québec, Canada.

Moretti, G. & Montanari, A. (2006) AFFDEF: a spatially distributed grid-based rainfall–runoff model for continuous time simulations of river discharge. *Environmental Modeling and Software* (in press).

Moussa, R. (1991) Variabilité spatio-temporelle et modélisation hydrologique. Application au bassin du Gardon d'Anduze. PhD Thesis, University of Montpellier II, France.

Orlandini, S. & Rosso, R. (1996) Diffusion wave modelling of distributed catchment dynamics. *J. Hydrol. Engng, ASCE* **1**(3), 103–113.

Perrin, C., Michel, C. & Andréassian V. (2003) Improvement of a parsimonious model for streamflow simulation. *J. Hydrol.* **279**(1–4), 275–289.

Rawls, W. & Brakensiek, D. (1989) Estimation of soil water retention and hydraulic properties. In: *Unsaturated Flow in Hydrologic Modeling: Theory and Practice* (ed. by H. J. Morel-Seytoux), 275–300. NATO ASI series. C. Kluwer Academic, Arles, France.

Refsgaard, J. (1997) Parameterisation, calibration and validation of distributed hydrological models. *J. Hydrol.* **198**, 69–97.

Singh, V. P. (1995) *Computer Models of Watershed Hydrology*. Water Resources Publications, Fort Collins, Colorado, USA.

Turcotte, R., Rousseau, A. N., Fortin, J. P. & Villeneuve, J. P. (2003) Development of a process-oriented, multiple-objective, hydrological calibration strategy accounting for model structure. In: *Calibration of Watershed Models* (ed. by Q. Duan, S. Sorooshian, H. Gupta, A. N. Rousseau & R. Turcotte), 153–163. American Geophysical Union (AGU), Washington, USA.

Turcotte, R., Lacombe, P., Dimnik, C. & Villeneuve, J. -P. (2004) Prévision hydrologique distribuée pour la gestion des barrages publics du Québec. *Can. J. Civil Engng* **31**(2), 308–320.

Flow simulation in an ungauged basin: an alternative approach to parameterization of a conceptual model using regional data

MONOMOY GOSWAMI & KIERAN MICHAEL O'CONNOR
Department of Engineering Hydrology, National University of Ireland, Galway, Ireland
kieran.oconnor@nuigalway.ie

Abstract An alternative approach to parameterization of the conceptual Soil Moisture Accounting and Routing (SMAR) model for continuous simulation of flow in an ungauged basin, using daily discharge data from neighbouring gauged catchments, is presented. The methods are suited for situations where adequate data on soil, geology, land use, etc., which are required for relating model parameters to catchment-specific hydrometeorological and physiographical characteristics, are not available. Four methods, namely: regional averaging of data, regional pooling of data, transposition of data, and averaging the parameters of the model calibrated for different gauged catchments, are applied to a group of 12 catchments in France. The concept of a pseudo-ungauged basin is used whereby each catchment is first considered as ungauged for the purpose of flow estimation, and only subsequently considered as gauged to evaluate the performance of the flow estimation procedures. Performance evaluation using the Nash-Sutcliffe efficiency index shows that pooling of data is the best in this approach.

Key words conceptual model; flow simulation; rainfall–runoff; regional analysis; ungauged basin

INTRODUCTION AND OBJECTIVE

Parameters of a conceptual model, when calibrated for a catchment, are generally regarded as a potential source of information about the physiographic and hydrometeorological characteristics of that catchment. Explicit relations between the descriptors of relevant features of the catchment and perceptibly predominant processes of the hydrometorological systems on one hand, and the calibrated values of parameters on the other, are often difficult to prescribe. Yet efforts are made to link the parameters of the model with some descriptors of catchment characteristics and hydrometeorological behaviour using physically plausible and usually empirical relations. This assumed capability of producing meaningful relationships for transferring information from the gauged to the ungauged basin is often considered a prospective tool for application in ungauged basins for rainfall–runoff simulation. However, in many situations, particularly where an ungauged catchment is located in the remote headwater region of a river, adequate and reliable data about soil, geology, land use, land cover, etc., may not be available. Although important topographical and hydrometeorological data can generally be obtained with some effort, lack of complete data sets representative of all major hydrological processes, and consequent difficulty in relating, empirically or otherwise, the parameters of a conceptual model to the catchment-characteristics greatly limit the application of the conceptual model for such purposes. In such situations, however, it may not be difficult to collect hydrometeorological data from a number of gauged catchments from a homogeneous region to which the ungauged catchment under study may be considered to belong. In this context, the term "region" may not be restricted to geographical proximity. Judging on the nature of the response behaviour to rainfall inputs, and broad hydrometeorological conditions, contiguous or "local" catchments may also be included in the "region". An approach of regionalization for evaluation of parameter values of the conceptual model for subsequent use in ungauged catchments in such cases may be useful. Three methods involving regionalization of discharge data of the gauged catchments, and one method in which parameters calibrated for each of the gauged catchments are combined are presented in this study. The physically-inspired lumped conceptual Soil Moisture Accounting and Routing (SMAR) model has been used for transferring the values of the parameters from gauged basins to the ungauged basin for this purpose.

Seven years of continuous daily hydrometeorological data from 12 French catchments were generously provided by Météo France and the Direction de l'Eau, through Dr Vazken Andréassian, of Cemagref, Paris, for application in the MOPEX (Model Parameter Estimation Experiment) research project, and made available to the present authors for their contribution to the 2004 MOPEX Workshop held in Paris. The Nash-Sutcliffe efficiency index (R^2) is used for assessing the performance of the SMAR model. Results are presented, and conclusions are drawn on the efficacy of the procedure.

METHODOLOGY AND REGIONALIZATION STRATEGIES

Data gaps in discharge series of six catchments in the sub-group are first filled synthetically by using rainfall, evaporation and available discharge records with the SMAR model in an iterative scheme. With the starting values of –9.99 for the missing data, the model is calibrated, and the iterations continued until the model performance in two successive iterations converged, the discharge data used for the gaps in each iteration being the corresponding estimates obtained in the previous iteration.

For assessing relative performance of the methods in the regionalization approach, each of the 12 gauged catchments considered in the region is used initially in turn as if it was ungauged, but subsequently, after simulation of its flows from the regional analysis, as gauged (using its measured flow data) for the purpose of evaluating the simulation efficiency of the procedure of flow simulation in an ungauged catchment in the region. The catchment, thus considered, is called "pseudo-ungauged" in this study.

With the underlying assumption that regionalization reflects the general characteristics of the region, which are also representative of the ungauged basin in the region, four regionalization methods are applied for flow estimation in pseudo-ungauged catchments. The methods are described below.

Regional averaging of data

In this method, the naïve "no-model" discharge estimate of a pseudo-ungauged (equally applicable to ungauged) catchment is obtained as the average of the synchronous discharges of N–1 number of gauged catchments excluding the pseudo-ungauged catchment in the selected region comprising N catchments. The average discharge series, thus generated, is used for rainfall–runoff simulation by calibration of the SMAR model. The rainfall and the evaporation data series, as required by a model, are those observed for the pseudo-ungauged catchment, i.e. it is assumed that rainfall and evaporation data are available for the pseudo-ungauged catchment and ultimately for any ungauged catchment in the region.

Regional pooling of data

In this method, the observed hydrological data series are combined by putting the m years of data from all gauged catchments (N–1 in a group of N) in a region, in series, end to end, and appending the data of the pseudo-ungauged (Nth) catchment as the last one, thereby making $N \times m$ years of data in all. Hydrological models are then calibrated by maximizing the combined R^2 value over the calibration period using the corresponding combined (end to end) input to the N–1 gauged catchments in a region as inputs to the "regional model" to simulate flows in this calibration period. Finally, considering the entire data set of the Nth pseudo-ungauged catchment as that belonging to the "validation" period of the "regional model", and using the rainfall and evaporation inputs to that catchment as inputs to the model, the discharge series for the pseudo-ungauged catchment is estimated.

Transposition of data

This "nearest neighbour" approach is used when very few catchments in a homogeneous region in the neighbourhood of an ungauged catchment are gauged. The flow data series of a gauged catchment measured in volume of flow per unit time, i.e. $m^3 s^{-1}$, are scaled up or down in the proportion of catchment areas depending on whether the ungauged (pseudo-ungauged in this study) catchment is larger or smaller in area than the gauged catchment considered. Noting that for the catchments used in the present study the rainfall values within each subgroup are similar, the mean annual rainfall data of the gauged and the ungauged catchments are not used for additional scaling. By scaling of flow rates from the gauged catchments by area, it is thus assumed that for all catchments in a homogeneous region, the flow depth over the catchment during any given time interval is uniform. Taking the rainfall and the evaporation data series as those observed for the ungauged catchment, and the flow data series as that obtained by scaling of the data series of the nearest-neighbour gauged catchment, i.e. of the "index basin", the SMAR model is calibrated. For validation, the output of the model is compared with the corresponding measured discharge of the pseudo-ungauged catchment.

Regional averaging of parameters

In this method, unlike the previous three, the SMAR model is first applied to N–1 gauged catchments individually in a group of N catchments in the region obtaining the best possible fit for each catchment. The weighted average of N–1 values of each parameter from the N–1 parameter sets is obtained. The weights used in this study are based on the R^2 efficiency values, which reflect the degree of fit of the model. These weighted-average parameters are then applied, without further calibration of the SMAR model, to the Nth catchment which is considered pseudo-ungauged, and the simulated discharge is compared with the discharge observed at the pseudo-ungauged catchment.

For simplicity, despite its shortcomings (Kachroo & Natale, 1992), only the dimensionless efficiency index R^2 (Nash & Sutcliffe, 1970) is used in this study for judging the relative performance of the SMAR model while using different methods of regionalization. Whereas R^2 = 100% would denote an ideal or "perfect" fit, it is generally agreed that $R^2 > 90\%$ is indicative of a very good model fit, while that in the range of 80–90% is a fairly good fit, and a range of 60–80% is unsatisfactory. For consistency, a "warm-up" period corresponding to first year's data is not considered for performance evaluation in calibration as well as in validation in all four methods.

ASSESSMENT OF REGIONAL HOMOGENIETY

Assessment of regional homogeneity is very important for reducing the uncertainty of estimation of flow in an ungauged catchment by regional analysis. In the group of 12 catchments, the three in the northwest are in the humid seaboard climatic zone, three in the southeast are in the Mediterranean zone, one in the northeast is in the semi-continental zone, and the remaining five are characterized by an intermediate climatic zone. A1522020 in the east is the wettest and H3613020 is the driest. The southeastern catchments, namely, V6035010, Y3514020 and Y5615030 have significant evaporation, with evaporation exceeding rainfall for almost 80% of the time. The northwestern catchments, namely, J2034010, J3024010 and J4124420 constitute a hydrometeorologically homogeneous region with very little variability in hydrometeorological data values. The contributing catchment at station K0744010 is contained within that of station K0753210, both stations being located on the same river. These two catchments are therefore considered as being hydrometeorologically homogenous for the purpose of this study.

Topographically, the mean altitude, the altitude at the highest point, and the altitude at the outlet of the three catchments in the west are of the same order of magnitude, whereas Y3514020 in the sub-group of three catchments in the southeast is located at a much lower altitude in comparison with V6035010 and Y5615030 in that sub-group. A1522020 is located at a significantly high altitude in the northeast France. Among the three catchments in the central region, H5723011 and H3613020 are located at nearly the same altitude, whereas H2001020 is at a higher altitude.

Characteristics of the hydrological data of all 12 catchments are presented in Table 1. It is seen from the table that the rainfall in three southeastern catchments, namely, V6035010, Y3514020 and Y5615030, exceeds the evaporation for only about 15–20% of the time. Although conscious that just seven years of data is hardly sufficient to adequately characterise the rainfall–runoff process in these drier catchments, this number was adopted for uniformity in application, the same number being used in the analysis for consistency with the lengths of data available for the other catchments in

Table 1 Characteristics of hydrological data (1 August 1995–31 July 2002) 2557 data values.

Station code no.	% days $R>E$	$R_{mean} - Q_{mean}$ (mm day^{-1})	$\frac{R_{mean} - Q_{mean}}{E_{mean}}$	$\frac{E_{mean}}{R_{mean}}$	$R_{mean} - E_{mean}$ (mm day^{-1})
J2034010	37.3	1.74	0.90	0.74	0.69
J3024010	38.7	1.33	0.71	0.68	0.90
J4124420	38.1	1.95	0.99	0.58	1.41
A1522020	40.0	2.52	1.25	0.44	2.55
H5723011	30.7	1.54	0.75	0.94	0.14
H3613020	31.6	1.87	0.93	0.91	0.20
H2001020	37.0	1.18	0.58	0.57	1.53
K0744010	34.1	1.66	0.83	0.74	0.69
K0753210	34.6	1.56	0.78	0.72	0.78
V6035010	20.1	2.07	0.70	1.02	–0.06
Y3514020	15.4	1.61	0.51	1.37	–0.86
Y5615030	17.1	1.59	0.52	0.97	0.10

the sample. It is also observed that the ratio of mean values of evaporation to rainfall is significantly greater than unity (at 1.37) for the Y3514020 catchment in the southeast, very close to unity for the two neighbouring catchments, i.e. V6035010 and Y5615030, and less than unity for the remaining nine catchments. These statistics, while indicating higher evaporation levels in the three catchments in the south-east of the country, also suggest that the catchment Y3514020 is an outlier within the group of 12 catchments. As a further indication of the heterogeneity effect of Y3514020, the mean evaporation for this catchment is seen to be higher than the mean rainfall by 0.86 mm day^{-1}, whereas it is either less than or very near to the mean rainfall in the case of the other eleven catchments. A comparison of the values in Table 1 of $[(R_{mean} - Q_{mean})/E_{mean}]$, i.e. the ratio of the difference between the mean values of rainfall and discharge to the corresponding value of evaporation, shows that, for catchment A1522020, this ratio is significantly greater than unity (at 1.25) whereas it is less than unity for the other 11 catchments. This suggests that, in the transformation of rainfall to precipitation, quite apart from evapotranspiration losses, there is substantial unaccounted-for loss in this catchment, so that the hydrological system, represented by the observed rainfall, evaporation, and discharge data, is apparently non-conservative.

On estimation of regionally averaged "no-model" discharge for each pseudo-ungauged catchment for regional analysis, it is observed that the ratio of the mean regionally averaged discharge to the mean observed discharge for H3613020 is very high (3.79) followed by H5723011 (1.99), Y3514020 (1.77) and V6035010 (1.51), whereas for other catchments it is near to 1. This shows that in each of these four catchments, actual discharge production is much less than the regionally averaged discharge, and there may have been some component of rainfall which is not accounted for in the water-balance expression dependent on the observed data series. In the light of the above comments, the six catchments namely, A1522020, H3613020, H5723011, Y3514020, Y5615030 and V6035010 are considered heterogeneous in the group of 12 catchments.

Although initial analyses were conducted using all 12 catchments without considering regional heterogeneity, in the light of the observations regarding homogeneity presented above, the catchments A1522020, H3613020, H5723011,

V6035010, Y3514020 and Y5615030 were excluded from the subsequent analyses having the objective of the parameterization of the SMAR model variant applicable to conservative systems. Thus the remaining six catchments: J2034010, J3024010, J4124420, H2001020, K0753210 and K0744010, having discarded the six apparently heterogeneous catchments, as indicated above, were used only for exploratory tests to assess the effect of homogeneity, as reflected in the sub-group within the whole sample of 12 catchments, on the overall performance of the chosen methods.

Clearly, it would be desirable from the perspective of drawing a generalized conclusion on the performance of the regional methods tested in this study to include more catchments in the whole sample and hence in the sub-group. However, for the purposes of this heuristic study, the number used was deemed sufficient and was also used to demonstrate the applicability of the methods, their relative efficiencies, and the importance of homogeneity in selection of catchments for the simulation of flow in the case of ungauged basins.

MODEL DESCRIPTION

Whereas the standard form of the model, generally indicated by SMARG and referred in this study by SMAR, was designed for conservative systems, modifications were incorporated in its structure to make it applicable to both conservative and "apparently non-conservative" systems. These variants are named SMAR-NC1 and SMAR-NC2, where "NC" denotes "non-conservative". Detailed description of the structure of the model, which is not included in this paper for brevity, may be found in Kachroo (1992) for the original version of SMAR, Goswami *et al.* (2002) for SMARG, and Goswami & O'Connor (2005) for SMAR-NC1 and SMAR-NC2.

Briefly, the SMAR model is a parsimonious nine-parameter lumped quasi-physical conceptual rainfall–evaporation–runoff model, with distinct water-balance and routing components. Using a number of empirical and intuitively assumed relations which are considered to be at least physically plausible, the nonlinear water balance (i.e. soil moisture accounting) component ensures satisfaction of the continuity equation over each time-step, i.e. it preserves the balance between the rainfall, the evaporation, the generated runoff, and the changes in the various elements (layers) of soil moisture storage. Five parameters, namely, Z (moisture holding capacity of soil layers), T (evapotranspiration conversion factor), H (fast response separation factor), Y (infiltration excess separation term), and C (factor for soil moisture depletion by evapotranspiration), control the overall operation of the water-budget component of the SMAR model. The four parameters G (saturation excess separation factor), N (shape factor of the Nash cascade model for surface water routing), NK (lag of the Nash cascade model) and K_g (linear reservoir constant for groundwater routing) control the operation of the routing component.

RESULTS AND DISCUSSION

A summary of the result of application of the SMAR model using different regionalization methods, first considering all 12 catchments and subsequently considering the sub-group of six catchments, are given in Tables 2 and 3. These tables also show the "no model" efficiencies, which correspond to the discharge series generated for a pseudo-ungauged catchment by averaging the discharges of all gauged catchments (except the pseudo-ungauged) in the region. The efficiency of the model, when applied to the actual observed discharge data of the pseudo-ungauged basin, considering this basin as gauged, is given for comparison. It is observed from Table 2 that, in comparison with the method of regional averaging of data, the regional pooling method performs significantly better in calibration and in 10 out of 12 catchments in validation. For the sub-group of six catchments, the pattern is similar, the R^2 efficiency in this case being significantly better in validation in five out of six catchments. The R^2 efficiency in validation, which really reflects the ability of the model to perform in an ungauged catchment, has a median value of 70.21% and 70.65% for all 12 catchments and for the sub-group of six catchments, respectively, which is reasonably good for an ungauged catchment. The improvement in performance in this method is attributed to the use of catchment-specific data series from each catchment in a group without

Table 2 R^2 efficiency values (%) by different regionalization methods considering all 12 catchments.

Catchment (station code) and area (km^2)	H2001020 (98)		J2034010 (125)		J3024010 (43)		J4124420 (32.1)		K0744010 (181)		K0753210 (371)	
	Calib.	Verif.	Calib.	Verif.	Calib.	Verif.	Calib.	Verif.	Calib.	Verif.	Calib.	Verif.
Regional averaging of data	68.92	17.89	60.01	63.81	51.33	65.68	55.80	66.15	62.01	64.97	61.86	67.82
Regional pooling of data	74.45	72.41	74.60	68.21	73.35	73.21	75.51	55.08	74.65	70.57	74.90	66.76
Regional averaging of parameters	---	67.86	---	49.75	---	73.57	---	45.97	---	72.96	---	72.30
"No model"	---	10.03	---	23.56	---	34.33	---	39.63	---	37.16	---	44.01
Best fit with actual observed discharge data of the pseudo-ungauged basin	---	73.42	---	83.07	---	83.43	---	87.69	---	75.76	---	75.09

Catchment and area (km^2)	A1522020 (68.1)		H3613020 (252)		H5723011 (104)		V6035010 (150)		Y3514020 (291)		Y5615030 (279)	
	Calib.	Verif.	Calib.	Verif.	Calib.	Verif.	Calib.	Verif.	Calib.	Verif.	Calib.	Verif.
Regional averaging of data	50.78	21.72	58.74	–1722	57.57	26.89	42.48	42.68	6.18	26.42	10.79	15.11
Regional pooling of data	74.58	74.42	75.01	54.57	73.77	64.82	74.91	69.85	74.31	82.37	72.17	72.68
Regional averaging of parameters	---	68.38	---	–685.31	---	54.43	---	72.15	---	68.88	---	78.34
"No model"	---	15.65	---	–2766	---	7.84	---	15.62	---	–110	---	11.89
Best fit with actual observed discharge data of the pseudo-ungauged basin	---	78.07	---	–78.23	---	68.06	---	77.50	---	79.14	---	89.30

Calib: efficiency in calibration with regionally derived data, Verif: efficiency (in verification, wherever applicable) when used for pseudo-ungauged basin

Table 3 R^2 efficiency values (%) by different regionalization methods considering six catchments.

Catchment (station code) and area (km^2)	H2001020 (98)		J2034010 (125)		J3024010 (43)		J4124420 (32.1)		K0744010 (181)		K0753210 (371)	
	Calib.	Verif.	Calib.	Verif.	Calib.	Verif.	Calib.	Verif.	Calib.	Verif.	Calib.	Verif.
Regional averaging of data	55.42	37.71	71.84	31.84	61.47	74.85	43.54	84.13	59.25	59.52	60.02	65.28
Regional pooling of data	74.45	65.69	72.05	75.04	71.58	77.75	73.55	58.26	72.09	71.42	72.61	69.87
Transposition												
Index basin												
J2034010	---	---	---	---	82.97	54.08	75.28	56.68	---	---	---	---
J3024010	---	---	76.22	43.12	---	---	76.78	82.53	---	---	---	---
J4124420	---	---	77.17	32.90	84.09	76.76	---	---	---	---	---	---
K0744010	---	---	---	---	---	---	---	---	---	---	55.90	35.75
K0753210	---	---	---	---	---	---	---	---	42.04	58.52	---	---
Regional averaging of parameters	---	63.37	---	60.00	---	75.75	---	59.39	---	69.34	---	69.29
"No model"	---	19.42	---	-9.75	---	39.90	---	44.20	---	36.00	---	50.47
Best fit with actual observed discharge data of the pseudo-ungauged basin	---	73.42	---	83.07	---	83.43	---	87.69	---	75.76	---	75.09

Calib: efficiency in calibration with regionally derived data, Verif: efficiency (in verification, wherever applicable) when used for pseudo-ungauged basin.

any averaging, and hence without dilution of the response characteristics of each of the catchments. Models calibrated to the data series generated by pooling simulate the response from an ungauged catchment in the region better because of the response of the ungauged catchment being similar to many, if not all, catchments of the region.

The method of transposition, being suitable for a small sub-group of catchments, was applied only to two relatively more homogeneous sub-groups, one comprised of the catchments J2034010, J3024010 and J4124420 in the northwest, and the other consisting of two catchments, K0744010 and K0753210. The results of the method of transposition are given in Table 3. From the results of transposition for catchment J4124420 considering J3024010 as the index basin, and *vice versa*, it may be observed that for catchments having areas of identical order of magnitude good results may be obtained by transposition. For a large difference in areas of the two catchments considered for transposition (one gauged and the other pseudo-gauged), the scaling of discharge series of the index catchment lowers the performance. This is seen from the performances of the pairs K0744010 and K0753210, J2034010 and J3024010, and J2034010 and J4124420, when considered for transposition.

Comparison of results of the method of regional averaging of parameters with those of the regional pooling method in Table 2 shows that, for the case with 12 catchments, the former performs better in five out of 12 catchments, and that the efficiency for the catchment H3613020 by the method of regional averaging of parameters is unacceptably negative. This reflects that the method of regional pooling is clearly best for the group of 12 catchments. It also shows that catchment H3613020 may have been an outlier in the group, and that the inclusion of this catchment in the regional analysis is likely to reduce the efficiency of flow modelling by any regionalization method. A similar comparison of the method of regional averaging of parameters with the regional pooling method given in Table 3 shows that, in the case of the sub-group of six catchments, the efficiencies in the pseudo-ungauged basins, although generally lower, are comparable. In the context of the regional averaging method, however, it may be noted that due to equifinality (Beven & Freer, 2001), the parameter values in the optimum parameter set, as obtained by the modelling exercise, may differ significantly in some cases from catchment to catchment in the group, and meaningful averaging of parameters may not be possible. A number of tests may be required in such a case to choose the appropriate set of parameter values giving the best fit of the model to be considered for averaging.

From Tables 2 and 3, it is found that in five out of the six catchments indicated in Table 3, the method of regional pooling of data, considering these six catchments as a homogeneous sub-group, generally performs better than when the whole group of 12 catchments is selected. Thus, although a larger volume of data is used in the case of 12 catchments, the homogeneity in the sub-group of six catchments yields better efficiency of the modelling method in the case of the test with the homogeneous sub-group of only six catchments in comparison with that with all 12 catchments. Similarly, in three out of these six catchments, the method of regional averaging of parameters, considering the six catchments in the homogeneous sub-group, performs better in comparison with the case when the whole group of 12 catchments is chosen.

From Tables 2 and 3 it is observed that, as expected, efficiency values obtained by all regionalization methods range between those achieved by "no model" and by using the model best fitted to the actual observed discharge series of the pseudo-ungauged basins except in the case of catchments H3613020 and Y3514020. In the case of these two catchments, the efficiencies in verification obtained by the method of regional pooling are higher than the corresponding values obtained by considering the observed discharge data series. This inconsistency is attributed to the peculiarity in hydrological characteristics of these two catchments as explained earlier in the section regarding assessment of regional homogeneity. As demonstrated in this study on regionalization approaches for continuous flow simulation in an ungauged basin, the conceptual SMAR model proved to be a suitable choice.

CONCLUSIONS

Assessing hydrometeorological and physiographical homogeneity is very important for reducing uncertainty of estimation of flow in an ungauged catchment by regional analysis. The method of regional pooling of data is considered to be the best when a number of gauged catchments are available in the region. When only a few gauged catchments are available, transposition of data from the gauged to the ungauged basin may be used provided the catchments are similar in characteristics and their areas are of similar order of magnitude. As expected, the lack of homogeneity in a larger sample of catchments generally reduces the efficiency of the regionalization methods for flow modelling in ungauged basins. The conceptual SMAR model is seen to be a suitable choice for regionalization studies in continuous flow simulation in an ungauged basin.

REFERENCES

Kachroo, R. K. (1992) River flow forecasting, Part 5: Applications of a conceptual model. *J. Hydrol.* **133**, 141–178.

Goswami, M., O'Connor, K. M. & Shamseldin, A. Y. (2002) Structures and performances of five rainfall–runoff models for continuous river-flow simulation. In: *Proc. 1st Biennial Meeting of Int. Env. Modeling and Software Soc.* (Lugano, Switzerland), 1, 476–481.

Goswami, M. & O'Connor, K. M. (2005) Application of a conceptual rainfall–runoff simulation model to three European catchments characterised by non-conservative system behaviour. In: *Proc. Int. Conf. Hydrological Perspectives for Sustainable Development* (Roorkee, India), 1, 117–130.

Beven, K. & Freer, J. (2001) Equifinality, data assimilation, and uncertainty estimation in mechanistic modelling of complex environmental systems using the GLUE methodology. *J. Hydrol.* **249**, 11–29.

Modelling ungauged basins with the Sacramento model

TERRI HOGUE[1], KORAY YILMAZ[2], THORSTEN WAGENER[3] & HOSHIN GUPTA[2]

1 *Dept of Civil and Environmental Engineering, University of California, Los Angeles, California, USA*
thogue@seas.ucla.edu

2 *Dept of Hydrology and Water Resources, University of Arizona, Tucson, Arizona, USA*

3 *Dept of Civil and Environmental Engineering, Pennsylvania State University, College Station, Pennsylvania, USA*

Abstract This paper evaluates two contrasting approaches to parameter estimation for ungauged basins using the US National Weather Service's SACramento Soil Moisture Accounting (SAC-SMA) model. An automatic calibration scheme (Multi-Step Automatic Calibration Scheme, MACS) provides deterministic parameter estimates using a three-step, multiple objective approach. The MACS estimates are then transferred to similar or "sister" watersheds for basins in the French MOPEX data set. Physically-based parameter estimates are also developed for the same basins based on the *a priori* approach of Koren *et al.* (2000). In general, the two methods, the transfer and the *a priori* approaches, show similar overall performance. Parameter estimates appear more consistent between basins using the *a priori* approach, but statistically the regionalized MACS parameters and the *a priori* parameters show very similar model performance for the three basins investigated in this study. Model simulated hydrographs are also very similar between the two methods, with both methods tending to underpredict most events (peak and volume) but matching the shape and pattern of flow well. However, both methods have worse performance than a calibrated model for the same basin, indicating the possibility for further refinement and adjustment of the techniques presented here.

Key words parameter estimation; rainfall–runoff modelling; regionalization; Sacramento model

INTRODUCTION

A priority of MOPEX is to improve the estimation of parameters, primarily through the development of *a priori* methods. The goal of this paper is to compare two fairly recent, but diverse, methods for estimating parameters for ungauged basins. One method is based on the transfer of parameters from calibrated, neighboring catchments using an automated procedure for estimation in the gauged basins, and the second method is based on the estimation of *a priori* estimates from relationships developed between soil physics and model parameters. This study uses the conceptual rainfall–runoff SACramento Soil Moisture Accounting (SAC-SMA) model used by the National Weather Service (NWS) for forecasting river flows in the United States. We test the *a priori* and regionalization methods on three basins in France. This work is directly tied to the goals of MOPEX, which are to advance the state of knowledge of parameter estimation techniques and provide guidance on procedures for improvement of *a priori* estimates of parameters for land surface and hydrological models.

METHODS

The two methods used here were developed for the case in which streamflow observations are not available for calibration of the SAC-SMA model (i.e. the basins under analysis are ungauged). These parameter estimation methods include: (1) a direct estimate of parameters for the ungauged basin using theoretical understanding of (small-scale) soil physics proposed by Koren *et al.*, (2000); and (2) transfer of parameters calibrated on neighboring basins to the ungauged basin. A third possible approach of parameter regionalization using statistical regression equations (e.g. Wagener *et al.*, 2004) was not tested here due to the small number of basins available. The second method (sister approach) involved calibrating the model to streamflow observations in each basin using an automated procedure (MACS; Hogue *et al.*, 2000, 2006), and applying the mean parameter values of the calibrated estimates to the ungauged basin. Ideally, and in our future work, one could use physically-derived relationships (i.e. watershed size, precipitation/evaporation (PE) ratio, mean flow ratios, etc.) to adjust the parameters before application to the ungauged basin. The sister approach assumes that there will be some regional basins with observational data for calibration, and that geographically proximal basins behave in a similar manner in terms of their hydrological response. The advantage of the *a priori* method lies in the "physical" relationship of the derived parameters to actual watershed properties. In principle, the *a priori* approach should result in more consistent parameter distributions within (distributed models) and between watersheds. A disadvantage of this method, however, is the difficulty of finding reliable and accurate watershed information (soils, land use, etc.). Other challenges include unexplored problems of scale between the watershed data and model parameters.

Study sites

Table 1 lists the watershed identifiers along with their soil and/or watershed characteristics for our three study basins. Each basin was treated as ungauged and parameters were derived using (1) *a priori* estimates, and (2) a mean parameter set from two regional sister-basins. Each of the methods, along with the model, is outlined in more detail below.

Table 1 Study basins with identifier, size, land cover and soil type.

Basin ID	Size (km^2)	Land cover	USDA soil texture class
J3024010	43	Agricultural (50%) Arable land (40%) Other (10%)	4 (100%)
V6035010	150	Forest (70%) Scrub and herbaceous (30%)	6 (80%) 4 (20%)
Y5615030	279	Forest (40%) Natural grassland (36%) Urban (10%) Other (14%)	4 (55%) 6 (45%)

Model

The SAC-SMA is a conceptual model using a two-layer soil moisture system to continuously account for storage and flow through the soil layers. The upper layer represents surface soil regimes and interception storage, while the lower layer represents deeper soil layers and groundwater storage (Brazil & Hudlow, 1981). Each layer consists of fast components (free water), driven mostly by gravitational forces, and slow components (tension water), driven by evapotranspiration and diffusion. The SAC-SMA, with a total of 16 parameters (Table 2), is a saturation excess model; when precipitation amounts exceed percolation and interflow capacities, upper zone storage will overflow and overland flow will occur. Inputs to the model are Mean Areal Precipitation and Potential Evapotranspiration (PET). An evapotranspiration demand curve (or adjustment curve) is used for estimating the potential evaporation for the watershed. Output from the model, channel inflow, is routed through a unit hydrograph for forecasting basins in the USA.

The model was run at the hourly time-step for each of the study basins. Channel routing was performed using a series of Nash-cascade linear reservoirs. A Monte Carlo based sensitivity analysis was used to determine the optimum number of reservoirs; for the basins under study five reservoirs (each using the same recession coefficient, k_route) was found to be optimal. In the MACS procedure, k_route was optimized along with 13 SAC-SMA parameters. In the *a priori* method, k_route was estimated by averaging the MACS obtained value for the other two basins.

A priori parameter estimation

Koren *et al.* (2000) present equations (hereafter called Koren equations) for deriving *a priori* estimates for the 11 major SAC-SMA parameters from soil texture, hydrological

Table 2 List of parameters within the SAC-SMA model and their description.

SAC-SMA	Parameter description
UZTWM	Upper zone tension water max. storage (mm)
UZFWM	Upper zone free water max. storage (mm)
LZTWM	Lower zone tension water max. storage (mm)
LZFPM	Lower zone free water primary max. storage (mm)
LZFSM	Lower zone free water suppl. max. storage (mm)
UZK	Upper zone free water lateral depletion rate (day^{-1})
LZPK	Lower zone prim. free water depletion rate (day^{-1})
LZSK	Lower zone suppl. free water depletion rate (day^{-1})
ADIMP	Additional impervious area (fraction)
PCTIM	Impervious fraction of the watershed (fraction)
ZPERC	Maximum percolation rate (dimensionless)
REXP	Exponent of the perco. equation (dimensionless)
PFREE	% of water percolating directly to lower zone free water storage
RIVA	Riparian vegetation (fraction)
SIDE	Ratio of deep recharge to channel baseflow (fraction)
RESERV	% of lower zone free water not transferable to lower zone tension water

soil group, soil depth and vegetation. The physical basis of the Koren equations and their derivation are given by Duan *et al.* (2001) and Koren *et al.* (2003), therefore only a brief overview is given here. The Koren equations assume that the SAC-SMA tension water storages are related to "available" soil water and that the free water storages are related to "gravitational" soil water. Available and gravitational soil water can be derived from soil properties (i.e. saturated moisture content, field capacity, and wilting point). Following Koren *et al.* (2003), regression equations derived by Cosby *et al.* (1984) were used together with Campbell's matric water potential equation to determine soil hydraulic properties from the USDA soil texture class information.

Soil texture classes in the French MOPEX Data set (available in GIS), were mapped to the USDA soil texture classes (Table 1). A soil depth of 2.5 m was assumed. Soil hydraulic properties for each basin were defined as the area average of soil property values of soil texture polygons in a basin. The combined thickness of the upper and lower layers was assumed to be equal to the soil profile depth. The thickness of the upper layer was estimated using a concept of initial rain abstraction based on the Curve Number (CN) classification system developed by The Natural Resources Conservation Service (NRCS) (McCuen, 1982). Following Koren *et al.* (2003), it was assumed that under the average soil moisture conditions stipulated by NRCS, the upper layer tension water storage is full and the free water storage is empty. In this case initial rain abstraction should satisfy the upper layer free water capacity (Koren *et al.* 2003). The upper layer thickness can then be calculated based on a CN for the soil profile. NRCS soil hydrological group was assumed to be "B" for all soils in the study basins. Information on land cover was available through the CORINE land cover project (http://reports.eea.eu.int/COR0-part1/en) launched by the Commission of the European Communities. For each soil texture class polygon, a dominant land use was assigned and the corresponding CN was estimated. For each basin an average CN was calculated as the area weighted average of the CNs in a specific basin.

Under these assumptions the Koren equations were used (see Koren *et al.*, 2003) to estimate the SAC-SMA storages (UZTWM, UZFWM, LZTWM, LSFSM, LZFPM) and runoff depletion rates (UZK, LZPK, LZSK). The impervious fraction (PCTIM) of the watersheds was estimated from the percent of watershed area with land cover types of urban and bare rocks. One of the major limitations of the *a priori* methods is the scale difference between the soil hydraulic properties (point measurements) and model parameters (representative of a spatially heterogeneous model grid area of several kilometres). This limitation is being investigated in ongoing research and is not a focus of the current study.

Automated parameter estimation

MACS is a departure from previously developed single-step, single-criterion automatic calibration techniques and is based on a progressive evaluation of objective function values throughout the optimization procedure. The procedure uses the Shuffle Complex Evolution-University of Arizona (SCE-UA) algorithm developed by Duan *et al.* (1992, 1993). The method has been tested in a wide variety of hydro-climatic regimes in the USA and has been shown to produce model simulations as good as, or better in some circumstances, to traditional manual calibration techniques. Details are presented in Hogue *et al*., (2000, 2006); and therefore only a brief overview is given here.

In *step one* of MACS, 13 of the parameters of the SAC-SMA and the linear reservoir parameter (routing parameter) are selected and optimized using the LOG criterion (see below). This first run places strong weighting on the low-flow portions of the hydrograph and gives good estimates of the lower zone parameters.

$$LOG = \sum (LOG_{Qsim,t} - LOG_{Qobs,t})^2$$

where $Q_{sim,t}$ = simulated flows, and $Q_{obs,t}$ = observed flows at time step t.

The *second step* of MACS emphasizes the estimation of parameters that influence higher flow events. Lower zone parameters estimated in the first step are held constant, and a second optimization is run using the RMSE function using the upper zone parameters and routing parameter:

$$RMSE = \sqrt{1/n(\sum_{t=1}^{n}(Q_{sim,t} - Q_{obs,t})^2)}$$

Step three is run using the LOG function to fine-tune the parameters which affect the lower zone processes (upper zone parameters from step two are held constant).

The MACS procedure was run for each basin to find a set of "calibrated" parameters for the SAC-SMA. A total of 14 parameters were estimated: 13 SAC-SMA and one routing parameter (RIVA, SIDE and RSERV are set to book values). A split-sample approach was undertaken with an initial period used for calibration and a separate evaluation period run with the calibrated parameters to assess calibration adequacy. Calibrated parameters from each of the two neighbouring basins were then averaged to find parameters for the third basin, assumed to be ungauged. In this study we had the benefit of comparing this "sister" parameter set with a parameter set calibrated specifically to the basin (theoretically our ungauged basin), along with the parameter set determined via the *a priori* method.

RESULTS

Three parameter sets were compared for each study basin, including *a priori* estimation, parameters calibrated specifically to the basin (MACS), and parameters estimated using the sister approach (mean of two neighboring basins). Figure 1 shows a normalized plot of the three parameter sets for each of the three basins (J302, V603 and Y561). In general, a larger variability is seen between the study basins for the MACS parameters (parameters specifically calibrated to each basin). There is slightly better consistency between basins with the *a priori* method. Intuitively this makes sense. The basin specific calibration would tend to pick up nuances and differences in flow regimes between basins and adjust parameters accordingly. As each of the basins have fairly similar soil types (all have type 6 and two have type 4), the *a priori* estimates should have some similarity and the estimated values reflect this. The mean values transferred to each basin (sister approach) lie somewhere in between these two extremes (slightly more consistent between basins than the MACS, but not as consistent as the *a priori* values). Also of interest, there is much better consistency between all three approaches in the lower zone parameters indicating similar baseflow patterns while the parameters controlling surface flow (UZTWM through REXP) show more variability between methods.

Statistics for the model simulations are presented in Figs 2 and 3. Statistical results from the calibration and evaluation period simulations for each study basin are shown in Fig. 2. Four statistics are presented: Root Mean Squared Error (RMSE), square root RMSE (SQRT RMSE), Nash-Sutcliffe Efficiency (NSE) and percent bias (%Bias). The MACS calibration tends to outperform the *a priori* during the calibration period, showing better statistics in most cases except for a slightly lower NSE for V603.

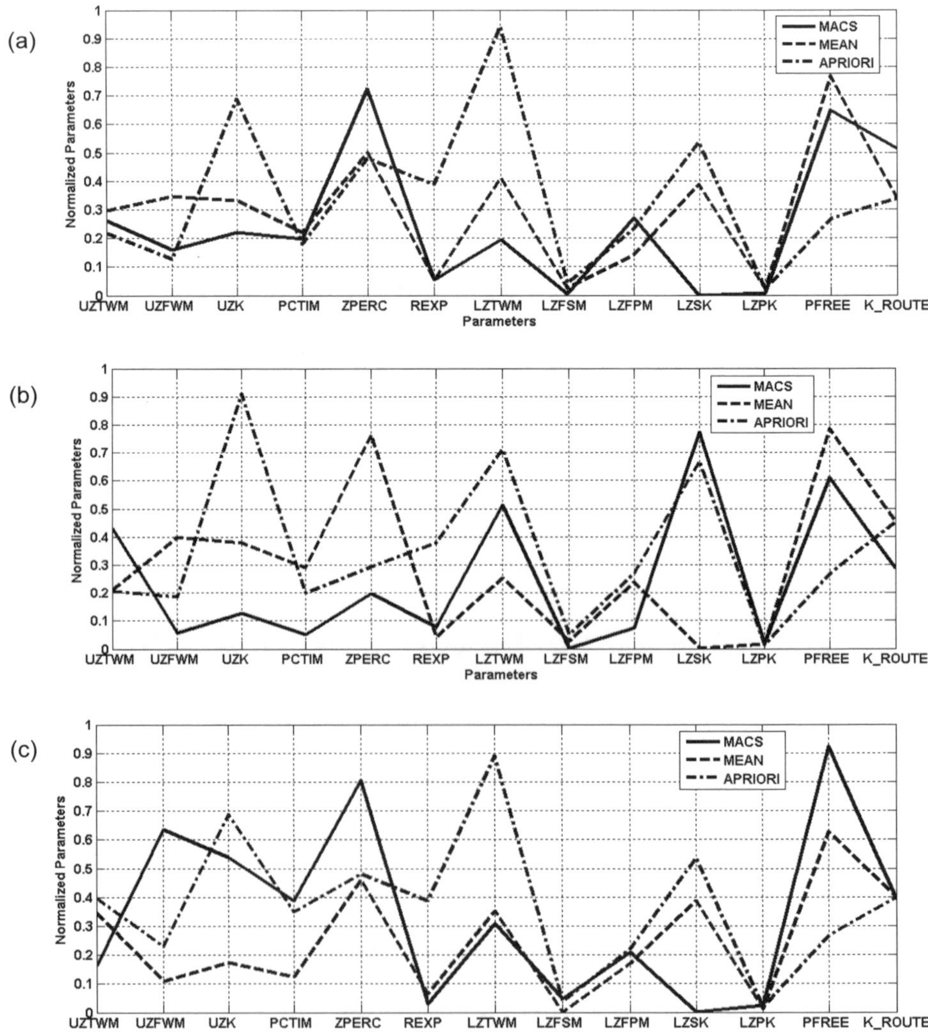

Fig. 1 Comparison of parameters from the *a priori* method, mean (or sister approach) and parameters calibrated at the specific basin (MACS) for each of the study basins: (a) J302, (b) V603, and (c) Y561.

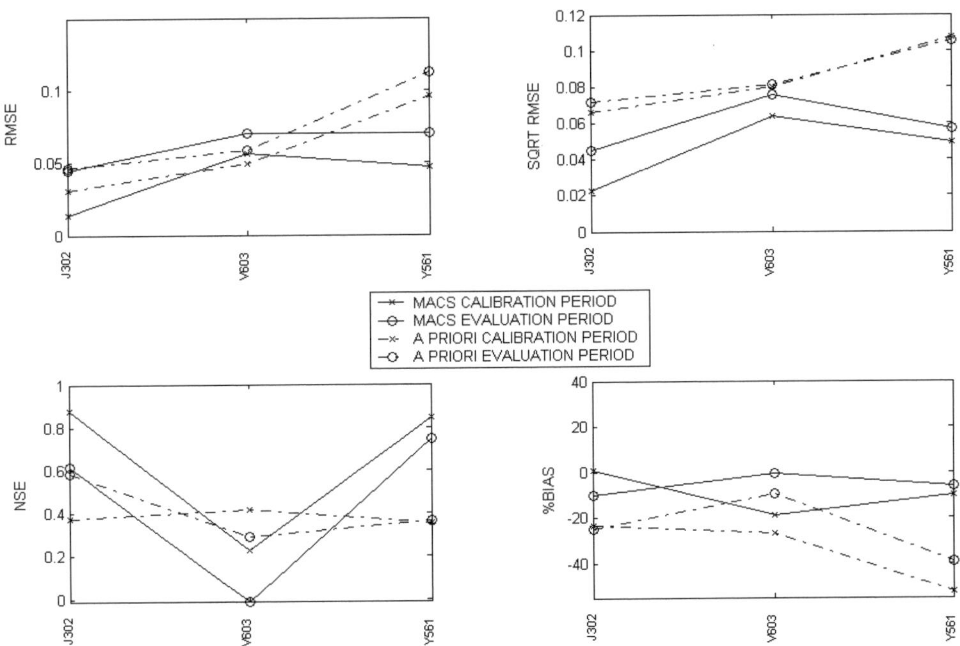

Fig. 2 Comparison of statistics for calibration and evaluation period for each study basin with MACS and *a priori* estimation methods (root mean square error (RMSE), square root RMSE (SQRT RMSE), Nash-Sutcliffe efficiency (NSE), and percent bias (%BIAS).

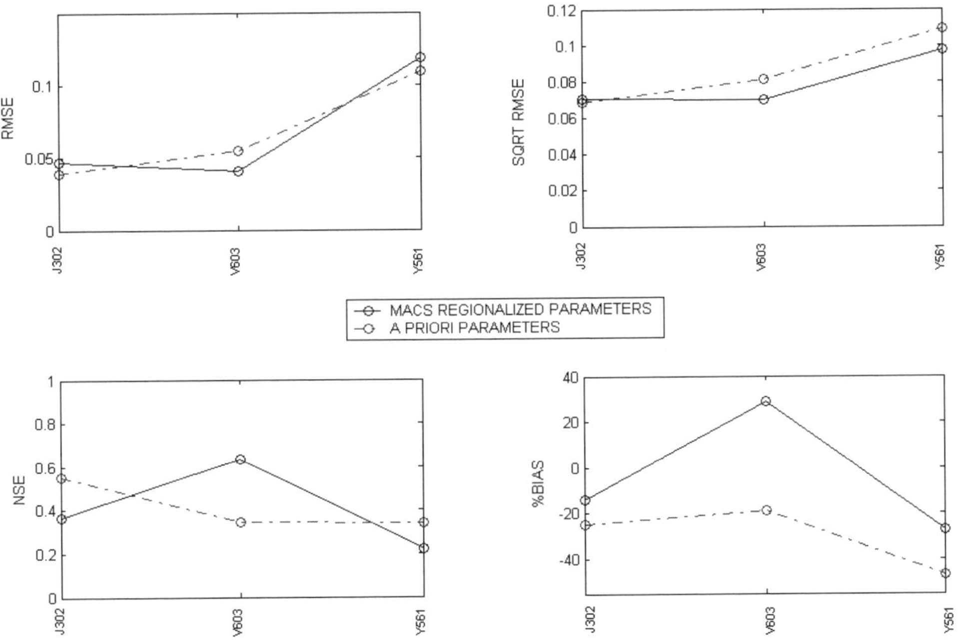

Fig. 3 Comparison of statistics for the three basin as "ungauged" using *a priori* and sister approach (MACS regionalized parameters). Same statistics as in Fig. 2.

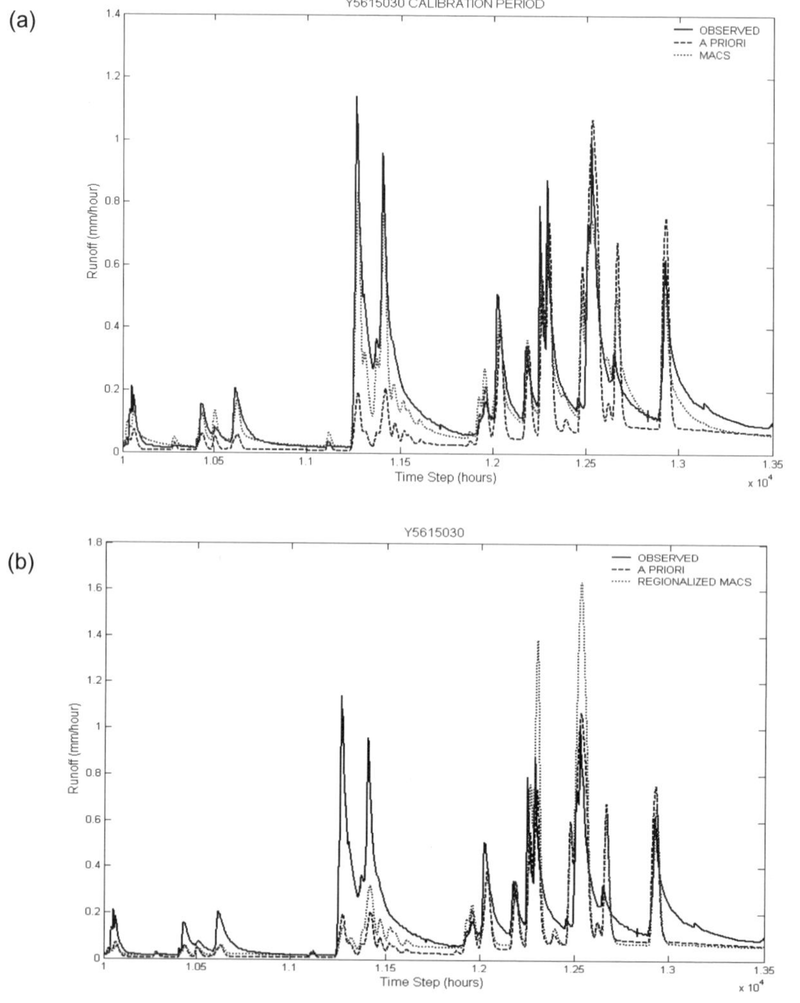

Fig. 4 Model simulations for basin Y561 as: (a) gauged basin (calibration period), and (b) ungauged basin with only *a priori* and regionalized parameters.

basin. However, the MACS performance declines somewhat during the evaluation period, showing more similarity to the *a priori* performance and even slightly worse for the V603 basin (NSE and %Bias). Figure 3 presents the same set of statistics for the basins treated as ungauged. Interestingly, performance is very similar between the two methods. *A priori* estimates tend to have a negative %Bias for all basins, while the transferred parameters show a positive percent bias only for the V603 basin. The *a priori* method does slightly better in the NSE statistic for two of the basins (J302 and Y561) while the MACS has a much better NSE in the V603 basin.

An example of model simulations for basin Y561 (our largest basin) is shown in the hydrographs presented in Fig. 4. Two simulations are shown against observed flows for the calibration period (i.e. gauged basin) in Fig. 4(a) (*a priori*, and MACS calibration) and the same period is shown in Fig. 4(b) with the basin treated as ungauged (showing observed, *a priori* and MACS regionalized parameters). In the gauged case, simulations from both methods, MACS and *a priori*, tend to fit the pattern fairly well. The *a priori* underestimates the first large event around time-step 11 200 but does better catching the peaks in the remaining period. The MACS catches most of the peaks very well and also matches the recessions and volume better than the *a priori*. In the ungauged case of the basin (Fig. 4(b)), both methods (*a priori* and transferred parameters) miss the same first large event (around time-step 11 200). Both methods do better in the remaining time period, however the regionalized parameters drastically overpredict two of the large events (around time-steps 12 200 and 12 500). Also, neither method matches the recessions during most of the simulations. In the case of the *a priori* method, this is possibly due to the lack of deep soil property information (i.e. groundwater).

SUMMARY AND CONCLUSIONS

In summary, it is observed (and is consistent with intuitive reasoning) that better model performance is obtained when the model is specifically calibrated to a basin with observational data. However, the goal of this paper was to evaluate and compare methods for estimating parameters when streamflow data are not available, and for further insight, to compare these methods to the ideal case of having calibration data. Interestingly, both the *a priori* and sister approach show similar performance for the case of the ungauged basins in our study. The simple method of taking the mean of calibrated parameters from two neighboring basins with observational data and directly applying these estimates within the ungauged basin model, showed surprisingly good results. The performance (statistics and hydrograph visualization) was nearly identical to the *a priori* estimates for two of the basins. However, this is not to say that both methods are ideal in their present form. Both have poorer statistics (and poorer hydrograph fit) than if observational data were available, indicating the need for improvement in both methods. Some of the issues to be overcome include the scaling of the point-scale soil physics to the watershed scale and the accuracy of watershed-scale soil surveys when using *a priori* methods. The sister approach could be improved by development of relationships to scale or relating calibrated parameters from basins with observational data to ungauged basins. We continue our work in both of these research areas and plan to provide more insight on these topics in future publications.

REFERENCES

Brazil, L. E. & Hudlow, M. D. (1981) Calibration procedures used with the National Weather Service river forecast system. In: *Water and Related Land Resource Systems* (ed. by Y. Y. Haimes & J. Kindler), 457–566. Pergamon Press, New York, USA.

Cosby, B. J., Hornberger, G. M., Clapp, R. B. & Ginn, T. R. (1984) A statistical extrapolation of the relationships of soil moisture characteristics to the physical properties of soil. *Water Resour. Res.* **20**(6), 682–690.

Duan, Q., Gupta, V. K. & Sorooshian, S. (1992) Effective and efficient global optimization for conceptual rainfall–runoff models. *Water Resour. Res.* **28**, 1015–1031.

Duan, Q., Gupta, V. K. & Sorooshian, S. (1993) A shuffled complex evolution approach for effective and efficient global minimization. *J. Optimization Theory Application* **76**, 501–521.

Duan, Q., Schaake, J. & Koren, V. (2001) *a priori* estimation of land surface model parameters. In: *Land Surface Hydrology, Meteorology, and Climate: Observation and Modeling*, (ed. by V. Lakshmi), 77–94. Water Science and Applications, vol. 3. AGU, Washington, DC, USA.

Hogue, T. S., Sorooshian, S., Gupta, H., Holz, A. & Braatz, D. (2000) A multi-step automatic calibration scheme for river forecasting models. *J. Hydrometeor.* **1**, 524–542.

Hogue, T. S., Gupta, H. V. & Sorooshian, S. (2006) A "user-friendly" approach to parameter estimation in hydrologic models. *J. Hydrol.* (in press).

Koren, V. I., Smith, M., Wang, D. & Zhang, Z. (2000) Use of soil property data in the derivation of conceptual rainfall–runoff model parameters. In: *Proc. 15th Conf. on Hydrology* (AMS, Long Beach, California, USA), 103–106.

Koren, V., Smith, M. & Duan, Q. (2003) Use of *a priori* parameter estimates in the derivation of spatially consistent parameter sets of rainfall–runoff models. In: *Calibration of Watershed Models, Water Science and Applications* (ed. by Q. Duan, S. Sorooshian, H. Gupta, H. Rosseau & R. Turcotte), vol. 6, 239–254. AGU, Washington, DC, USA.

McCuen , R. H. (1982) *A Guide to Hydrologic Analysis using SCS Methods*. Prentice Hall, Englewood Cliffs, New Jersey, USA.

Wagener, T., Wheater, H. S. & Gupta, H. V. (2004) *Rainfall–Runoff Modelling in Gauged and Ungauged Catchments*. Imperial College Press, London, UK.

A parameter estimation scheme of the land surface model VIC using the MOPEX databases

ZHENGHUI XIE & FEI YUAN

Institute of Atmospheric Physics, Chinese Academy of Sciences, Beijing 100029, China

zxie@lasg.iap.ac.cn

Abstract A parameter estimation scheme for the land surface model VIC-3L is discussed. VIC-3L parameters, i.e. the variable infiltration curve parameter (B), the baseflow parameters (D_m and D_s) and the depth of the second soil layer (d_2), are chosen for calibration by a systematic manual calibration approach. The VIC-3L parameters are calibrated in the first half of the period of record for 12 MOPEX watersheds in France and used to predict the streamflow for the second half of the record. Comparisons of the simulated results using the *a priori* parameters with those using calibrated parameters show that the calibrated parameters, when evaluated against the *a priori* parameter estimates, are able to reduce the model bias and increase the Nash-Sutcliffe efficiency coefficient of the streamflow simulation. The averaged hourly Nash-Sutcliffe model efficiency coefficient for the 12 watersheds increases from 0.463 to 0.623, and that for the calibrated parameters in predicting the hourly streamflow for the validation period is 0.553. A sensitivity analysis on the calibrated parameters mentioned above shows that the variable infiltration curve parameter (B) and the depth of the second soil layer (d_2) are more sensitive than the other two parameters (D_m and D_s). Therefore, suitable calibration for the parameters of the land surface model VIC-3L, especially for the variable infiltration curve parameter (B) and the depth of the second soil layer (d_2) is very important for simulating land surface behaviour in a specific region.

Key words land surface model; parameter calibration; sensitivity; VIC

INTRODUCTION AND MODEL DESCRIPTION

Land surface models have been widely used for a variety of applications including hydrological forecasting, water resources management, and climate change studies (Dickinson *et al.*, 1993; Zeng *et al.*, 2002; Dai *et al.*, 2003). To properly simulate land surface behaviour in a specific region, the model parameters for the region must be specified *a priori*. Studies have shown that land surface models could perform well if their model parameters are appropriately estimated on the basis of calibration with observations but perform poorly if their model parameters are not calibrated properly. Therefore, a practical parameter estimation method is critical to hydrological modelling (Duan *et al.*, 1992; Huang *et al.*, 2003). There are two main approaches to estimating the model parameters. The first (*a priori*) approach estimates model parameters by relying on theoretical or empirical relationships that relate such parameters to observable characteristics of the region, such as soil and vegetation properties, regional geomorphology, topographical features, and more. The second approach (model calibration) adjusts model parameter values, so that the model input-output response closely matches the observed input-output response of the region for some historical period in which data have been collected (Nijssen *et al.*, 2001). In this work, we discuss parameter estimation of the land surface model VIC-3L using the MOPEX databases. Four VIC-3L parameters, i.e. the variable infiltration curve parameter (B), the baseflow parameters (D_m and D_s) and the depth of the second soil layer (d_2) are chosen to be calibrated by using a systematic manual calibration approach with 12 MOPEX watersheds' data. The VIC-3L parameters are calibrated in the first half of the period of record for 12 MOPEX watersheds in France and are used to predict the hourly streamflow for the second half of the record.

The variable infiltration capacity model (VIC), called VIC-2L, was developed by Liang *et al.* (1994), which includes two different time scales (i.e. fast and slow) for runoff to capture the dynamics of runoff generation. The upper soil layer of the model is designed to represent the dynamic response of the soil to rainfall events, and the lower layer is used to characterize the seasonal soil moisture behaviour. The VIC model uses physically-based formulations for the calculation of the sensible and latent heat fluxes, but uses the conceptual ARNO baseflow model (Franchini *et al.*, 1991; Todini, 1996) to simulate runoff generation from the deepest soil layer. To better represent quick bare soil evaporation following small summer rainfall events, a thin soil layer is included in VIC-2L (Liang *et al.*, 1996), and VIC-2L becomes VIC-3L. Soil moisture diffusion processes between the three soil layers are considered in VIC-3L. Cherkauer & Lettenmaier (1999) improved the representation of processes for cold climates within VIC. Liang & Xie (2001) developed a parameterization to represent the infiltration excess runoff mechanism in VIC-3L and combined it effectively with the original representation of the saturation excess runoff mechanism (Zhao, 1992). Xie *et al.* (2003) developed a surface runoff parameterization with the Philip infiltration formulation as the time compression analysis (TCA) that dynamically represents both the Horton and Dunne runoff generation mechanisms within a model grid cell. In this paper, the VIC-3L with the new runoff parameterization is applied to simulate runoff for the 12 MOPEX watersheds.

STUDY DOMAIN AND DATA SETS

Study domain

Twelve MOPEX watersheds in France are chosen for the calibration. Table 1 shows the basic basin information for the 12 MOPEX watersheds. The drainage areas of those basins range from 43.0 to 371.0 km^2. All the watersheds are humid watersheds, with the annual mean precipitation ranging from about 1000 to 2000 mm. The period from November to May of the next year is the wet season, with 70% to 80% of rainfall occurring, while it is hot and dry in the summer and early autumn. Therefore, these watersheds have the typical characteristics of a Mediterranean climate.

Vegetation data

Vegetation related parameters such as architectural resistance, minimum stomatal resistance, leaf-area index, albedo, roughness length, zero-plane displacement, and

Table 1 Watershed information.

Code	Name	Area (km²)
J3024010	Le Guillec à Trézilidé	43.0
V6035010	Le Toulourenc à Malaucène [Veaux]	150.0
Y5615030	Le Loup à Villeneuve-Loubet [Moulin du Loup]	279.0
A1522020	La Lauch à Guebwiller	68.1
H2001020	L'Yonne à Corancy	98.0
H3613020	Le Lunain à Épisy	252.0
J2034010	Le Guindy à Plouguiel	125.0
J4124420	La Rivière de Pont-l'Abbé à Plonéour-Lanvern [Tremillec]	32.1
K0744010	L'Anzon à Débats-Rivière-d'Orpra [Cotes]	181.0
K0753210	Le Lignon du Forez à Boën	371.0
Y3514020	Le Vistre à Bernis	291.0

fraction of root depth of each soil layer are based on the University of Maryland's (UMD) 1 km global land cover classification with 14 unique vegetation types in total (Hansen *et al.*, 2000). For each type of vegetation, the vegetation parameters mentioned above are derived from the literature and the land data assimilation system (LDAS). Since the MOPEX database provides the vegetation data based on the Corine land cover classification for the 12 French watersheds, we regroup it into the UMD classification to use the related vegetation parameters. As shown in Table 2, the area covered with the Corine-based coniferous forest, broad-leaved forest, mixed forest or transitional woodland-scrub is assigned with the corresponding vegetation parameters for the UMD-based deciduous needle-leaf forest, deciduous broadleaf forest, mixed forest, and woodland, respectively. In the same way, the region with the cover of moor and heathland, natural grassland or agricultural area in the Corine land cover classification is assigned with the vegetation parameter values for the open shrubland, grassland and crop land in the UMD classification. Table 2 lists the vegetation parameters in VIC-3L for various vegetation types (Su & Xie, 2003).

Soil data

The soil parameters in VIC-3L, such as, porosity θ_s ($m^3 m^{-3}$), saturated soil potential ψ_s (m), saturated hydraulic conductivity K_s ($m s^{-1}$), and the exponent parameter b, are derived according to Cosby *et al.* (1984) and Rawls *et al.* (1993) based on the Food and Agriculture Organization (FAO) soil classification (FAO, 1998). In this study, the soil data provided by MOPEX is based on the French soil texture classification, which are regrouped into the FAO soil classification. Table 3 shows the soil classification and the corresponding values of soil parameters used in the VIC-3L model (Su & Xie, 2003). The French soil texture classification consists of five soil types, which are coarse soil, moderate soil, moderately fine soil, fine soil and very fine soil. The area with the coverage of those five French soil types are set with the corresponding soil parameters for the FAO-based sandy loam, silt loam, silty clay loam, silty clay and clay, respectively.

Table 2 Vegetation-related parameters in the VIC-3L model.

UMD classification	Corine land cover legend	Albedo	Minimum stomatal resistance (sm^{-1})	Leaf-area index	Roughness length (m)	Zero-plane displacement (m)
Evergreen needleleaf forest		0.12	250	3.40~4.40	1.476	8.04
Evergreen broadleaf forest		0.12	250	3.40~4.40	1.476	8.04
Deciduous needleleaf forest	3.1.2 Coniferous forest	0.18	150	1.52~5.00	1.23	6.7
Deciduous broadleaf forest	3.1.1 Broad leaved forest	0.18	150	1.52~5.00	1.23	6.7
Mixed forest	3.1.3 Mixed forest	0.18	200	1.52~5.00	1.23	6.7
Woodland	3.2.4 Transitional woodland-scrub	0.18	200	1.52~5.00	1.23	6.7
Wooded grassland		0.19	125	2.20~3.85	0.495	1
Closed shrubland		0.19	135	2.20~3.85	0.495	1
Open shrubland	3.2.2 Moor and heathland	0.19	135	2.20~3.85	0.495	1
Grassland	3.2.1 Natural grassland	0.2	120	2.20~3.85	0.0738	0.402
Crop land (corn)	2 Agricultural area	0.1	120	0.02~5.00	0.006	1.005

Table 3 Soil-related parameters in the VIC-3L model.

FAO soil texture	French soil texture	θ_s (m^3m^{-3})	ψ_s (m)	K_s (mm day^{-1})	$2b$+3	Bulk density (kg m^{-3})
Sand		0.445	0.069	92.45	11.2	1490
Loamy sand		0.434	0.036	1218.24	10.98	1520
Sandy loam	1 Coarse	0.415	0.141	451.87	12.68	1570
Silt loam	2 Moderate	0.471	0.759	242.78	10.58	1420
Silt		0.523	0.759	242.78	9.1	1280
Loam		0.445	0.355	292.03	13.6	1490
Sandy clay loam		0.404	0.135	384.48	20.32	1600
Silty clay loam	3 Moderately fine	0.486	0.617	176.26	17.96	1380
Clay loam		0.467	0.263	211.68	19.04	1430
Sandy clay		0.415	0.098	623.81	29	1570
Silty clay	4 Fine	0.497	0.324	115.78	22.52	1350
Clay	5 Very fine	0.482	0.468	84.15	27.56	1390

Forcing and streamflow data

In this study, the forcing data were provided by the Meteo-France, which include the hourly precipitation, atmospheric pressure, water vapour pressure, wind speed, short wave radiation and long wave radiation in the period from 1 August 1995 to 31 July 2002. Hourly streamflow data were also provided for the same period.

PARAMETER ESTIMATION SCHEME

Although most of vegetation and soil parameters in VIC-3L can be estimated according to the literature, some soil parameters are subject to calibration based on the agreement between simulated and observed hydrographs. These include the infiltration parameter (B), which controls the amount of water that can infiltrate into the soil; the depths of the three soil layers d_i (i = 1, 2, 3), which affect the maximum storage available for transpiration; the three parameters in the baseflow scheme including the maximum velocity of baseflow (D_m), the fraction of maximum baseflow (D_s), and the fraction of maximum soil moisture content of the third layer (W_s) at which a nonlinear baseflow response is initiated, which determines how quickly the water stored in the third layer is depleted. *A priori* estimates for those parameters over a humid region, the Huaihe River basin in China, are presented in Su & Xie (2003). In this study, the related parameters for the Huaihe River basin are used as the *a priori* values for the 12 humid French watersheds, which are shown in Table 4. The hourly streamflow simulation is conducted for the whole period of record (1 August 1995–31 July 2002), called the ungauged simulation mode. Among the seven VIC-3L parameters, the variable infiltration capacity curve parameter (B), baseflow parameter (D_m and D_s) and the depth of the second soil layer (d_2) are the more sensitive, and hence they are chosen for calibration by a systematic manual calibration approach. The other three parameters are assigned with the *a priori* values mentioned above. Calibration is made manually and the Nash-Sutcliffe model efficiency coefficient (Nash & Sutcliffe, 1978) is used as the objective function, which describes the matching extent of the hydrograph between the simulated and observed values:

$$C_e = \frac{\sum (Q_{i,o} - \overline{Q}_o)^2 - \sum (Q_{i,c} - Q_{i,o})^2}{\sum (Q_{i,o} - \overline{Q}_o)^2}$$

where $Q_{i,o}$ is the observed streamflow (m^3 s^{-1}), $Q_{i,c}$ is the simulated streamflow (m^3 s^{-1}), and $\overline{Q}_o$ is the mean observed streamflow (m^3 s^{-1}).

Table 4 *A priori* parameter values of the VIC-3L model for the 12 MOPEX French watersheds.

Parameter	Physical meaning	Value
B	Variable infiltration curve parameter	0.3
D_m	Maximum velocity of baseflow (mm day^{-1})	10
D_s	Fraction of D_m where non-linear baseflow begins	0.02
W_s	Fraction of maximum soil moisture where non-linear baseflow occurs	0.8
D_1	Thickness of the first soil layer (m)	0.1
D_2	Thickness of the second soil layer (m)	0.5
D_3	Thickness of the third soil layer (m)	1.5

Model calibration is performed using the following procedures:

(1) Set the estimated value for the depth of the second soil layer (d_2), commonly with a deeper depth for arid regions and a lower depth for humid regions.

Table 5 Calibrated parameter values of the VIC-3L model for the 12 MOPEX French watersheds.

Gauge	B	D_s	D_m (mm day^{-1})	d_2 (m)
J3024010	0.3	0.02	10	1.7
V6035020	0.3	0.02	8	0.48
Y5615030	0.28	0.01	4	0.55
A1522020	0.45	0.035	6	0.45
H2001020	0.25	0.001	20	0.31
H3613020	0.15	0.02	2	10
J2034010	0.3	0.05	2	0.5
J4124420	0.35	0.35	2	2
K0744010	0.38	0.0001	20	0.45
K0753210	0.35	0.0002	20	0.45
Y3514020	0.21	0.01	0.95	0.5
A5723010	0.25	0.01	0.8	1.1

Table 6 Hourly Nash-Sutcliffe efficiency in ungauged and gauged modes.

Gauge	*Priori* (ungauged) Mode (1 Aug. 1995–31 July 2002) (1)	Gauged mode			*Improvement*
		Calibration period (1 Aug. 1996–31 July 1999)	Validation period (1 Aug. 1999–31 July 2002)	Entire period (1 Aug. 1995–31 July) (2)	(2) – (1)
J3024010	0.663	0.528	0.728	0.682	0.019
V6035020	0.424	0.492	0.361	0.423	–0.001
Y5615030	0.774	0.826	0.788	0.788	0.014
A1522020	0.515	0.551	0.519	0.546	0.031
H2001020	0.697	0.761	0.819	0.778	0.081
H3613020	–7.511	0.284	0.317	0.452	7.963
J2034010	–0.107	0.501	0.292	0.389	0.496
J4124420	0.192	0.454	0.556	0.565	0.373
K0744010	0.584	0.722	0.59	0.653	0.069
K0753210	0.634	0.785	0.63	0.691	0.057
Y3514020	0.582	0.762	0.616	0.67	0.088
A5723010	0.136	0.468	0.189	0.342	0.206
Mean	0.463	0.623	0.553	0.593	0.130

Note: The Nash-Sutcliffe efficiency for Gauge H3613020 is excluded from of the mean statistics.

(2) Calibrate the ARNO model parameters (D_m and D_s) to fit the low flow.
(3) Adjust the infiltration parameter B to match the observed flow peaks, with a higher value to increase the peak and a lower value to lower the peak.
(4) Make a fine adjustment to these parameters to get the best simulation results.

The four VIC-3L parameters are calibrated in the first half of the period of record (1 August 1995–31 July 1999) for each of the 12 MOPEX watersheds in France and are used to predict the hourly streamflow for the second half of the period of record (1 August 1999–31 July 2002), which is called gauged simulation mode. The first year is considered as a warm-up period and is not used in the error criteria calculations. Table 5 shows the calibrated parameter values for the 12 French watersheds.

Since the drainage area of the watersheds is not large, ranging from 43.0 to

371.0 km^2, each watershed is treated as a computational unit of the VIC-3L model. The VIC-3L is used to simulate the water balance for the 12 French watersheds for ungauged and gauged modes. In this study, only those vegetation types whose proportions over the computational grid cell are greater than 10% are involved in computing the water and energy balances, and we use the parameter values of the soil type with the highest proportion over the watershed as the parameters for the whole watershed.

Table 6 shows the hourly Nash-Sutcliffe efficiency in ungauged and gauged modes for the 12 French watersheds. Figure 1 shows the comparison of the observed daily streamflow with the simulated in the ungauged and gauged modes at Gauge Station Y5615030. For the ungauged mode, the VIC-3L model with the *a priori* parameters can basically simulate the hourly runoff for the French watersheds. Among the 12 watersheds, the hourly Nash-Sutcliffe efficiency for eight watersheds is larger than 0.40, and the mean value for eleven watersheds (except H2001020) is 0.463. For gauged mode, the VIC-3L model with the calibrated parameters simulates better than that with the *a priori* parameters. For the calibration period (1 August 1996–31 July 1999), the hourly Nash-Sutcliffe efficiency for most of watersheds is larger than 0.50, and the mean value of hourly Nash-Sutcliffe efficiency for the 12 watersheds is 0.623. The mean value of hourly Nash-Sutcliffe efficiency for the 12 watersheds for the validation period is 0.553.

The hourly Nash-Sutcliffe efficiency cumulative distribution and the simulated annual runoff with the observed from 1998 to 2002, are shown in Figs 2 and 3 respectively, which imply that the gauged mode is superior in simulating streamflow for the 12 French watersheds.

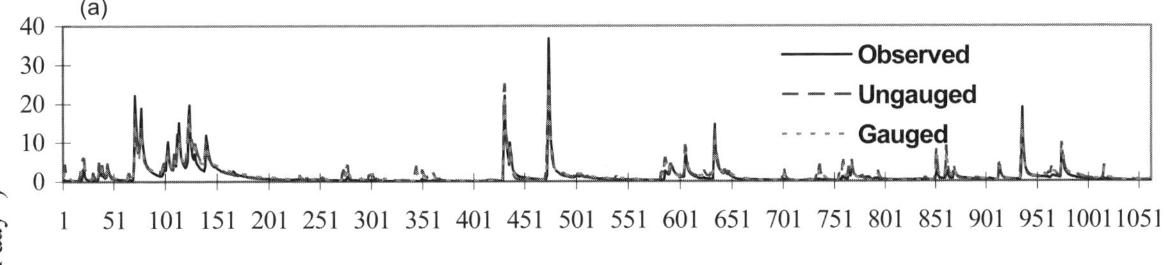

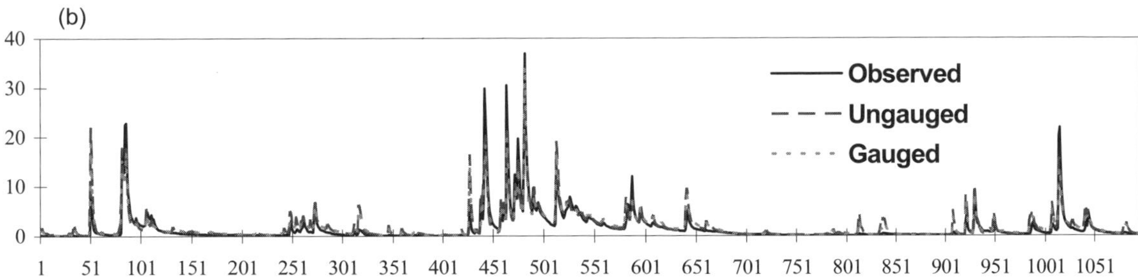

Fig. 1 Comparison of the observed daily streamflow with the simulated in the ungauged and gauged modes at Gauge Y5615030. (a) Calibration period (1 August 1996–31 July 1999); (b) validation period (1 August 1999–31 July 2002).

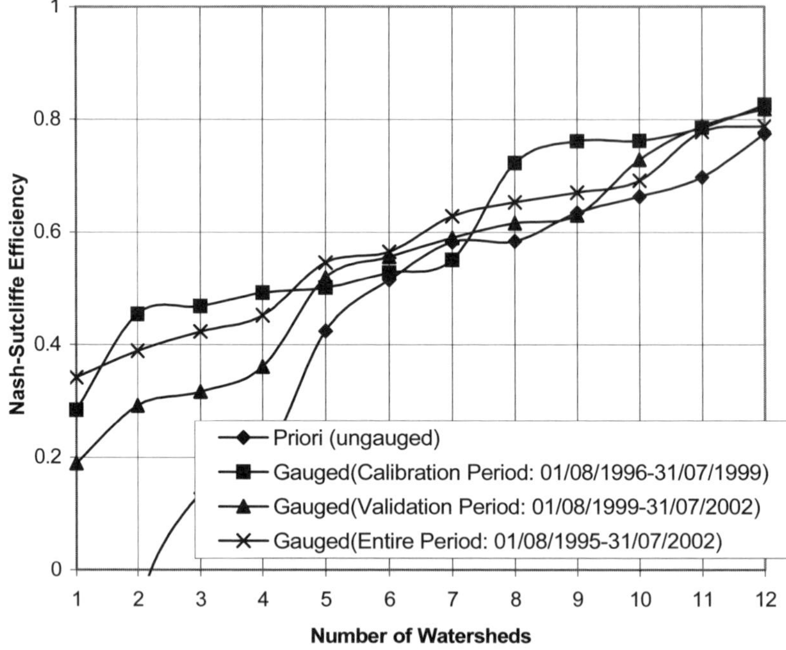

Fig. 2 Hourly Nash-Sutcliffe efficiency cumulative distribution.

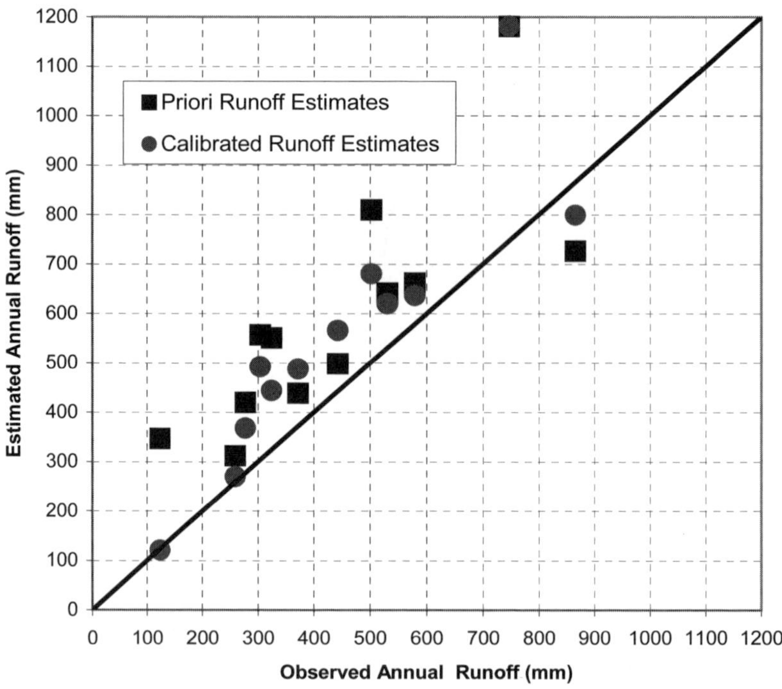

Fig. 3 Comparison of the simulated annual runoff with the observed during 1998–2002.

PARAMETER SENSITIVITY

A sensitivity analysis on the four parameters mentioned above is presented by varying each parameter value around its best estimate by ±10% and ±25% and comparing the Nash-Sutcliffe model efficiency coefficient and runoff for the different parameters. The sensitivities of the Nash-Sutcliffe efficiency coefficient and runoff to the four parameters for calibration are presented in Figs 4 and 5, respectively, which show that B and d_2 are more sensitive than the other two parameters (D_m and D_s). When B is altered by +25%, +10%, –10% and –25%, the mean variation of the Nash-Sutcliffe coefficient is 0.018, 0.007, 0.009 and 0.029, and the mean variation of runoff volume

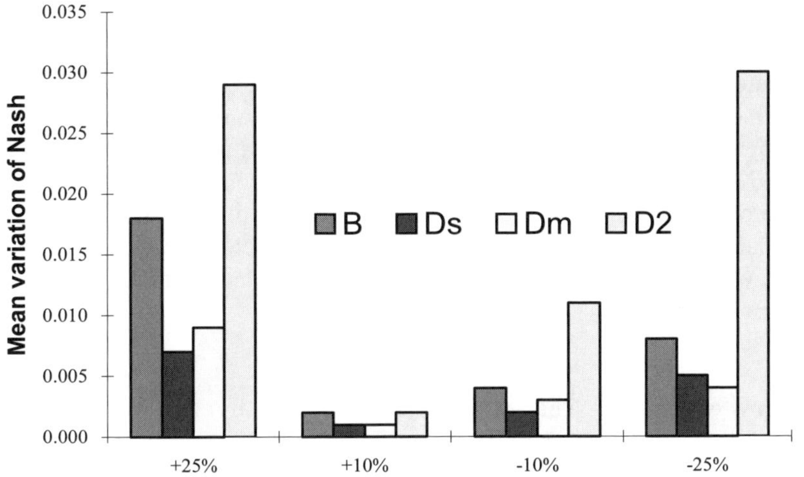

Fig. 4 Sensitivities of the Nash-Sutcliffe efficiency coefficient to the four model parameters for calibration.

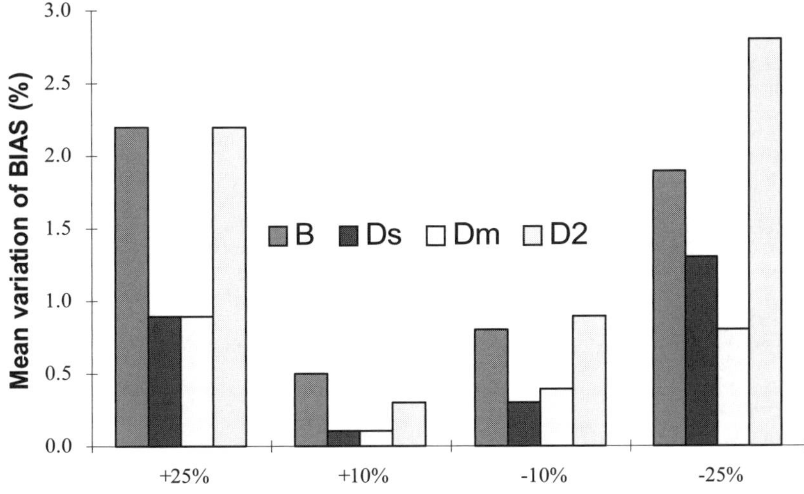

Fig. 5 Sensitivities of runoff to the four model parameters for calibration.

is 2.2%, 0.9%, 0.9% and 2.2%, respectively. By altering d_2 in the same way, the mean variation of the Nash-Sutcliffe coefficient is 0.008, 0.005, 0.004 and 0.030, and the mean variation of runoff volume is 1.9%, 1.3%, 0.8% and 2.8%, respectively. However, the same alteration of d_m and d_s produces much smaller variation in the Nash-Sutcliffe coefficient and runoff volume.

CONCLUSIONS

In this study, a parameter estimation scheme for the VIC-3L land surface model is discussed and applied to the 12 MOPEX watersheds in France, in which four VIC-3L parameters, i.e. the variable infiltration curve parameter (B), the baseflow parameters (D_m and D_s) and the depth of the second soil layer (d_2) are chosen for calibration. The VIC-3L with the calibrated parameters, when evaluated against the *a priori* parameter estimates, is able to reduce the model bias and increase the Nash-Sutcliffe efficiency coefficient of the streamflow simulation. Also, a sensitivity analysis on the calibrated parameters shows that the variable infiltration curve parameter (B) and the depth of the second soil layer (d_2) are most sensitive among the model parameters for calibration. Therefore, suitable calibration for the parameters of the land surface model VIC-3L, especially for the variable infiltration curve parameter (B) and the depth of the second soil layer (d_2) is very important for simulating land surface behaviour in a specific region.

Acknowledgements This work was supported by the National Natural Science Foundation of China under Grant no. 90411007, the National Basic Research Program under the Grant no. 2005CB321703, and CAS International Partnership Creative Group "The Climate System Model Development and Application Studies".

REFERENCES

Cherkauer, K. A. & Lettenmaier, D. P. (1999) Hydrologic effects of frozen soils in the upper Mississippi River basin. *J. Geophys. Res.* **104**, 599–610.

Cosby, B. J., Hornberger, G. M., Clapp, R. B. & Ginn, T. R. (1984) A statistical exploration of the relationships of soil moisture characteristics to the physical properties of soils. *Water Resour. Res.* **20**, 682–690.

Dai, Y., Zeng, X., Dickinson, R. E., Baker, I., Bonan, G. B., Bosilovich, M. G., Denning, A. S., Dirmeyer, P. A., Houser, P. R., Niu, G., Oleson, K. W., Schlosser, C. A. & Yang, Z. (2003) The Common Land Model. *Bull. Am. Met. Soc.* **84**(8), 1013–1023.

Dickinson, R. R., Henderson-Sellers, A. & Kennedy, P. J. (1993) Biosphere–Atmosphere Transfer Scheme (BATS) version 1 as coupled to the NCAR Community Climate Model. *NCAR technical note NCAR/TN-387+STR*. National Center for Atmospheric Research, Boulder, Colorado, USA.

Duan, Q., Sorooshian, S. & Gupta, V. K. (1992) Effective and efficient global optimization for conceptual rainfall–runoff models. *Water Resour. Res.* **28**, 1015–1031.

Food and Agriculture Organization (FAO) (1998) Digital soil map of the world and derived soil properties (CD-ROM), Land Water Digital Media Ser., vol. 1, Rome, Italy.

Franchini, M. & Pacciani, M. (1991) Comparative analysis of several conceptual rainfall runoff models. *J. Hydrol.* **122**, 161–219.

Hansen, M., DeFries, R., Townshend, J. R. G. & Sohlberg, R. (2000) Global land cover classification at 1km resolution using a decision tree classifier. *Int. J. Remote Sens.* **21**, 1331–1365.

Huang, M. & Liang, X. (2003) A transferability study of model parameters for the variable infiltration capacity land surface scheme. *J. Geophys. Res.* **108**(D22), 8864, doi:10.102.

Liang, X., Lettenmaier, D. P., Wood, E. F. & Burges, S. J. (1994) A simple hydrological based model of land surface water and energy fluxes for general circulation models. *J. Geophys. Res.* **99**(D7), 14 415–14 428.

Liang, X., Wood, E. F. & Lettenmaier, D. P. (1996) Surface soil moisture parameterization of the VIC-2L model: Evaluation and modifications. *Global Planetary Change* **13**, 195–206.

Liang, X. & Xie, Z. (2001) A new surface runoff parameterization with subgrid-scale soil heterogeneity for land surface models. *Adv. Water Resour.* **24**, 1173–1193.

Nash, J. E. & Sutcliffe, J. V. (1970) River flow forecasting through conceptual models. Part I: A discussion of principles. *J. Hydrol.* **10**, 282–290.

Nijssen, B., O'Donnell, G. M., & Lettenmaier, D. P. (2001) Predicting the discharge of global rivers. *J. Clim.* **14**, 3307–3323.

Rawls, W. J., Ahuja, L. R., Brakensiek, D. L. & Shirmohammadi, A. (1993) Infiltration and soil water movement. In: *Handbook of Hydrology* (ed. by D. R. Maidment), 5.1–5.51. McGraw-Hill Inc., New York, USA.

Su, F. & Xie Z. (2003) A model for assessing effects of climate change on runoff of China. *Prog. Nat. Sci.* **13**(9), 701–707.

Todini, E. (1996) The ARNO rain-runoff model. *J. Hydrol.* **175**, 339–382.

Xie Z., Su. F., Liang, X. *et al.* (2003) Applications of a surface runoff model with Horton and Dunne runoff for VIC. *Adv. Atmospheric Sci.* **20**(2), 165–172.

Zeng, X., Shaikh, M., Dai, Y., Dickinson, R. E. & Myneni, R. (2002) Coupling of the common land model to the NCAR community climate model. *J. Clim.* **15**(14), 1832–1854.

Zhao, R. J. (1992) The Xianjiang model applied in China. *J. Hydrol.* **135**, 371–381.

Application of global 1-degree data sets to simulate runoff from MOPEX experimental river basins

OLGA N. NASONOVA & YEUGENIY M. GUSEV
Institute of Water Problems, Russian Academy of Sciences, Gubkina St.3, 119991 Moscow, Russia
nasonova@aqua.laser.ru

Abstract Global 1-degree data sets provided within the framework of the Second Global Wetness Project (GSWP-2) were applied for the simulation of river flow from 12 MOPEX (Model Parameter Estimation Experiment) experimental river basins (with an area of the order of 10^3 km^2) to reveal the applicability of global forcing data and land surface parameters for regional runoff predictions. The simulations were performed at 3-h time steps for a 10-year period (1986–1995) using the SWAP land surface model (Soil Water–Atmosphere–Plants). The results of simulations were compared with the observed streamflow and with analogous simulations based on the regional data set. The comparison allowed us to reveal the influence of uncertainties in the forcing data and the land surface parameters on runoff prediction.

Key words global data; GSWP-2; land surface modelling; MOPEX; river runoff; uncertainty analysis

INTRODUCTION

At the moment there are a lot of global data sets containing hydrometeorological data, land use information, and soil and vegetation characteristics with 1-degree spatial resolution. Global data sets are widely used for global and macroscale runoff simulations (e.g. Oki *et al.*, 1995; Jayawardena & Mahanama, 2001), while their resolution is supposed to be crude for regional and local applications. In the latter case the downscaling procedure is usually applied to span the gap between large-scale forcing data generated by general circulation models (GCMs) and regional or local hydrological simulations (e.g. Salathé, 2003). Downscaling techniques are generally divided into statistical and dynamical, and spatial and temporal ones; for a review see e.g. Wilby & Wigley (1997). Each of them has its own advantages and disadvantages and is more or less costly. Also, different techniques result in different spatial/temporal patterns of downscaled data that may necessitate the investigation of a number of different downscaling techniques before a suitable methodology is identified. All these circumstances have led us to the following questions. Is it possible to simulate runoff at regional scale (for the basins, comparable in size with a 1-degree grid cell) without using the spatial downscaling technique? What will be the reliability of such simulations? Uncertainties in what type of data will be crucial for the final results? The present work is aimed at the investigation of these issues.

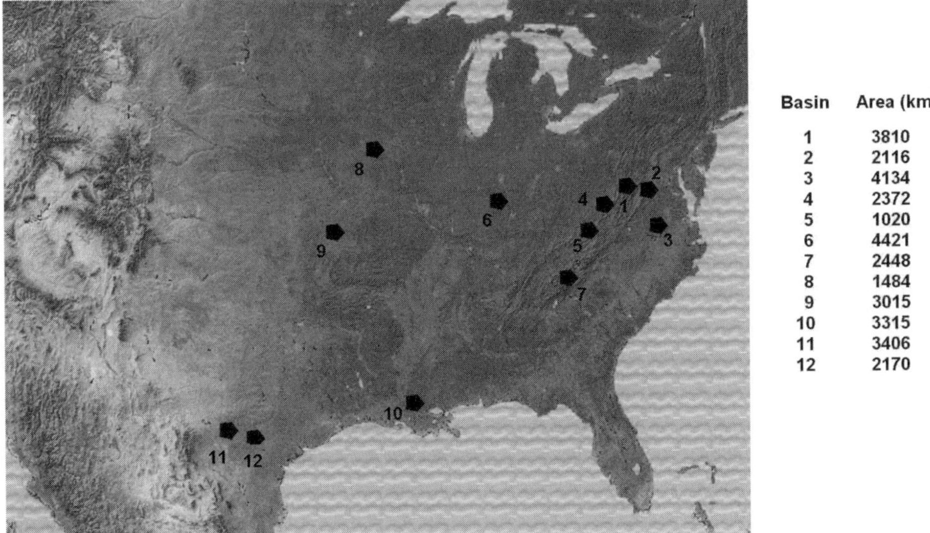

Basin	Area (km^2)
1	3810
2	2116
3	4134
4	2372
5	1020
6	4421
7	2448
8	1484
9	3015
10	3315
11	3406
12	2170

Fig. 1 Location of the 12 MOPEX river basins.

DESCRIPTION OF STUDY SITES AND DATA SETS

Twelve river basins (with an area ranging between 1020 and 4421 km^2), which were selected within the framework of the Second MOPEX (Model Parameter Estimation Experiment) Workshop (Duan *et al.*, 2006), were used for the given investigation (Fig. 1). The basins are located within the southeastern part of the United States and characterized by a wide range of hydrological and climatic conditions varying from desert conditions to very wet ones. The dominant vegetation types represent deciduous broadleaf (basins: 1, 2, 4, 5, 9), evergreen needleleaf (basin 10) and mixed (basins: 3, 7) forests, as well as cropland (6, 8) and grassland (basins: 11, 12).

Regional and global data sets were used for model simulations. Both data sets include near-surface meteorological (forcing) data and land surface parameters (soil and vegetation characteristics).

The regional data set was described in Gusev & Nasonova (this issue). The forcing data were provided by the Second MOPEX Workshop organizers. The values of model parameters were partly provided by the organizers, and partly derived by the authors (see the set of model parameters referred to as "CAL1-MOP" in Gusev & Nasonova (this issue)).

The applied global data represent 1-degree data sets (including forcing data and land surface parameters) produced within the framework of the Second Global Wetness Project (GSWP-2) (Dirmeyer *et al.*, 2002; Zhao & Dirmeyer, 2003). The global forcing data are based on re-analyses and gridded observational data used in ISLSCP (the International Satellite Land-Surface Climatology Project) Initiative II. ISLSCP-II global database includes two versions of meteorological data representing the products of NCEP/DOE (the National Center for Environmental Prediction—Department of Energy) and ERA-40 (ECMWF Re-analysis-40, European Centre for Medium-Range Weather Forecasts) re-analysis and meteorological data sets based on observations. The former contain systematic errors, the latter are characterized by low temporal resolution (as a rule, one month) and cannot be used directly for modelling. That is why corrections to the systematic biases in the re-analysis fields were made by hybridization of the 3-hourly re-analysis data with global observation-based gridded data sets within the framework of GSWP-2 project. As a result, several alternative data sets with different interpretations of global meteorological fields were suggested (http://www.iges.org/gswp/):

B0 Baseline data set (*SRB downward shortwave and longwave radiation*, CRU near-surface air temperature and humidity, ECOR surface pressure, NCEP wind speed, *hybrid NCEP/DOE precipitation* with GPCC and GPCP precipitation data sets).
M1 All *NCEP/DOE forcing data* (no hybridization with observational data).
M2 All *ECMWF forcing data* (no hybridization with observational data).

All the other global forcing data sets are the same as B0, but either precipitation or radiation is replaced by an alternative one:

PE *Hybrid ERA-40 precipitation* (instead of NCEP/DOE).
P1 *ERA-40 precipitation without hybridization* (pure re-analysis).
P2 *Hybrid NCEP/DOE precipitation, but without the relaxation to GPCP* (satellite-estimated) precipitation.
P3 *Hybrid precipitation as in P2, but without correction for gauge undercatch.*
P4 *NCEP/DOE precipitation without hybridization* (pure re-analysis).
R1 *NCEP/DOE downward shortwave and longwave radiation* (instead of SRB).
R2 *ERA-40 downward shortwave and longwave radiation* (instead of SRB).
R3 *ISCCP downward shortwave and longwave radiation* (instead of SRB).

Following the GSWP-2 strategy (Dirmeyer *et al.*, 2002), we used the described data sets to perform different sensitivity experiments.

To reveal the influence of model parameters on the runoff simulations, in addition to MOPEX and GSWP-2 parameter data sets, we produced two alternative global data sets of soil parameters on the basis of their relationships with the clay and sand contents in a soil. The relationships were derived from generalized tables published in Clapp & Hornberger (1978) and Dunne & Willmott (1996) for one soil data set and in Cosby *et al.* (1984) for another.

RESULTS

The described data sets were applied for the simulations of river runoff for the 12 basins at 3-hour time steps for a 10-year period (1986–1995) using the SWAP land surface model (Soil Water–Atmosphere–Plants) (Gusev & Nasonova, 2003). Some results are shown in Fig. 2, where annual observed runoff from each basin and for each year is compared with runoff simulated using the MOPEX regional data and 11 global forcing data sets. Different statistics for validation of the simulated annual runoff against observations (the ratio between modelled and observed runoff (*rat*), the Nash-Sutcliffe efficiency (*eff*) and the coefficient of correlation (*corr*)) are also given in the panels. Analysis of different chains of experiments allows us to reveal how uncertainties in forcing data influence the final results. Thus, following the sequences

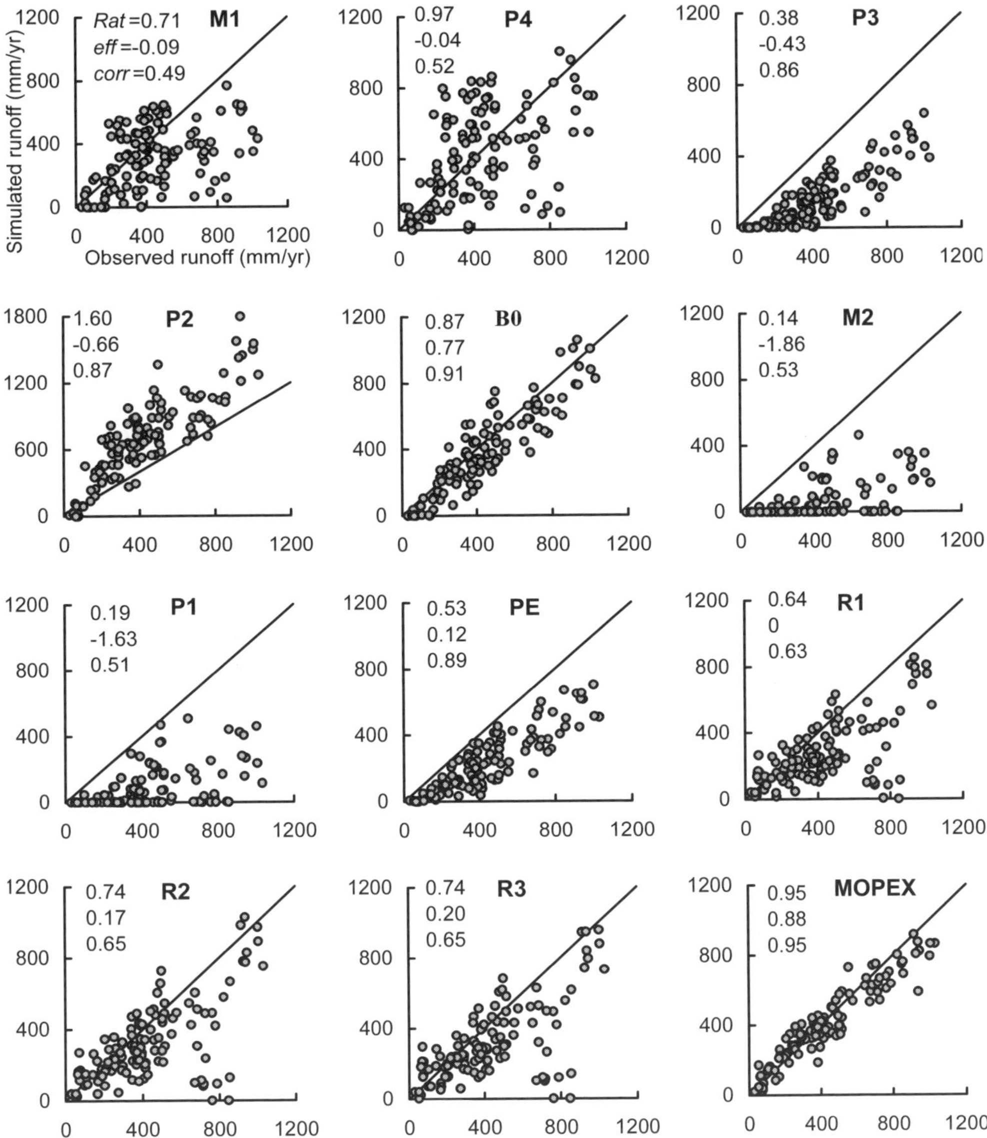

Fig. 2 Comparison of simulated and observed annual streamflow for each of the 12 basins over the 1986-1995 period. All designations are placed on the first panel.

of experiments M1→P4→P3→P2→B0 and M2→P1→PE, we move step by step from pure re-analysis NCEP/DOE (M1) or ERA-40 (M2) products to fully hybridized ones. The best result among the experiments with global data is in the case of B0 experiment: *rat* = 0.87, *eff* = 0.77 and *corr* = 0.91. For this experiment the simulated monthly and annual runoff from each basin was validated against observed runoff. The results of validation are shown in Table 1, which also contains, for comparison, the results from the MOPEX experiment simulation. As seen from the Table, for some

Table 1 Comparison of modelled and observed runoff from the twelve basins on monthly (mm month^{-1}) and annual (mm year^{-1}) basis for 1986–1995. The numerator refers to the B0-experiment results, the denominator represents the MOPEX-experiment results.

Basin	Mean observed	Mean simulated	Bias	RMSD	Nash-Sutcliffe efficiency	Correlation
Monthly statistics						
1	25	32.4/26.8	7.5/1.8	26.9/20.7	0.01/0.51	0.82/0.77
2	32.9	26.4/33.1	–6.6/0.2	22.3/19.8	0.57/0.66	0.86/0.83
3	29.7	30.4/27.5	0.8/–2.1	27.3/17.2	–0.07/0.57	0.77/0.77
4	59.6	43.6/59.1	–16.0/–0.54	34.5/33.6	0.5/0.53	0.81/0.74
5	34.5	38.7/31.5	4.3/–2.9	26.5/24.1	0.45/0.55	0.84/0.75
6	31.0	28.8/34.1	–2.2/3.1	25.8/13.3	0.07/0.75	0.74/0.88
7	65.5	58.6/59.3	–7.0/–6.2	39.6/19.8	–0.33/0.67	0.82/0.84
8	24.2	18.8/24.4	–5.4/0.2	22.5/19.7	0.58/0.68	0.78/0.83
9	33.1	24.7/28.2	–8.5/–4.9	33.3/34.0	0.46/0.44	0.71/0.68
10	60.1	54.9/53.1	–5.1/–7.0	36.7/34.7	0.62/0.66	0.84/0.85
11	12.1	4.4/10.6	–7.7/–1.5	15.8/9.2	0.37/0.79	0.73/0.89
12	16.4	6.6/12.7	–9.8/–3.7	18.8/13.3	0.33/0.67	0.76/0.83
Mean	35.3	30.7/33.4	–4.6/–2.0	27.5/21.6	0.30/0.62	0.79/0.80
Annual statistics						
1	299.6	389.3/321.6	89.8/22.0	122.1/39.2	–1.17/0.78	0.92/0.92
2	395.3	316.6/397.8	–78.7/2.5	110.1/71.3	0.21/0.67	0.88/0.83
3	356.1	365.4/330.3	9.3/–25.8	79.7/63.7	0.44/0.64	0.85/0.84
4	715.2	522.9/708.7	–192.3/–6.5	206.9/103.6	–0.30/0.67	0.93/0.87
5	413.6	464.9/378.6	51.3/–35.1	112.4/81.7	–0.23/0.35	0.86/0.70
6	371.5	345.6/408.7	–25.9/37.2	66.2/57.3	0.76/0.82	0.89/0.95
7	786.5	702.6/711.8	–83.8/–74.6	145.5/123.5	0.60/0.71	0.93/0.93
8	290.1	225.5/292.9	–64.6/2.7	93.8/133.2	0.87/0.75	0.97/0.95
9	397.3	295.9/338.5	–101.4/–58.7	123.0/122.0	0.52/0.53	0.97/0.85
10	720.9	659.1/637.0	–61.8/–83.9	123.6/101.3	–0.06/0.29	0.76/0.89
11	145.2	53.1/127.3	–92.1/–17.9	107.5/45.9	0.24/0.86	0.91/0.94
12	196.7	79.2/152.4	–117.5/–44.3	133.6/90.8	0.21/0.64	0.91/0.90
Mean	424.0	368.3/400.5	–55.6/–23.5	118.7/86.1	0.17/0.64	0.90/0.88

RMSD; root mean square deviation.

basins the values of the coefficient of correlation in the B0 experiment are even higher than in the MOPEX experiment, while the values of bias and RMSD are greater in B0. No wonder that the efficiency of simulations in B0, as a rule, is lower, than in the MOPEX run.

All experiments with different precipitation data sets are generalized in Fig. 3(a),(b) where annual precipitation values (averaged over 10 years and 12 basins) are sorted in increasing order. As seen, they vary among the experiments greatly: from 924 (experiment P1) to 1732 (experiment P2) mm year^{-1}. The appropriate values of modelled runoff (Fig. 3(a)) and runoff ratio (Fig. 3(b)) are within the ranges 80–680 mm year^{-1} (i.e. modelled runoff may differ by an order of magnitude in dependence of precipitation) and 0.09–0.39, respectively. In the case of the lowest precipitation, annual runoff is underestimated by 343 mm year^{-1} (or 81%). In the case of the largest precipitation, annual runoff is overestimated by 256 mm year^{-1} (or 60%).

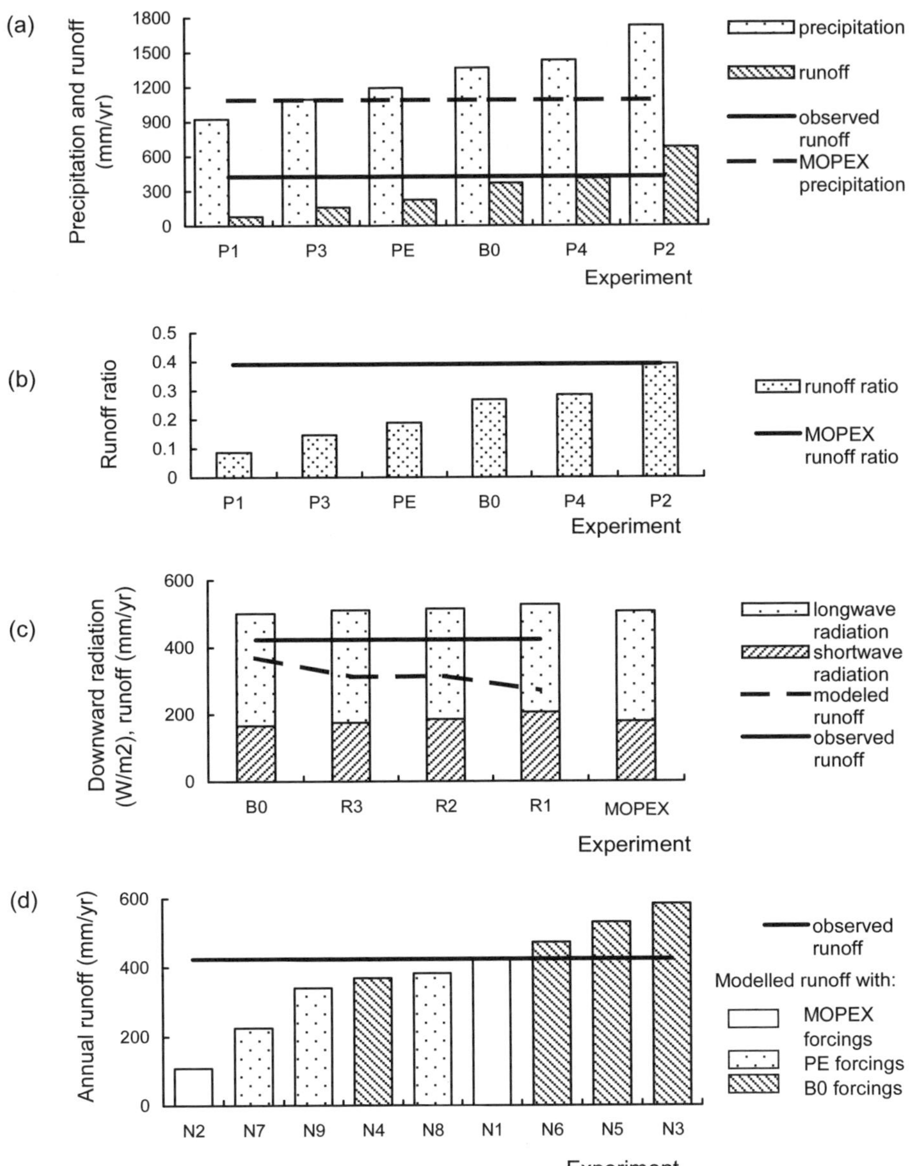

Fig. 3 Generalized results from the sensitivity experiments with different precipitation (a, b), downward radiation (c), and soil parameter data sets (d). The annual values of precipitation, runoff and radiation averaged over 10-year period (1986–1995) and over 12 basins are shown.

Four experiments with downward radiation $R\downarrow$ are generalized in Fig. 3(c). Mean annual $R\downarrow$ with partitioning between shortwave and longwave radiation is shown, being also sorted in increasing order, MOPEX radiation is given for comparison. The largest difference, equalled to 27 W m^2 (≈5%), is between SRB and NCEP/DOE radiation. This corresponds to a difference in simulated annual runoff equal to 98 mm year^{-1} (≈27%). Compared to the observed runoff, the simulated annual runoff is underestimated in all these experiments (by 13% for the lowest $R\downarrow$ and by 36% for the highest $R\downarrow$).

Different combinations of forcing data (MOPEX forcings and fully hybridized global forcing data sets: B0 and PE) and land surface parameters (from MOPEX and GSWP-2, as well as two soil data sets derived by us) allowed us to perform the following sensitivity experiments:

- N1 MOPEX forcing data and MOPEX land surface parameters.
- N2 MOPEX forcing data and GSWP-2 land surface parameters.
- N3 B0 forcing data and MOPEX land surface parameters.
- N4 B0 forcing data and GSWP-2 land surface parameters.
- N5 B0 forcing data and global soil parameters derived from Clapp & Hornberger (1978) and Dunne & Willmott (1996), the other land surface parameters are from GSWP-2.
- N6 B0 forcing data and global soil parameters derived from Cosby *et al.* (1984), the other land surface parameters are from GSWP-2.
- N7 PE forcing data and GSWP-2 parameters.
- N8 PE forcing data and global soil parameters derived from Clapp & Hornberger (1978) and Dunne & Willmott (1996), the other land surface parameters are from GSWP-2.
- N9 PE forcing data and global soil parameters derived from Cosby *et al.* (1984), the other land surface parameters are from GSWP-2.

The results from these experiments with different sets of parameters are presented in Fig. 3(d). The simulated annual runoff is sorted in increasing order and compared to the observed one. As seen, the impact of uncertainties in parameters on the simulated runoff is of the same order of magnitude as the impact of uncertainties in precipitation. The simulated annual runoff, averaged over 10 years and over all basins, varies from 109 to 584 mm year^{-1}. In experiment N2, being the lowest runoff is underestimated by 315 mm year^{-1} (74%), while in experiment N3 it is overestimated by 160 mm year^{-1} (38%). Combination of B0 forcings with soil parameters from GSWP-2 (N4 experiment) and PE forcings with soil parameters derived by using relationships reported by Clapp-Hornberger and Dunne-Willmott (N8 experiment) show the best results among simulations with different global 1-degree parameter data sets. The statistics for the simulated annual runoff in both cases are nearly the same: for PE, *rat* = 0.90, *eff* = 0.77 and *corr* = 0.90; for B0, they are 0.87, 0.77 and 0.91, respectively. Further improvement of these results may be achieved by means of model calibration by tuning some model parameters (e.g. the hydraulic conductivity at saturation) (Gusev & Nasonova, this issue).

The results obtained have shown that uncertainties in the two primary factors that force hydrological processes—precipitation and incoming radiation, as well as in the land surface parameters may cause large differences in simulated runoff. That is why global data sets should be tested for a wide range of river basins located all over the world (because data quality may vary spatially) before application on ungauged basins. It should also be noted that more attention should be paid to the development of alternative global data sets with the land surface (especially soil) parameters since their impact on runoff simulations is comparable to that of the main forcing factors.

Acknowledgements This work was supported by the Russian Foundation for Basic Research (Grant 05-05-64161). We acknowledge the MOPEX and GSWP-2 experiment organizers for providing us with the data to run the model.

REFERENCES

Clapp, R. B. & Hornberger, G. M. (1978) Empirical equations for some soil hydraulic properties. *Water Resour. Res.* **14**, 601–604.

Cosby, B. J., Hornberger, G. M., Clapp, R. B. & Ginn, T. R. (1984) A statistical exploration of the relationships of soil moisture characteristics to the physical properties of soils. *Water Resour. Res.* **20**(6), 682–690.

Dirmeyer, P., Gao, X. & Oki, T. (2002) The second global soil wetness project. Science and implementation plan. *IGPO Publication Series*, 37.

Duan, Q., Schaake, J., Andreassian, V., Franks, S., Goteti, G., Gupta, H. V., Gusev, Y. M., Habets, F., Hall, A., Hay, L., Hogue, T., Huang, M., Leavesley, G., Liang, X., Nasonova, O. N., Noilhan, J., Oudin, L., Sorooshian, S., Wagener, T. & Wood, E. F. (2006) Model Parameter Estimation Experiment (MOPEX): An overview of science strategy and major results from the second and third workshops. *J. Hydrol.* **320**, 3–17.

Dunne, K. A. & Willmott, C. J. (1996) Global distribution of plant-extractable water capacity of soil. *Int. J. Climatol.* **16**, 841–859.

Gusev, Ye. M. & Nasonova, O.N. (2003) Modelling heat and water exchange in the boreal spruce forest by the land-surface model SWAP. *J. Hydrol.* **280**, 162–191.

Gusev, Ye. M. & Nasonova, O. N. (2006) Simulating runoff from MOPEX experimental river basins using the SWAP land surface model and different parameter estimation techniques. In: *Large Sample Basin Experiments for Hydrological Model Parameterization: Results of the Model Parameter Experiment–MOPEX* (ed. by V. Andréssian, A. Hall, N. Chahinian & J. Schaake), 188–195. IAHS Publ. 307. IAHS Press, Wallingford, UK (this issue).

Jayawardena, A. W. & Mahanama, S. P. P. (2001) Daily river discharge prediction using GCM generated atmospheric data. In: *Soil-Vegetation–Atmosphere Transfer Schemes and Large-Scale Hydrological Models* (ed. by A. J. Dolman, A. J. Hall, M. L. Kavvas, T. Oki & J. W. Pomeroy), 159–166. IAHS Publ. 270. IAHS Press, Wallingford, UK.

Oki, T., Musiake, K., Matsuyama, H. & Masuda, K. (1995) Global atmospheric water balance and runoff from large river basins. *Hydrol. Processes* **9**, 655–678.

Salathé, E. P. (2003) Comparison of various precipitation downscaling methods for the simulation of streamflow in a rainshadow river basin. *Int. J. Climatol.* **23**, 887–901.

Wilby, R. L. & Wigley, T. M. L. (1997) Downscaling general circulation model output: a review of methods and limitations. *Progr. Phys. Geogr.* **21**, 530–548.

Zhao, M. & Dirmeyer, P. A. (2003) Production and Analysis of GSWP-2 near-surface meteorology data sets. *COLA Technical Report 159.*

Simulating runoff from MOPEX experimental river basins using the SWAP land surface model and different parameter estimation techniques

YEUGENIY M. GUSEV & OLGA N. NASONOVA

Institute of Water Problems, Russian Academy of Sciences, Gubkina St. 3, 119991 Moscow, Russia

gusev@aqua.laser.ru

Abstract Different techniques for the *a priori* estimation and calibration of parameters in the SWAP land surface model for the 12 MOPEX experimental river basins were investigated. Two approaches for *a priori* parameter estimation were applied: (1) derivation of parameters from spatial distribution of different soil and vegetation classes within a basin; (2) application of a global parameter data set provided within the framework of the Second Global Wetness Project (GSWP-2). Two parameter calibration techniques were also used: (1) manual calibration of hydraulic conductivity at saturation that is the most crucial for runoff generation; (2) automatic optimization of several soil parameters. The results of river flow simulations with all the derived sets of parameters were compared with observations to reveal optimal values of soil parameters. Since the optimal values differed from *a priori* estimations using the USDA soil classification, an attempt was made to connect the optimal parameter values with the USDA soil texture classes.

Key words *a priori* parameter estimation; land surface modeling; MOPEX; parameter calibration; river runoff

INTRODUCTION

Model simulation of river flow requires data on soil and vegetation parameters for a river basin. In physically-based models, some parameters have a physical meaning and can be measured directly, some can be estimated from the other measured characteristics, and some represent empirical coefficients obtained for a particular case and cannot be applied for an object with substantially different natural properties. In addition, most parameters are characterized by spatial variability while measurements have a point character. Besides that, these measurements, as a rule, are not carried out on a regular basis and do not cover all the watersheds of the Earth. As a result, there are many locations in the world where data on soil and vegetation parameters are not available. This raises the problem of *a priori* parameter estimation.

A priori estimation of the model parameters is usually based on relationships between the parameters and soil, vegetation, topographic and climatic characteristics of the object under study (Schaake *et al*., 2001). Some examples of such estimations for conceptual and physically-based hydrological models can be found in Koren *et al.* (2000) and Kuchment & Gelfan (2005), respectively. In the land surface models (LSMs), *a priori* parameters are usually derived from the relationships between model parameters and soil texture and vegetation classes. The shortcomings of existing *a priori* parameter estimation procedures are shown in (Duan *et al.*, 2006). The Model Parameter Estimation Experiment (MOPEX) was initiated in 1996 to improve this situation by developing enhanced *a priori* parameter estimation techniques for hydrological and land surface models (Schaake *et al.*, 2001). According to the MOPEX strategy the improved *a priori* parameters may be estimated using the results of model calibration, in particular, by application of new relationships between the calibrated parameters and basin characteristics.

In the present paper, following the MOPEX strategy, investigation of two *a priori* model parameter estimation techniques was undertaken using the SWAP land surface model (Soil Water–Atmosphere–Plants) (Gusev & Nasonova, 1998, 2002, 2003) for river runoff simulations. Another goal of this study was investigation of how different calibration techniques can improve simulation of river flow based on *a priori* estimated parameters. An attempt was made to use the results of calibration for construction of a new set of *a priori* parameters, which can be treated as effective parameters for river basins.

HYDROLOGICAL BASINS AND DATA

The hydrological objects under study represent 12 river basins (with an area of the order of 10^3 km^2), which were selected within the framework of the Second MOPEX Workshop (Duan *et al.*, 2006). The basins are located within the southeastern part of the United States and characterized by a wide range of hydrological and climatic conditions, ranging from arid to a very wet regime. The main characteristics of the basins are given in Table 1, a full description can be found in Duan *et al.* (2006).

The data set for each of the 12 basins includes: (1) 1-hourly near-surface meteorological data (downward shortwave and longwave radiation, air temperature and humidity, atmospheric precipitation, air pressure and wind speed) for a 39-year period (1960–1998), and (2) basin land surface characteristics (basin boundary, elevation, spatial distribution of different soil and vegetation classes within a basin, soil texture,

Table 1 Characteristics of the 12 MOPEX river basins.

Basin	Basin ID	Area (km^2)	Elevation (m)	Latitude	Longitude	Dominant soil type	Dominant vegetation type	Clay in soil (%)	Sand in soil (%)
1	01608500	3810	171	39.45	–78.65	Loam	Dec. broadleaf	43.5	30.6
2	01643000	2116	71	39.39	–77.38	Silt loam	Dec. broadleaf	17.1	23.9
3	01668000	4134	17	38.32	–77.52	Clay loam	Mixed forest	29.4	26.4
4	03054000	2372	390	39.15	–80.04	Loam	Dec. broadleaf	32.0	26.0
5	03179000	1020	465	37.54	–81.01	Si cl loam/loam	Dec. broadleaf	24.6	20.5
6	03364000	4421	184	39.20	–85.93	Si loam/cl loam	Croplands	25.0	24.6
7	03451500	2448	594	35.61	–82.58	Loam	Mixed forest	30.9	35.9
8	05455500	1484	193	41.47	–91.72	Clay loam	Cropland	32.6	10.8
9	07186000	3015	254	37.25	–94.57	Si loam/cl loam	Dec. broadleaf	27.4	17.5
10	07378500	3315	0	30.46	–90.99	Silt loam	Ever. needleleaf	20.5	21.0
11	08167500	3406	289	29.86	–98.38	Clay	Grassland	40.0	14.8
12	08172000	2170	98	29.67	–97.65	Clay	Grassland	49.1	23.5

monthly surface albedo, roughness length and greenness fraction). All these data were provided by the Second MOPEX Workshop organizers and described in detail in Duan *et al.* (2006).

Evidently, the information provided on soil and vegetation characteristics is rather general. Application of the SWAP model requires soil hydrophysical parameters including field capacity W_{fc}, wilting point W_{wp}, soil porosity W_{sat}, and relations of soil hydraulic conductivity K and soil matric potential ϕ with soil moisture W, i.e. $K(W)$ and $\phi(W)$. In SWAP, Clapp & Hornberger (1978) parameterizations for $K(W)$ and $\phi(W)$ are used. These parameterizations require information on soil matric potential at saturation ϕ_0, the B parameter, W_{sat}, and soil hydraulic conductivity at saturation k_0. Besides that, snow-free albedo of bare soil α_{soil} and soil depth h_{soil} are also required. As to vegetation parameters, along with provided ones, the SWAP model needs monthly values of the leaf area index LAI, zero plane displacement height d_0, snow-free albedo of vegetation α_{veg}, root depth h_{root}. Since these parameters were not available for the selected basins, we had to estimate them ourselves by means of the techniques described in the next section.

A PRIORI PARAMETER ESTIMATION AND CALIBRATION TECHNIQUES

Two approaches for *a priori* parameter estimation were applied. First, soil and vegetation parameters for each of the 12 basins were derived by us using available data on spatial coverage of the USDA soil texture classes and of the University of Maryland vegetation types. As to soil parameters, experimentally determined soil textural and bulk density data (corresponding to each soil class) were used as predictors for the calculation of hydraulic and soil water retention parameters using the Rosetta code that implements pedotransfer functions based on an artificial neural network predictions (Schaap *et al.*, 1998). Thus, the values of W_{sat} and k_0 and Van Genuchten's (1980) shape parameters for the $K(W)$ and $\phi(W)$ functions were determined. Then, the values of ϕ_0 and B were obtained on the basis of interpolation of the estimated $\phi(W)$ by the Clapp and Hornberger equation within the range –33 to –1500 kPa. The values of W_{wp} and W_{fc} were determined as soil moisture at ϕ = –1500 kPa and at ϕ = –33 kPa, respectively. In such a manner, hydrophysical parameters were estimated for all soil classes with the exception of "bedrock" that is absent in the USDA classification. For bedrock, we roughly estimated the hydrophysical parameters on the basis of our experience. For each basin, soil parameters were calculated as weighted average values with accounting for the spatial coverage of each soil class within a basin. For k_0, logarithms of k_0 were averaged. As to vegetation parameters, we assigned their typical values for each vegetation class and then calculated their weighted average values for each basin. The constructed *a priori* parameters set hereafter will be referred to as the MOPEX *a priori* data set ("APR_MOP").

The second set of *a priori* parameters was constructed from the global parameter data set provided within the framework of the Second Global Wetness Project (GSWP-2) (Zhao & Dirmeyer, 2003). Hereinafter, this *a priori* parameter set will be referred to as GSWP-2 *a priori* data set ("APR_GSW")

To improve runoff simulations based on *a priori* parameters, some parameters related to soil, which are more important from the viewpoint of runoff generation in the SWAP model, were calibrated by means of minimization of the root-mean-square-deviation (RMSD) between simulated and measured daily total runoff from a basin. Two parameter calibration techniques were applied: (1) manual calibration of hydraulic conductivity at saturation (referred to as "CAL1") which is the most crucial for runoff generation, and (2) automatic procedure for optimization of six parameters (W_{fc}, W_{wp}, W_{sat}, k_0, h_{soil}, h_{root}) (referred to as "CAL6"). The latter is based on a stochastic or Monte-Carlo technique. The first 20 years (1960–1979) were used for calibration in both cases. Each calibration procedure was applied for each of the two *a priori* parameters sets. As a result, we obtained four sets of calibrated parameters, related to the MOPEX (referred to as "CAL1_MOP" and "CAL6_MOP") and GSWP-2 (referred to as "CAL1_GSW" and "CAL6_GSW") data sets.

RESULTS

Simulations of river flow from the 12 MOPEX basins for a 39-year period (1960–1998) were performed using the land surface model SWAP with two sets of *a priori* estimated and four sets of calibrated parameters. The results of the simulations were compared with each other and with measured daily streamflow. The results are given in Figs 1–3.

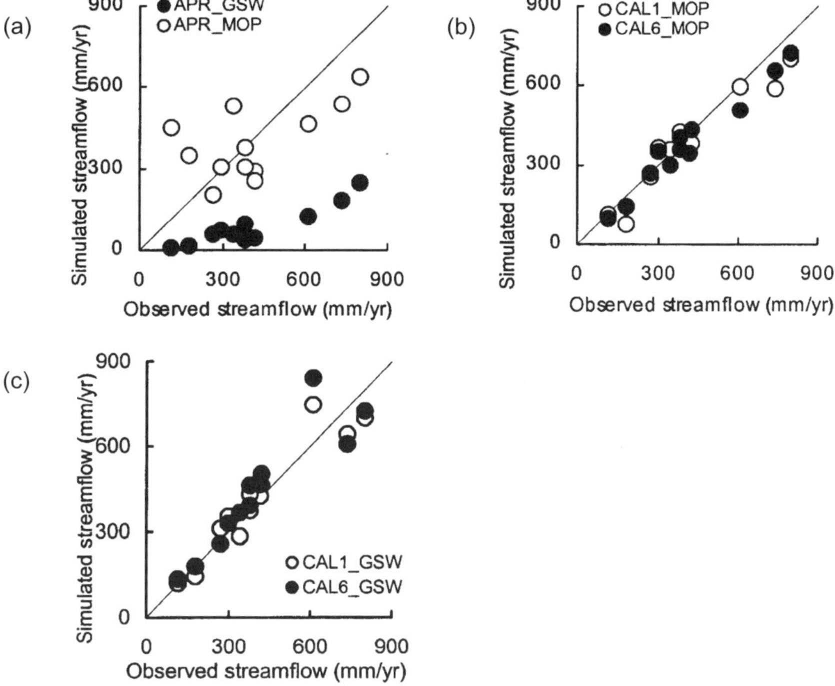

Fig. 1 Comparison of simulated and observed annual streamflow averaged over the 1960–1998 period for 12 basins when two sets of *a priori* estimated parameters (a), and one/six calibrated parameters from MOPEX (b) and GSWP-2 (c) parameter sets are used.

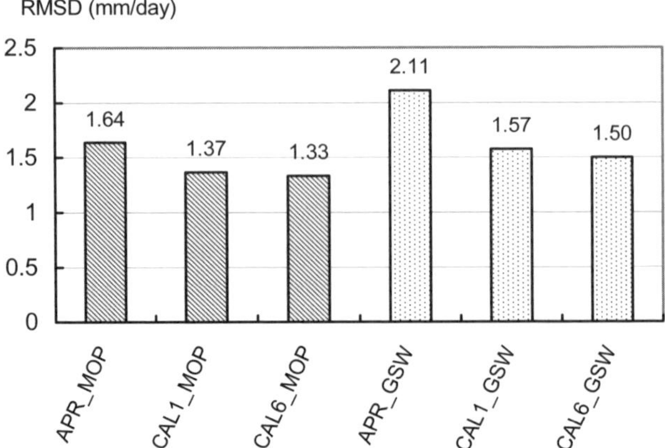

Fig. 2 The root-mean-square deviation between simulated (with six parameter sets) and observed daily streamflow, for 1960–1998, averaged over the 12 basins.

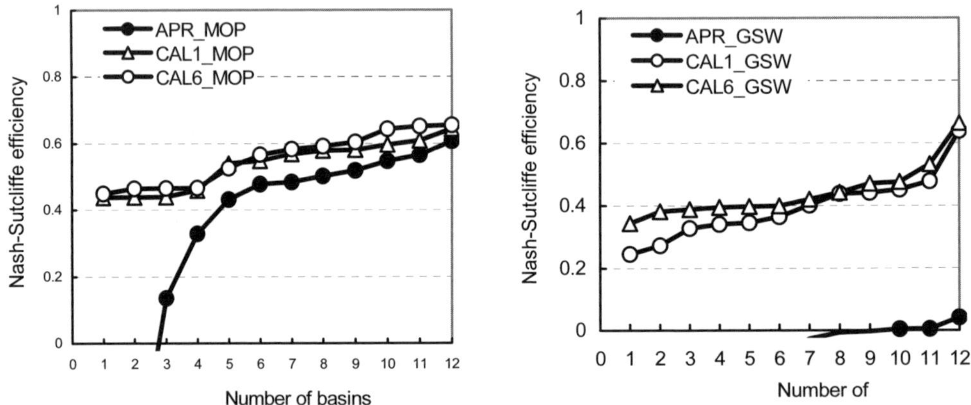

Fig. 3 Daily Nash-Sutcliffe efficiency of runoff simulation for the 12 MOPEX basins with *a priori* estimated and calibrated parameters sorted in increasing order.

Figure 1(a) shows the comparison between observed and simulated annual runoff (averaged over the 39-year period) with *a priori* parameters from the 12 basins. As seen, in the case of GSWP-2 *a priori* parameters, the simulated runoff is greatly underestimated (by 80%), however, the coefficient of correlation (r) is very high (0.91). For the MOPEX *a priori* parameters, runoff is underestimated by only 5%, while the scatter of points is much greater ($r = 0.58$).

Application of the calibrated parameters sets allowed us to substantially improve the results (Fig. 1(b),(c)). This is also confirmed by the values of RMSD between simulated and observed daily runoff (Fig. 2) and by the daily Nash-Sutcliffe efficiency of runoff simulations with different parameters sets (Fig. 3). As seen from these figures, the values of simulated runoff in the case of one calibrated parameter are close to those when six parameters were calibrated (both for MOPEX and GSWP-2 data sets) and the agreement between simulations and observations is better for the MOPEX parameters sets. The former means that soil hydraulic conductivity at saturation k_0 is the main soil parameter effecting runoff generation. The latter means that non-calibrated parameters derived from the MOPEX data are more adequate than those taken from the global GSWP-2 data set. Consequently, the main focus of the rest of this paper is on k_0 and on the MOPEX data set.

As seen from Fig. 4, *a priori* estimated values of k_0 differ from the optimal values obtained when six parameters from the MOPEX data set were calibrated. Again, application of the MOPEX *a priori* parameters shows better results compared to the GSWP-2 *a priori* parameters. So, it is reasonable to concentrate only on the MOPEX data set and to use the optimal values of k_0 to improve the procedure of estimation of *a priori* values of k_0 for ungauged basins. We attempted to do this in the following manner.

As mentioned above, the MOPEX *a priori* values of k_0 for each basin were estimated with the help of the soil texture classes provided (from the USDA classification) and the values of k_0, assigned for each class by the US Salinity Laboratory on the basis of investigation of different soil patterns of local character. At the same time, k_0 is known to be highly variable in space. The optimal values of k_0 obtained by us on the basis of model calibration may be treated as effective values for inhomogeneous areas and may be used for the calculation of new (effective) values of k_0 for the USDA soil classes. For this purpose, we derived nonlinear regression equations connecting the optimal k_0 for each basin with the content of clay and sand in a soil averaged over each basin (the average values of clay and sand are given in Table 1):

$$\mathrm{Ln}(k_0) = a_0 + a_1 \cdot \mathrm{CLAY} + a_2 \cdot \mathrm{SAND} + a_3 \cdot \mathrm{CLAY}^2 + a_4 \cdot \mathrm{SAND}^2 + a_5 \cdot \mathrm{CLAY} \cdot \mathrm{SAND} \quad (1)$$

Using equation (1) it is easy to calculate new values of k_0 for the USDA soil textural classes because the average content of clay and sand for each class is known. However, the basins under study are very few in number and the average contents of clay and sand in their soils are within narrow ranges: 17% < clay < 43%, and 10% < sand < 35% (Table 1). For these ranges, the coefficients in equation (1) were obtained. Figure 5 shows the location of each basin (according to the averaged-over-a-basin

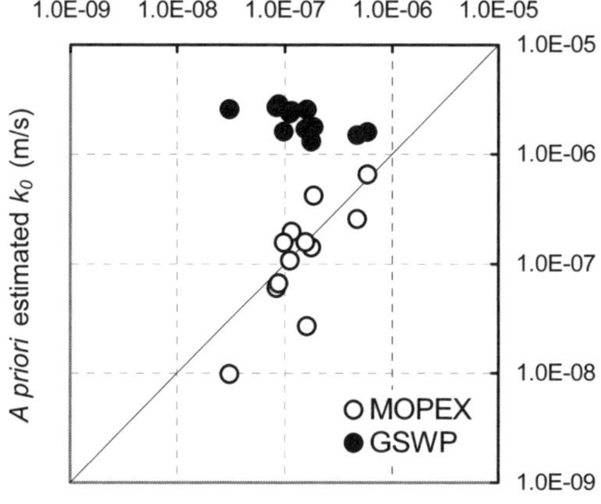

Fig. 4 Comparison of *a priori* estimated and calibrated hydraulic conductivity at saturation (in logarithmic scale).

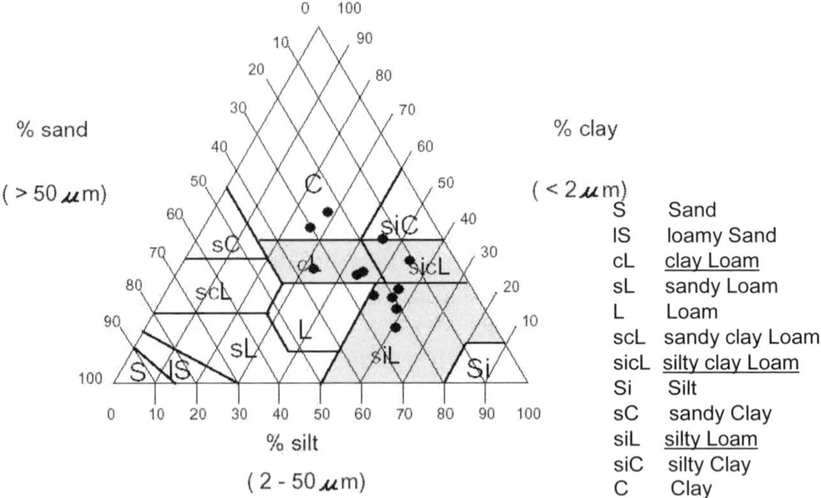

Fig. 5 Location of the 12 MOPEX basins (black circles) within the USDA-SCS soil textural triangle.

clay/sand content in a soil) within the USDA-SCS soil textural triangle. As shown, in the main, three soil classes correspond to the clay and sand ranges mentioned: clay loam, silty clay loam, and silty loam. This means that we can derive the new values of k_0 only for these three classes. These values were calculated by equation (1) using average contents of clay and sand for each soil class (Table 2). Increasing the number of basins will allow the calculation of the new values of k_0 for all soil classes and the construction of a new set of *a priori* parameters.

Table 2 Optimized values of hydraulic conductivity at saturation k_0 compared to initial values from the Rosetta code for the three USDA soil textural classes.

Soil class	Average content of clay (%)	Average content of sand (%)	k_0 (cm h^{-1}) (from the Rosetta code)	k_0 (cm h^{-1}) (calculated with equation (1))
Clay loam	34	32	0.038	0.006
Silty clay loam	34	10	0.047	0.079
Silty loam	14	18	0.053	0.127

CONCLUSIONS

1. When using the SWAP model on the 12 MOPEX basins, the application of *a priori* parameters estimated from the MOPEX database, results in better simulations of river flow than *a priori* parameters obtained from the GSWP-2 database.
2. Analysis of the soil parameter calibration results shows that: (a) the main soil parameter (in the model SWAP) controlling streamflow is the soil hydraulic

conductivity at saturation, and (b) an increase in the number of calibrated parameters only slightly improves the quality of streamflow simulation.

3. Following the MOPEX strategy, it seems possible to link the optimal soil hydrophysical parameters (from the viewpoint of streamflow simulation by SWAP) to more integral pedological characteristics (e.g. soil texture). For this purpose, it is necessary to increase the number of basins under study in order to cover all existing soil types.

Acknowledgements This work was supported by the Russian Foundation for Basic Research (Grant 05-05-64161). We acknowledge the MOPEX and GSWP-2 experiment organizers for providing us with the data to run the model.

REFERENCES

Clapp, R. B. & Hornberger, G. M. (1978) Empirical equations for some soil hydraulic properties. *Water Resour. Res.* **14**, 601–604.

Duan, Q., Schaake, J., Andreassian, V., Franks, S., Goteti, G., Gupta, H. V., Gusev, Y. M., Habets, F., Hall, A., Hay, L., Hogue, T., Huang, M., Leavesley, G., Liang, X., Nasonova, O. N., Noilhan, J., Oudin, L., Sorooshian, S., Wagener, T. & Wood, E. F. (2006) Model Parameter Estimation Experiment (MOPEX): An overview of science strategy and major results from the second and third workshops. *J. Hydrol.* **320**, 3–17.

Gusev, Ye. M. & Nasonova, O. N. (1998) The Land Surface Parameterization scheme, SWAP: description and partial validation. *Global Planet. Change.* **19**(1–4), 63–86.

Gusev, Ye. M. & Nasonova, O. N. (2002) The simulation of heat and water exchange at the land–atmosphere interface for the boreal grassland by the land-surface model SWAP. *Hydrol. Processes* **16**(10), 1893–1919.

Gusev, Ye. M. & Nasonova, O. N. (2003) Modelling heat and water exchange in the boreal spruce forest by the land-surface model SWAP. *J. Hydrol.* **280**, 162–191.

Koren, V. I., Smith, M., Wang, D. & Zhang, Z. (2000) Use of soil property data in the derivation of conceptual rainfall–runoff model parameters. In: *Preprints, 15th Conf. on Hydrol.* (Long Beach, California, January 2000), Paper 2.16. Am. Met. Soc.

Kuchment, L. S. & Gelfan, A. N. (2005) Toward parameter estimation of physically-based models of river flow formation under insufficient hydrological observations. *Meteorology & Hydrology* **12**, 77–87 (in Russian).

Nash, J. E. & Sutcliffe, J. V. (1970) River flow forecasting through conceptual models: 1. A discussion of principles. *J. Hydrol.* **10**(3), 282–290.

Schaake, J., Duan, Q., Koren V. & Hall A. (2001) Toward improved parameter estimation of land surface hydrology models through the Model Parameter Estimation Experiment (MOPEX). In: *Soil–Vegetation–Atmosphere Transfer Schemes and Large-Scale Hydrological Models* (ed. by A. J. Dolman, A. J. Hall, M. L. Kavvas, T. Oki & J. W. Pomeroy), 91–97. IAHS Publ. 270. IAHS Press, Wallingford, UK.

Schaap, M. G., Leij, F. J. & van Genuchten, M. Th. (1998) Neural network analysis for hierarchical prediction of soil water retention and saturated hydraulic conductivity. *Soil Sci. Soc. Am. J.* **62**, 847–855.

van Genuchten, M. Th. (1980) A closed-form equation for predicting the hydraulic conductivity of unsaturated soils. *Soil Sci. Soc. Am. J.* **44**, 892–898.

Zhao, M. & Dirmeyer, P. A. (2003) Production and analysis of GSWP-2 near-surface meteorology data sets. *COLA Technical Report 159.*

Large Sample Basin Experiments for Hydrological Model Parameterization: Results of the Model Parameter Experiment–MOPEX. IAHS Publ. 307, 2006.

Performance comparison of a complex physics-based land surface model and a conceptual, lumped-parameter hydrological model at the basin-scale

THIAN YEW GAN[1], YEUGENIY GUSEV[2], STEPHEN J. BURGES[3] & OLGA NASONOVA[2]

1 *Department of Civil and Environmental Engineering, University of Alberta, Edmonton, Alberta T6G 2G7, Canada*
tgan@ualberta.ca

2 *Institute of Water Problem, Russian Academy of Sciences, Russia*

3 *Department of Civil and Environmental Engineering, University of Washington, Seattle, Washington 98195-2700, USA*

Abstract The Soil–Water–Atmosphere-Plants (SWAP) model of Gusev & Nasonova (2003) is a one-dimensional, land-surface model describing heat and water exchanges between the land surface and the atmosphere in a physics-based, analytical manner. The Sacramento model (SAC-SMA) is a deterministic, lumped-parameter, conceptual, rainfall–runoff model that does not explicitly consider spatial variability of terrain features, land-use, soil heterogeneities, and energy fluxes. SWAP is compared to SAC-SMA and both are tested using data from 12 MOPEX river basins located in the middle to southern eastern USA. Two model simulation comparisons are made: one using broad classifications of soil and vegetation to derive parameters (base case), and the second using calibrated parameters derived from measured data from the 12 MOPEX sites during the 20-year period (1960–1979). Only six of the SWAP parameters were automatically calibrated using a stochastic or Monte-Carlo technique. SAC-SMA was calibrated using manual and automatic means (Duan *et al.*, 1992). In terms of the goodness-of-fit between simulated and observed hydrographs, for the base case SAC-SMA was slightly better than SWAP. Both yielded results were unsuitable for any scientific inference. For the calibrated case SAC-SMA out-performed SWAP when comparing simulated hydrographs, but the simulations are still unsatisfactory. Analysis of the different behaviour of the models is presented.

Key words *a priori* parameter estimation; hydrological model; land surface model; MOPEX basins; parameter calibration

INTRODUCTION

As remotely sensed and digital elevation model (DEM) data have become more readily available, hydrological models used for simulating basin hydrological processes have been made more complex. More complex models may not necessarily produce better or more representative basin-scale simulations: simpler conceptual, lumped-parameter models can out-perform complex models, particularly when there is insufficient high quality data to support the complex models. The objective of this paper is to compare the performance of a relatively complex, physics-based model called SWAP (Gusev & Nasonova, 2003) with that of a simpler, conceptual, lumped-parameter model, the so-called Sacramento Soil Moisture Accounting model (SAC-SMA) (Burnash *et al.*,

1973). We tested model performance over a variety of hydroclimatic conditions using data from the 12 river basins of MOPEX (Model Parameter Experiment) (Duan *et al.*, 2005) in the southeastern United States (Fig. 1). The MOPEX basins represent a number of soil and vegetation types range from desert to very wet climatic conditions. Two model simulation comparisons are made: one using broad classifications of soil and vegetation to derive parameters (base case), and the second using calibrated parameters derived from measured data from the 12 MOPEX sites.

MOPEX RIVER BASINS

A full description of the 12 MOPEX basins is given in (Duan *et al.*, 2005); Fig. 1 and Table 1 provide summary information. All the basins are located below latitude 40°N, and from Texas to the east. Consequently, there is minimal to no snow or frozen ground influence. Different hydrological and climatic conditions are well represented in these basins, ranging from desert conditions (basins 11 and 12) to very wet conditions (basins 1, 4, 5, 7 and 10) (Table 2). The catchment areas of these MOPEX basins range from about 1000 to 4400 km^2. (Table 1), the annual average precipitation ranges from 764 mm to 1564 mm, while the annual Runoff to Precipitation (*QP*) and Precipitation to Potential Evapotranspiration (*PPE*) ratios range from 0.09 to 0.59, and 0.45 to 2.34, respectively (Table 2).

MODEL DESCRIPTION

Land surface model SWAP

SWAP is a physically based land-surface model (LSM) describing heat and water exchange between the land surface and the atmosphere throughout a year at different

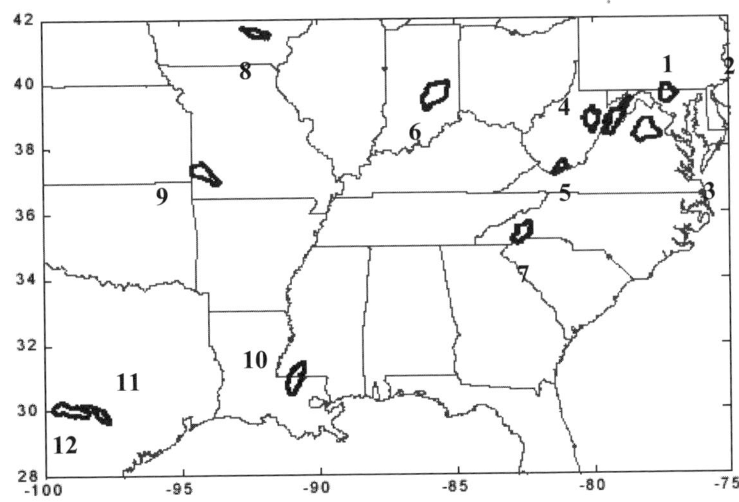

Fig. 1 Locations of 12 MOPEX river basins.

Table 1 Twelve MOPEX river basins.

No	Station#	Longitude, Latitude (°W, °N)	Area km^2	Description of river basin
1	1608500	–78.65, 39.45	3810	South Branch Potomac River Nr Springfield, WV
2	1643000	–77.38, 39.39	2120	Monocacy R At Jug Bridge Nr Frederick, MD
3	1668000	–77.52, 38.32	4130	Rappahannock River Near Fredericksburg, VA
4	3054500	–80.04, 39.15	2370	Tygart Valley River At Phillipi, WV
5	3179000	–81.01, 37.54	1020	Bluestone River Near Pipestem, WV
6	3364000	–85.93, 39.2	4420	East Fork White River At Columbus, IND
7	3451500	–82.58, 35.61	2450	French Broad River At Asheville, N. C.
8	5455500	–91.72, 41.47	1480	English River At Kalona, IA
9	7186000	–94.57, 37.25	3020	Spring River Near Waco, MO
10	7378500	–90.99, 30.46	3320	Amite River Near Denham Springs, LA
11	8167500	–98.38, 29.86	3410	Guadalupe River Nr Spring Branch, TX
12	8172000	–97.65, 29.67	2170	San Marcos River At Luling, TX

Table 2 Annual water budget and 8 major USDA soil texture classifications of the 12 MOPEX river basins.

	Annual water budget				8 Major USDA soil texture[b] (%)							
No.	*PPE*	*QP*	*AEPE*	*P* [a] mm	SL	SIL	L	SICL	CL	SIC	C	Br
1	1.64	0.47	0.86	1040	19	13	23	3	0	0	0	36
2	1.15	0.34	0.76	1040	9	64	10	6	0	0	4	0
3	1.2	0.36	0.77	1030	5	35	4	9	36	0	5	4
4	1.76	0.5	0.87	1170	0	25	49	9	0	0	0	17
5	1.5	0.45	0.83	1020	0	16	32	34	0	0	6	1
6	1.21	0.36	0.77	1020	0	31	16	20	31	0	0	0
7	2.34	0.59	0.96	1380	11	3	48	1	17	0	15	5
8	0.89	0.25	0.67	890	0	7	0	92	1	0	0	0
9	0.96	0.28	0.69	1080	0	29	10	29	14	14	0	0
10	1.46	0.44	0.83	1560	1	58	4	21	8	0	0	0
11	0.45	0.09	0.41	770	0	0	13	0	14	3	24	17
12	0.56	0.13	0.49	830	1	0	17	4	23	2	43	9

[b] SL, sandy loam; SIL, silt loam; L, loam; SICL, silty clay loam; CL, clay loam; SIC, silty clay; C, clay; BR, bedrock.
PPE, Precipitation/Potential ET; *QP*, Annual runoff/Precipitation; AEPE, actual ET/PET
[a] *P*, average annual precipitation in mm.

scales (from local to global). It was developed to use atmospheric forcing data from the lowest atmospheric layer of GCMs (Global Circulation Models) or from any reference height (Gusev & Nasonova, 2000, 2003). The main distinctive feature of SWAP is a combination of its physically based treatment of the main processes and rationality of modelling technique used. The latter is provided by application of analytical methods (contrary to the usual practice of application of numerical ones) to solve the systems of equations and by a relatively small number of model parameters (compared to other LSMs). This allows one to avoid many problems associated with solving numerical equations (such as instability, great consumption of computer resources and calculation time) and parameters estimation.

The direct use of analytical methods has led to a non-traditional model structure for SWAP. Thus, in SWAP, a calendar year is divided into two seasons: warm and cold. For each season, a separate submodel was developed. These two submodels were linked into one general model, named SWAP. SWAP operates at different time steps (from 30 minutes to 1 day), depending on available forcing data and includes the following processes: interception of rainfall/snowfall by the canopy; evapotranspiration (including transpiration by plants, soil/snow evaporation, canopy evaporation); formation of snowpack on the ground and on the trees' crowns (including snow accumulation, snow evaporation, snowmelt, water yield of snow cover, refreezing of meltwater); formation of surface runoff and drainage; water infiltration into soil; water exchange between soil layers; interaction between soil water and groundwater; formation of the energy balance at the land surface; soil freezing/thawing.

Input data for the model includes atmospheric forcings (incoming shortwave and longwave radiation, air temperature and humidity, surface air pressure, wind speed, precipitation) and land surface parameters. Model outputs include all the energy and water balance components, all evapotranspiration components, different surface and subsurface state variables, and cold season characteristics.

SWAP was one of the LSMs used in different international experiments, including PILPS, RhoneAGG, SnowMIP, GSWP-2.

Hydrological model-Sacramento Model (SAC-SMA)

The SAC-SMA is a deterministic, conceptually-based rainfall–runoff model with spatially lumped parameters (Burnash *et al.*, 1973). The model does not explicitly consider spatial variability of the land surface, soil heterogeneities and energy fluxes. The Sacramento model is used world wide to estimate streamflow for river basins. It is one of the most extensively studied conceptual rainfall–runoff models, e.g. Gan & Burges (1990a,b). The SAC-SMA model is generally applied by the US National Weather Service to river basins ranging from 300 km^2 to 5000 km^2. It can run at daily, 6-hourly, hourly or a smaller time step. Input to the model includes precipitation, temperature, pan evaporation data and lumped parameters of the basin physical characteristics. Full details for how we used this model for the 12 MOPEX sites are given in Gan & Burges (2005)

RESEARCH METHODOLOGY

Atmospheric forcing data to drive the models were provided by the Second MOPEX Workshop organizers and described in details in Duan *et al.* (2005). Additionally, some common basin characteristics (e.g. spatial distributions of different soil and vegetation classes within a basin), which can be used to derive the required model parameters, were provided. Two sets of parameters were obtained for each model: *a priori* estimated (base case) and calibrated (CAB) parameters. Concurrent 20-year long calibration data sets (January, 1960–December, 1979) were used to calibrate the models for each of the 12 MOPEX river basins (Table 1). Calibration was affected by attempting to match the daily simulated and recorded streamflow time series. The calibrations for each basin were tested (or "validated") using 19 years of data independent of the calibration experience (January 1980–December 1998). We prefer to use the term "tested"; strict requirements for model "validation" are rarely met.

The base case parameters for SAC-SMA were derived using a set of physically based relationships between SAC-SMA parameters and United States Department of Agriculture (USDA) soil properties developed by Koren *et al.* (2000). SAC-SMA was calibrated using a combination of the global optimization algorithm, the shuffle complex evolution method, SCE-UA, of Duan *et al.* (1992) and manual effort. Duan *et al.* showed that SCE-UA has overcome calibration problems in five areas that most, if not all local calibration algorithms suffer from: (i) regions of attraction—where more than one main convergence region exists; (ii) minor local optima—where there are small "pits" in each region; (iii) roughness—when the response surface is rough with discontinuous derivatives; (iv) sensitivity–poor and varying sensitivity of the response surface in the region of optimum, and nonlinear parameter interaction; and (v) shape—resulting from a non-convex response surface with long curved ridges. After calibrating SCA-SMA using SCE-UA, the resulting model parameters were fine tuned manually to obtain final calibrated model parameters (referred to as CAB) for each of the 12 MOPEX river basins. SAC-SMA was calibrated using 16 parameters: 11 "land parameter" parameters, plus three unitgraph ordinates (channel features), a precipitation scaling factor (that helps to account for bias—see e.g. Burges (2003)) and the potential evapotranspiration (PET) adjustment factor.

The base case data for SWAP were derived using the University of Maryland vegetation classification, the USDA soil texture classification, and some information from the International Satellite Land-Surface Climatology Project 1 (ISLSCP1) database. The calibrated parameters were obtained by means of automatic procedure for optimization (by minimization of root-mean-square-deviation between simulated and measured daily total runoff from a basin) based on a stochastic or Monte-Carlo technique, which as far as we know, also does not suffer from the five shortcomings identified by Duan *et al.* (1992); more extensive research is needed to clarify this issue. Only the six SWAP soil parameters that influence runoff (saturated hydraulic conductivity k_0, porosity W_{sat}, field capacity W_{fc}, wilting point W_{wp}, the depth of soil column h_{soil}, root zone depth h_{root}) were calibrated.

DISCUSSION OF RESULTS

The performance of SAC-SMA, when applied to the 12 MOPEX river basins using both base case and calibrated parameters, was compared with that of SWAP.

SAC-SMA model performance

Model parameters from optimization and manual calibration (denoted CAB) are shown in Table 3. The SAC-SMA model represents basin hydrologic response based on storage and release of water from five conceptual storages. The sum of these five storages should be consistent with physical soil water holding capability (Gan & Burges, 1990b). We show these values as "ΣStorage" in Table 3. The CAB based values for ΣStorage range from 328 mm to 710 mm. The latter value requires a deep equivalent soil column. The closest matches in this ratio were 0.92 for basins 1 and 6. We expect the conceptual storages obtained from calibration to reflect the soil storages of each basin, which is likely the case since the range of "ΣStorage" between basins varies widely.

Table 3 Comparisons of calibrated (CAB) SAC-SMA parameters and their respective performance, in terms of coefficient of efficiency (E_f) and Bias obtained at the calibration (E_f CAB) (1960–1979) and validation (E_f VAL) stages (1980–1998), and base case (no calibration), for the twelve MOPEX river basins.

	MOPEX River Basin #											
	1	2	3	4	5	6	7	8	9	10	11	12
Parameter	SCE–UA Optimized SAC–SMA Parameters with Manual Refinements											
UZTWM (mm)	5.4	39.5	3.4	6.3	20.2	49.	7.6	24.8	15.7	71.8	4.1	3.3
UZFWM (mm)	18.1	30	49.4	21.7	19.3	25.7	48.5	22.5	37.6	58.7	40.4	36.6
UZK	0.18	0.06	0.30	0.51	0.40	0.38	0.17	0.53	0.07	0.3	0.06	0.60
ZPERC	128.7	58.3	59.3	13.1	112.9	98.2	24.3	93.9	129.1	143.5	6.0	69.5
REXP	3.1	3.32	1.84	2.90	3.15	2.69	3.07	3.0	1.81	3.15	0.30	2.76
LZTWM (mm)	149.7	105	155.9	13.4	222.2	206.1	68.5	195.5	230.2	260.1	26.6	147.7
LZFSM (mm)	6.3	5.8	21.3	38.8	9.1	26.1	32.4	29.4	9.75	10.53	6.25	40.5
LZFPM (mm)	148.6	148.2	107.0	357.7	258.2	137.0	450.	80.0	30.	308.9	265.	147.3
LZSK	0.1	0.09	0.17	0.21	0.016	0.197	0.15	0.15	0.19	0.064	0.10	0.06
LZPK	0.01	0.015	0.007	0.013	0.010	0.008	0.01	0.018	0.004	0.005	0.01	0.011
PFREE	0.12	0.0	0.42	0.43	0.03	0.32	0.45	0.264	0.22	0.245	0.17	0.38
ΣStorage (mm)*	328.1	328.5	337	437.9	529	443.9	607	352.2	323.3	710.0	342.4	375.4
Ratio**	0.92	0.74	0.80	1.47	1.45	0.92	1.38	0.85	0.83	1.38	1.24	1.30
	Calibration (Jan. 1960–Dec. 1979) and Validation (Jan. 1980–Dec. 1998) Results											
E_fCAB	55.9	79.5	72.7	35.8	68.0	75.7	85.2	53.4	76.9	77.5	51.2	56.5
E_fVAL	60.1	63.5	66.4	39.0	60.8	79.6	84.6	61.7	81.6	80.6	51.5	78.3
Bias CAB	–0.9	0.6	–4.2	–13.8	–2.2	2.9	1.6	–1.8	–9.7	–5.9	0.8	3.3
Bias VAL	1.0	–0.2	–7.6	–13.6	2.3	8.9	2.1	20.8	–11.3	–7.8	–4.9	1.5
	A priori CAB (Jan. 1960–Dec. 1979) and VAL (Jan. 1980–Dec. 1998) Results											
E_fBase case C	48.9	69.8	23.3	50.3	55.4	75.1	40.2	46.9	75.0	72.2	10.2	16.2
E_f Base case V	45.1	55.1	19.4	53.6	45.3	79.4	29.9	55.0	80.5	81.0	49.6	66.4
Bias Base Case C	–5.0	–7.5	–71.6	–31.3	–0.4	1.0	–5.1	–8.7	–21.3	–5.7	–64.1	–69.5
Bias Base Case V	–2.3	–7.8	–72.0	–29.9	4.2	7.2	–6.3	14.4	–19.3	–7.7	–57.5	–63.3
Correlation CAB–BaseCase[a]	0.13	0.68	–0.16	–0.3	–0.36	–0.18	0.18	0.12	–0.07	0.05	–0.21	0.37

a = Correlation between each set of SAC-SMA CAB and Base Case parameters normalized by their respective median values; * ΣStorage = Sum of conceptual storages UZTWM, UZFWM, LZTWM, LZFPM, and LZFSM; **Ratio = CAB ΣStorage /BaseCase ΣStorage.

Table 3 shows for the calibration stage (1960–1979) two goodness-of-fit statistics between simulated and observed daily average flow rates for each river basin: the coefficient of efficiency (E_f) (also called the Nash and Sutcliffe coefficient of efficiency) and bias (Bias). None of these modelled hydrographs could be judged "satisfactory" given the stringent demonstrated requirements of Burges (2003); high quality calibrations should have values of E_f in excess of 95%. The median E_f for the 12 CAB cases is over 70% (Fig. 2). The CAB based simulations are generally satisfactory except for the two driest river basins, Guadalupe River and San Marcus River located in Texas (MOPEX Basins 11 and 12), and for Rappahannock River of VA (Basin 3) where E_f drops to less than 25%.

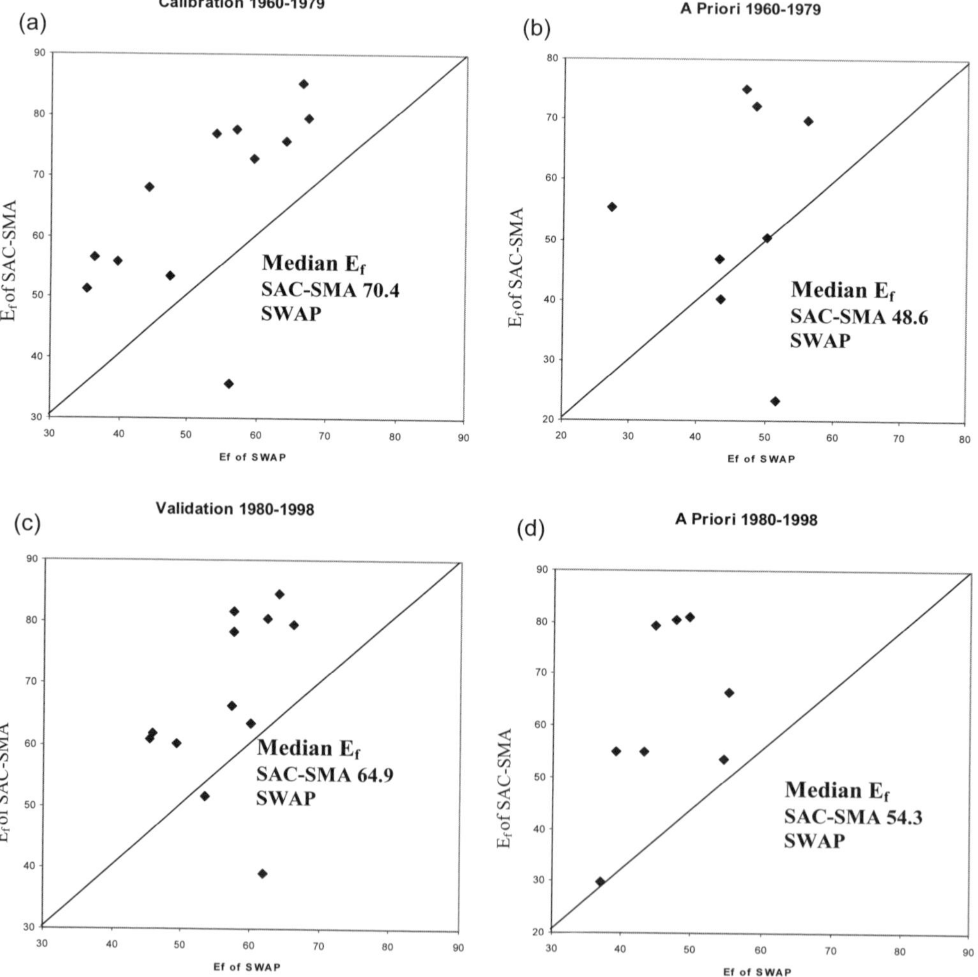

Fig. 2 Scatterplots of the coefficient of efficiency (E_f) of daily simulated streamflow for the 12 MOPEX river basins using SAC-SMA and SWAP for: (a) calibration stage (1960–1979), (b) base case stage (1960–1979), (c) validation stage (1980–1998), and (d) base case stage (1980–1998).

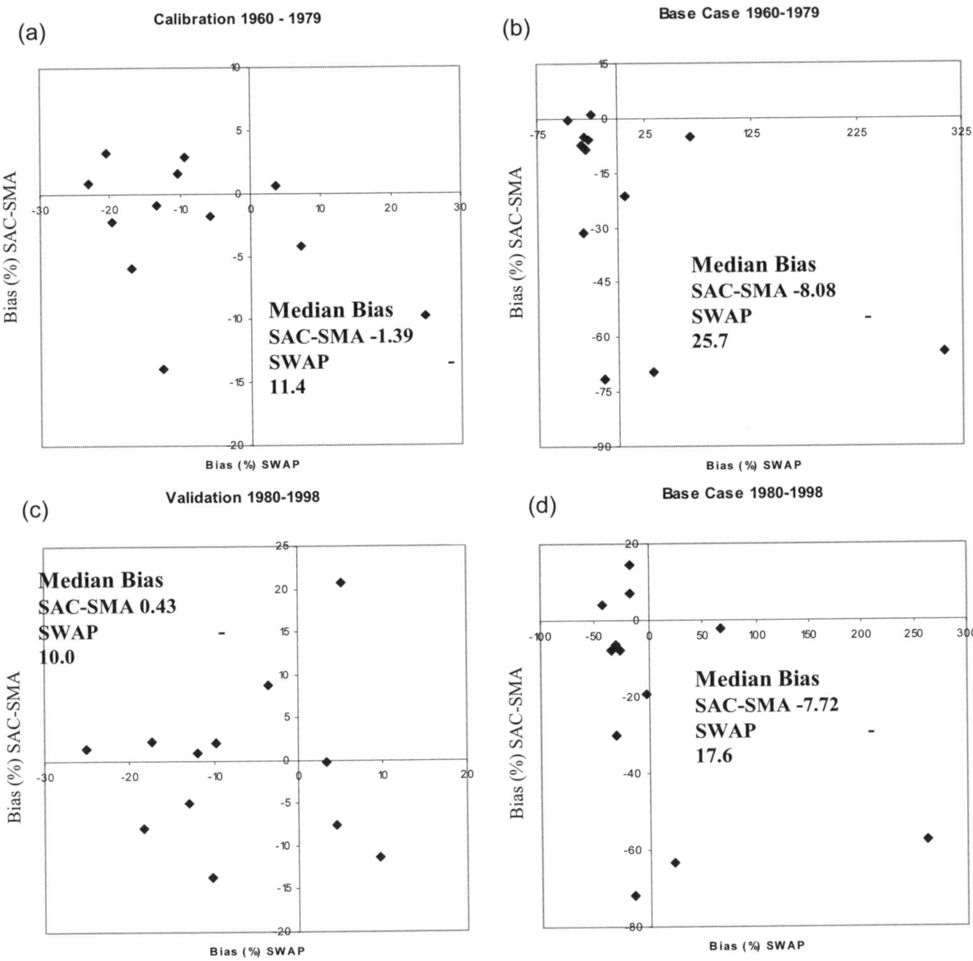

Fig. 3 Scatterplots of the bias (BIAS) of the 12 MOPEX river basins using SAC-SMA and SWAP for: (a) calibration stage (1960–1979), (b) base case stage (1960–1979), (c) validation stage (1980–1998), and (d) base case stage (1980–1998).

At the testing (validation) stage (1980–1998), the relative modelled hydrograph performances based on calibrated parameters (denoted as VAL) are similar, with the former mostly better than the latter except for Basin 4. We suspect that data inadequacies or errors in Basin 4 data make a meaningful calibration ($E_f = 35.8\%$) an almost impossible task. The median E_f for VAL drops to 64.5%. It is to be expected that, moving from the calibration to the validation stages, the performance of calibrated parameters (CAB) to generally decrease because the validation (VAL) data used are independent of the calibration experience (see Gan & Biftu (1996), and Gan & Burges (1990a,b)).

Among the 12 sets of CAB parameters, the standard deviations of the normalized moisture storage parameters (UZTWM, LZFSM and LZFPM) are 1.88, 0.83 and 0.84, respectively, while the corresponding standard deviations of the three normalized extraction parameters (UZK, LZSK and LZPK) are 0.84, 0.64, and 0.50, respectively. Physical reasoning supports our finding for CAB parameters that the standard deviation of UZK is higher than LZSK and LZPK. UZK controls release from the upper zone storage (related to surface runoff) while LZPK controls storage release from the lower zones (related to sub-surface runoff). Wetter basins that are more dominated by surface runoff have a wider range of upper zone storages than dry basins that are dominated by sub-surface runoff. Given that the 12 MOPEX basins represent a wide range of hydrologic conditions, from wet (Basin 7), to medium, to very dry (Basins 11 and 12), we expect their dominant hydrologic processes to differ widely, which should be more reflected in surface runoff parameters like UZK than in sub-surface parameters like LZPK or LZSK, or similarly in moisture storages UZTWM than in LZTWM (Table 3).

Performance of SWAP Model

The results of SWAP model simulations are given in Table 4. In the case of *a priori* (base case) parameters, the efficiency of simulated runoff is negative for basins 11, 12 and 1 (for 1960–1979): the *a priori* estimated parameters for these basins are inadequate. This could be expected because *a priori* parameters values for SWAP were derived using the USDA soil texture classification where class “bedrock” (Br) is absent; Br soil parameters could only be estimated very roughly. Bedrock covers 9, 17, and 36 % of the area of basins 12, 11 and 1, respectively. Consequently, the error of *a priori* estimation of model parameters for these basins may be large. Basins 11 and 12 are the driest of the 12 MOPEX basins. The influence of this factor may be evident when comparing the results for basins 4 (E_f = 50 and 55% for 1960–1979 and 1980–1998 years, respectively) and 11 (E_f <0). The bedrock coverage is the same (17%) in basins 11 and 14; basin 11 is the driest and basin 4 is the second wettest.

Table 4 Comparisons of calibrated SWAP parameters and their respective performance, in terms of coefficient of efficiency (E_f) and Bias obtained at the calibration (E_fCAB) (1960–1979) and validation (E_fVAL) stages (1980-1998), and Base Case (no calibration), for the twelve MOPEX river basins.

	MOPEX River Basin #											
	1	2	3	4	5	6	7	8	9	10	11	12
Parameter	Normalized by median Optimized SWAP Parameters											
Wfc (m^3 m^3)	1.11	0.82	0.91	0.90	0.99	0.97	1.06	1.03	1.15	1.01	1.04	0.97
Wwp(m^3 m^3)	0.51	0.77	0.76	1.14	0.80	1.20	1.12	0.88	1.25	1.49	1.02	0.98
Por (m^3 m^3)	1.01	1.10	1.13	1.02	1.13	0.95	0.81	1.13	0.93	0.99	0.89	0.97
Ko (m s^{-1})	1.45	0.735	0.90	0.14	0.68	0.78	0.86	1.61	1.79	1.10	5.82	7.52
hroot (m)	1.06	0.75	0.63	0.54	0.57	1.37	0.58	1.11	1.05	1.46	1.32	0.95
ho (m)	1.29	1.23	1.06	0.96	1.06	0.69	1.01	0.97	0.85	1.0	0.91	1.13
	Calibration (Jan. 1960–Dec. 1979) and Validation (Jan. 1980–Dec. 1998) Results											
E_fCAB	39.6	67.2	59.4	56.1	44.4	64.2	66.5	47.4	54	57	35.3	36.3
E_fVAL	49.9	60	57.2	61.8	45.4	66	64	45.8	57.4	62.3	53.4	57.5
Bias CAB	–13.4	3.5	7.1	–12.5	–19.7	–9.4	–10.3	–5.8	24.8	–16.8	–23.0	–20.5
Bias VAL	–11.9	3.2	4.5	–10.2	–17.3	–3.5	–9.8	5.1	9.7	–18.3	–12.9	–25.0
	A priori CAB (Jan. 1960–Dec. 1979) & VAL (Jan. 1980–Dec. 1998) Results											
E_f Base Case C	–36.4	56	51.6	50.1	27.1	46.9	43.3	43.1	46.8	48.4	–367	–75
E_f Base Case V	20.2	39.2	43.5	54.6	28.2	44.6	36.9	43.1	47.7	49.5	–227	55.3
Bias Base Case C	67.8	–34.0	–13.8	–32.9	–45.5	–24.4	–31.5	–30.3	6.1	–26.9	307	31.2
Bias Base Case V	67.7	–33.8	–14.7	–30.4	–41.9	–17.7	–30.4	–17.5	–1.9	–26.3	262	22.7

Application of the calibrated parameters resulted in better performance of SWAP compared to the base case. For the calibration period, E_f varies among the basins between 35 and 67%, Bias varies from –23 to 25%; for the validation period, E_f and bias range from 45–66%, and –25 to 10%, respectively. The median E_f for 12 basins increased by 10% for 1960–1979 and 14% for 1980-1998; the median Bias reduced by 14 and 8%, respectively. Basins 11 and 12 (with large bedrock coverage and driest conditions) are among the worse with respect to both Bias and E_f. For 8 basins the simulated runoff is underestimated. According to our experience underestimation of runoff modelled by SWAP usually results from underestimation of precipitation.

Comparing SAC-SMA with SWAP model performance

SAC-SMA and SWAP are comparable in their performance for the base case parameter simulations (without calibration), though SAC-SMA has a slight edge over SWAP. For the 12 MOPEX sites the median E_f for simulated daily streamflow was 49% (SCA-SMA) and 45% (SWAP) for the calibration period (1960–1979), and, respectively, 54% and 43% for the "test period" (1980–1998). The corresponding median bias values were –26% (1960–1979) and –18% (1980–1998) for SWAP, which are higher than that –8.1% and –7.7% for SAC-SMA. Why did SAC-SMA appear to perform better than SWAP? We offer several possibilities. A hydrological model developed for streamflow simulations should reproduce streamflow better than a land surface model which places emphasis on heat and water exchange processes occurring in a complex and multifactor soil-vegetation/snow cover-atmosphere system. A second explanation may be connected with better estimation of *a priori* parameters values for SAC-SMA. It should be noted, however, that both models produce simulations that are completely inadequate for any question of science and may have limited use for hydrological decision making. This is disappointing in that the model parameters are derived as if "ungauged catchments" were being modelled.

The second set of comparisons involved simulations from calibrated models of each of the 12 MOPEX catchments. The simple Sacramento model consistently out-performed the much more complex SWAP model when considering both the calibration period and the testing "validation" period (Tables 3 and 4, Figs 2 and 3).

The first possible reason that SAC-SMA outperformed SWAP in both calibration and validation stages could be because SWAP was only calibrated in terms of six soil parameters whereas SAC-SMA was calibrated with the 11 parameters shown in Table 3, plus the three unitgraph ordinates, one precipitation scaling factor (that helps account for bias—see e.g. Burges (2003)) and one potential ET adjustment factor, a total of 16 parameters. This was partly why after calibration (20 years of data), the median streamflow E_f of SAC-SMA was 70%, *vs* 55% for SWAP while for the test (validation) period (19 years), the gap between them is more modest, 65% of SAC-SMA over 58% for SWAP. For SWAP, calibration reduced the Bias relative to the base case parameters by about 8 to 14% (–26 to –11% for 1960–1979 and –18 to –10% for 1980–1998). However, calibration further reduced the Bias of SAC-SMA to –1.4% (1960–1979) and 0.4% (1980–1998), respectively, i.e. for SAC-SMA, the bias was reduced by 7–8%.

One more additional explanation why the site specific calibrated SAC-SMA model produced better simulated streamflow than the SWAP model may be related to the different calibration techniques used. SWAP parameter calibration is based on a stochastic or Monte-Carlo technique which may not be as effective as that of SCE-UA. More complete testing of the stochastic optimization scheme is needed before this issue can be resolved.

CONCLUSIONS

On the basis of results obtained from driving a complex, physics-based land surface model (SWAP) and a simple, conceptual lumped-parameter hydrological model (SAC-SMA) on 12 MOPEX river basins ranging from 1000 to 4400 km^2 in area, and representing a wide range of hydrological and climatic conditions (dry to wet), it seems that a land surface model may not necessarily reproduce streamflow better than a simple, hydrological model. This could be expected at the first stage of this study (when *a priori* parameters were used) because the model designed for streamflow simulation should perform better than the model that emphasizes heat and water exchange between the land surface and the atmosphere. Better performance (with respect to streamflow simulation) of SAC-SMA at the second stage (when calibrated parameters were used) may have resulted because SWAP was not as completely calibrated as SAC-SMA (6 versus 16 parameters). More extensive tests involving more models and basins are needed to confirm the finding of this study, that conceptual hydrologic models might model basin-scale hydrological processes more reliably than complex models.

Acknowledgements This research was partly funded by the Natural Science and Engineering Research Council (NSERC) of Canada. All the MOPEX data were provided jointly by the National Oceanic and Atmospheric Administration/National Weather Service, Silver Spring, and SHARA/University of Arizona.

REFERENCES

Burges, S. J. (2003) Process representation, measurements, data quality, and criteria for parameter estimation of watershed models. In: *Calibration of Watershed Models* (ed. by Q. Duan, H. Gupta, S. Sorooshian, A. N. Rousseau & R. Turcotte), 283–299. AGU Monograph, Water Science and Application 6.

Burnash, R. J. C., Ferral, R. L. & McGuire, R. A. (1973) A generalized streamflow simulation system-conceptual modeling for digital computers. US Department of Commerce, National Weather Service and State of California, Department of Water Resources, California, USA.

Clapp, R. B. & Hornberger, G. M. (1978) Empirical equations for some soil hydraulic properties. *Water Resour. Res.* **14**(4), 601–604.

Duan, Q., Sorooshian, S. & Gupta, V. K. (1992) Effective and efficient global optimization for conceptual rainfall runoff models. *Water Resour. Res.* **28**(4), 1015–1031.

Duan, Q., Schaake, J., Andreassian, V., Franks, S., Gupta, H. V., Gusev, Y. M., Habets, F., Hall, A., Hay, L., Hogue, T., Huang, M., Leavesley, G., Liang, X., Nasonova, O. N., Noilhan, J., Oudin, L., Sorooshian, S., Wagener, T. & Wood, E. F. (2005) Model parameter estimation experiment (MOPEX): overview and summary of the second and third workshop results, MOPEX Special Issue. *J. Hydrol.* **320**, 3–17, doi:10.1016/j.jhydrol.2005.07.031.

Gan T. Y. & Burges, S. J. (1990a) An assessment of a conceptual rainfall–runoff model's ability to represent the dynamics of small hypothetical catchments, 1: models, model properties, and experimental design. *Water Resour. Res.* **26**(7), 1595–1604.

Gan T. Y. & Burges, S. J. (1990b) An assessment of a conceptual rainfall–runoff model's ability to represent the dynamics of small hypothetical catchments, 2: hydrologic responses for normal, extreme rainfall. *Water Resour. Res.* **26**(7), 1605–1619,.

Gan, T. Y. & Biftu, G. F. (1996) Automatic calibration of conceptual rainfall–runoff models: optimization algorithms, catchment conditions, and model structure. *Water Resour. Res.*, AGU, Dec. **32**(12), 3513–3524.

Gan, T. Y. & Burges, S. J. (2005) Assessment of soil-based and calibrated parameters of the Sacramento model and parameter transferability. *J. Hydrol.* **320**, 117–131, doi:10.1016/j.hydrol.2005.07.008.

Gusev, Ye. M. & Nasonova, O. N. (2000) An experience of modeling heat and water exchange at the land surface on a large river basin scale. *J. Hydrol.* **233**(1–4), 1–18.

Gusev, Ye. M. & Nasonova O. N. (2003) Modelling heat and water exchange in the boreal spruce forest by the land-surface model SWAP. *J. Hydrol.* **280**(1–4), 162–191.

Koren, V. I., Smith, M., Wang, D. & Zhang, Z. (2000) Use of soil property data in the derivation of conceptual rainfall–runoff model parameters. In: *15th Conf. Hydrology* (Long Beach, California, USA), Am. Meteor. Soc., 10–14 January, 103–106.

Lohmann, D., Lettenmaier, D. P., Liang, X., Wood, E. F., Boone, A., Chang, S., Chen, F., Dai, Y., Desborough, C., Dickinson, R. E., Duan, Q., Ek, M., Gusev, Y. M., Habets, F., Irannejad, P., Koster, R., Mitchell, K. E., Nasonova, O. N., Noilhan, J., Schaake, J., Schlosser, A., Shao, Y., Shmakin, A. B., Verseghy, D., Warrach, K., Wetzel, P., Xue, Y., Yang, Z. -L. & Zeng, Q. -C. (1998) The project for intercomparison of land-surface parameterization schemes phase-2(c) Red-Arkansas River basin experiment. 3. Spatial and temporal analysis of water fluxes. *Global Plan. Change* **19**(1–4), 161–179.

Luo, L., Robock, A., Vinnikov, K. Y., Schlosser, A. C., Slater, A. G., Boone, A., Braden, H., Cox, P., de Rosnay, P., Dickinson, R. E., Dai, Y. -J., Duan, Q., Etchevers, P., Henderson-Sellers, A., Gedney, N., Gusev, Ye. M., Habets, F., Kim, J., Kowalczyk, E., Mitchell, K., Nasonova, O. N., Noilhan, J., Pitman, A. J., Schaake, J., Shmakin, A. B., Smirnova, T. G., Wetzel, P., Xue, Y., Yang, Z. -L. & Zeng, Q.-C. (2003) Effects of frozen soil on soil temperature, spring infiltration, runoff: results from the PILPS 2(d) experiment at Valdai, Russia. *J. Hydrometeorol.* **4**(2), 334–352.

4 Regionalization and parameterization studies based on large basin samples

A bounded version of the Nash-Sutcliffe criterion for better model assessment on large sets of basins

THIBAULT MATHEVET[1,2], CLAUDE MICHEL[1], VAZKEN ANDRÉASSIAN[1] & CHARLES PERRIN[1]

1 *Cemagref, Parc de Tourvoie, BP 44, F-92163 Antony cedex, France*
thibault.mathevet@edf.fr

2 *now at: EDF-DTG, 21, Avenue de l'Europe, BP 41, F-38040 Grenoble cedex 09, France*

Abstract Rainfall–runoff models are useful tools for hydrological research, water engineering and environmental applications. Given the large number of available rainfall–runoff models, many comparative studies have been done to compare models performances, to specify their domain of application and provide guidance to end-users. Although most existing comparative studies tested models only on a few basins, we believe that effective model evaluation requires large samples of test catchments. However, large test samples raise the issue of appropriate criteria to quantify model performances. This paper shows that the widely used Nash and Sutcliffe criterion may be difficult to apply for large test samples and that a bounded version of this criterion (called C_{2M}) is better suited for extensive model assessment.

Key words large sample of basins; Nash-Sutcliffe criterion; rainfall–runoff modelling

INTRODUCTION

Since the 1960s, hydrologists have developed a large number of more or less complex rainfall–runoff (RR) models. As a consequence of this proliferation, the need for comparative studies appeared quite early (WMO, 1975). Linsley (1982) suggested that *"because almost any model with sufficient free parameters can yield good results when applied to a short sample from a single basin, effective testing requires that models be tried on many basins of widely differing characteristics, and that each trial cover a period of many years"*. Few modellers have, however, followed these recommendations, and most of the RR modelling studies reported in the literature present the performances of one RR model on a single basin (or on a small number of similar basins). If studies include too few watersheds, the validity of their conclusions will be limited to the hydro-climatic domain of the test sample. Conversely, a large set of basins provides a general overview of the efficiency of one or several models, in a wide range of hydro-meteorological conditions and catchment physical characteristics (geology, soil, vegetation, topography, land use, etc.).

With large basins sets, however, assessing RR model performances becomes complex since a complete distribution of results is obtained, often over a large range of performances. In such cases, one must find ways to compare these distributions, or to summarize them into proper statistics, provided that the formulation of the selected criterion allows deriving such a summary.

In this paper we propose a criterion formulation suitable for comparing model performances on large basin samples. An application is made on a sample of 313 basins. We also show the usefulness of large basin sets for model assessment.

In the first section we discuss the need for using large basin samples. Then we introduce a new formulation of the classical Nash & Sutcliffe (1970) criterion, better suited to the assessment of models on large samples. In the following section we present the basin sample, the tested models and the assessment methodology. Last, we present the results of model tests and discuss the usefulness of this new criterion formulation for large basin samples.

WHY ARE LARGE BASIN SAMPLES NEEDED TO ASSESS RR MODELS EFFICIENCIES?

Most existing comparative studies rely only on a small number of watersheds. Among exceptions are the studies by Vandewiele *et al.* (1992), Makhlouf & Michel (1994) and Xu & Vandewiele (1995) at the monthly time-step, and by Perrin *et al.* (2001) at the daily time-step. Nash & Sutcliffe (1970), Linsley (1982) and Klemeš (1986) stressed that generality should be a fundamental requirement for any RR model. Linsley (1982) argued that *"it seems axiomatic that the fundamental processes of hydrology are the same in all catchments.[...] In some cases a process may not be present.[...] However, these differences do not mean that a single model cannot be applied in all cases. The model must represent the various processes with sufficient fidelity so that irrelevant processes can be "shut off" or will simply not function. Differences [...] must be represented by model parameters which can be preset to represent these characteristics.* (p. 14)*"*.

Thus, to assess model generality, it seems obvious that model assessment cannot rely only on a single or few basins. Given the large differences that exist between basins, a large number of basins is required to judge of the actual adaptability of a RR model to different conditions.

However, large sets of data raise some problems. It is difficult to systematically check data quality since data come from many different sources and result from different collection practices, so errors may affect data. There may also be influences (e.g. due to human activities) difficult to detect by visual inspection of time series. For these reasons, there will inevitably be a number of basins in the test set where the models will fail. These basins should not be excluded from the test set since, as argued by Linsley (1982), all models will suffer equally from these problems and this will not bias the comparative model assessment. However, these model failures may cause problems in the quantification of the average model performance over the test sample if appropriate criteria are not chosen. This need for appropriate criteria is discussed in the next section.

AN APPROPRIATE CRITERION FOR MODEL ASSESSMENT OVER LARGE BASIN SAMPLES

The Nash & Sutcliffe (1970) criterion is widely used in hydrological modelling. However, for model assessment on large basins sets, this criterion raises some problems.

The Nash & Sutcliffe criterion: advantages and drawbacks

The approach followed by Nash & Sutcliffe (1970) is to build a relative index of agreement (or disagreement) between observed and computed runoff that could be used to compare model performances between periods or basins.

They start from the sum of square errors given by:

$$F = \sum_{i=1}^{n} \left(Q_{obs,i} - Q_{sim,i}\right)^2 \tag{1}$$

where F is the index of disagreement, $Q_{obs,i}$ and $Q_{sim,i}$ are the observed and simulated discharges at time step i, the sum being taken over n time steps of a pre-selected period. F is analogous to the residual variance of a regression analysis. The initial variance F_0 is given by:

$$F_0 = \sum_{i=1}^{n} \left(Q_{obs,i} - \overline{Q_{obs}}\right)^2 \tag{2}$$

where $\overline{Q_{obs}}$ is the mean of the observed discharge over the pre-selected period. Nash & Sutcliffe (1970) then define the efficiency of the model E as the proportion of the initial variance accounted for by the model:

$$NS = 1 - \frac{F}{F_0} \tag{3}$$

NS can take values between $-\infty$ and 1. A value of 1 indicates a perfect agreement and a value of zero indicates that the model does not explain any part of the initial variance.

The Nash & Sutcliffe criterion can also be interpreted as a criterion that determines the improvement made by a given model in simulating flows in comparison with a reference model that would simulate a flow equal to $\overline{Q_{obs}}$ at each time step. The value of zero for the criterion therefore means that the model is not better than this basic one-parameter model, and a negative value indicates that the model is worse than this basic model.

This criterion is very useful in model assessment since its adimensional form is hoped to allow comparison of performances on different catchments or periods. Several authors mention, however, that this criterion has some drawbacks. Garrick *et al.* (1978) and Martinec & Rango (1989) showed that the NS criterion may produce relatively high values, even for quite poor models. This is mainly due to the fact that the basic model ($Q_i = \overline{Q_{obs}}$ for all i) can be very primitive in some instances, so it becomes easy to be better than this basic reference ($F \ll F_0$). To overcome this problem, Garrick *et al.* (1978) proposed to use another reference model, such as a "seasonal model", to compare in F_0 the measured runoff to the long-term average measured runoff for each Julian day. Conversely, it is difficult to obtain high values for periods or basins where flow does not vary much in time.

These considerations indicate that this criterion may not be similarly demanding in all circumstances and that it can yield wide ranges of performances when models are assessed on large basin samples that include many different characteristics. This is all the more true as this criterion has no lower bound and may give strongly negative values when the model fails.

The problem of criteria formulation for extensive model assessment

If one focuses on comparative assessments, the simplest case is to test two models on one basin. There are numerous ways of comparing performances in such a simple case. However, when the number n of models and the number m of test basins increase, the classification of models becomes more complex. Indeed, one has to compare n lists of m values of the NS criteria. To summarize such a large amount of information, Perrin *et al.* (2001) used:

- the mean value of the m NS criteria,
- the distribution of NS criteria obtained by each model over the basin sample and the percentiles (0.1, 0.5, 0.9) of this distribution.

The mean performance is actually the best measure to have a synthetic overview of model performance. Unfortunately, the mean of NS criteria can be heavily influenced by a few strongly negative values obtained on a small number of basins. Therefore the mean value can be artificially biased, which may impair the conclusions of the comparison exercise. The use of distribution percentiles (e.g. the median value) could be a remedy to this problem. Unfortunately the distributions of NS criteria may cross each other and therefore the relative value of two models will depend on the selected percentile. It could also be argued that basins where some models do not work should be discarded. However, this would bias the comparison in favour of the models used for this prior screening.

C_{2M}, a bounded formulation for the Nash & Sutcliffe criterion

To avoid these problems, we propose to adopt a new formulation of the NS criterion to make it vary between –1 and +1, like a correlation coefficient. This will generate less skewed criteria distributions and it will be possible to compute significant mean values. To keep the same zero value as the NS criterion, we propose the following formulation, called C_{2M}:

$$C_{2M} = \frac{1 - \frac{F}{F_0}}{1 + \frac{F}{F_0}} \tag{4}$$

The NS criterion is related to C_{2M} as follows (Fig. 1):

$$NS = \frac{2.C_{2M}}{1 + C_{2M}} \tag{5}$$

$$C_{2M} = \frac{NS}{2 - NS} \tag{6}$$

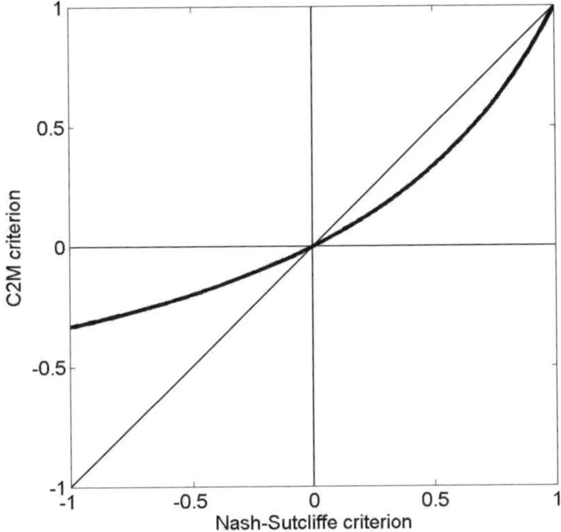

Fig. 1 Relation between the Nash-Sutcliffe and the C_{2M} criteria.

Note that the C_{2M} is less optimistic than the NS criterion for positive values, which partially provides an answer to the criticism made by Garrick *et al.* (1978) who argued that the NS criterion produces too high values. Since the criteria distributions are now bounded, it should be possible to derive more meaningful statistics to summarize model performances.

DATA AND METHODS

Basin sample test

To test the usefulness of this new criterion formulation, we used a sample of 313 basins (see characteristics in Table 1). Most basins are in France (227) and in the USA (70), some in Australia (12), Spain (2) and Slovenia (2). Hydro-climatic conditions are varied: semiarid, Mediterranean, oceanic, temperate, mountainous and continental. Snowmelt influences are generally limited. Data were collected at the hourly time step.

Rainfall–runoff models

We used three model structures derived from existing RR models (see Table 2). All three are lumped and continuous, and were run at the hourly time step. More details can be found in Mathevet (2005). As the purpose of this study is to discuss methodological aspects and not to compare original models, the models are called M1, M2 and M3 hereafter. The models were fed with exactly the same input data, i.e. rainfall time-series and potential evapotranspiration estimates, and their free parameters were calibrated against observed runoff.

Table 1 Minimum–mean–maximum characteristics of the 313 basins.

Country	France	USA	Australia	Spain	Slovenia
Number of watersheds	227	70	12	2	2
Annual runoff (mm)	*35*–44–*1655*	*0*– 279–*1612*	*9*–35–*96*	429–736	1024–1182
Annual rainfall (mm)	*403*–963–*2067*	*193*–1163–*2996*	*569*–674–*1025*	1183	1384
Annual PE (mm)	*595*–791–*1252*	*1104*–1545–*2085*	1226	639	735
Annual runoff–rainfall ratio(%)	*0.05*–0.44–*2.59*	*0*– 0.20 –*0.81*	*0.01*–0.05–*0.13*	0.36–0.62	0.74–0.85
Type of climate	Temperate, Mediterranean, oceanic, continental	Temperate, oceanic, semiarid	Semiarid	Mountainous Mediterranean	Mountainous
Watershed area (km²)	*1.1*–280–*4978*	*1.2*–33–*334*	*2.7*–48–*2538*	0.56–4.17	457–1385
Length of the time series (year)	*3*–8–*34*	*3*–11–*43*	6	3–4	5

Table 2 Characteristics of the three RR models used in this study. The details of the modified versions tested here can be found in Mathevet (2005).

Tested model	Number of free parameters	Number of reservoirs	Original model	Reference of original model
XINANJ	8	4	XINANJIANG	Zhao *et al.* (1995)
IHAC	6	3	IHACRES	Jakeman *et al.* (1990)
GR4H	4	2	GR4J	Perrin *et al.* (2003)

Assessment methodology

To assess the performances of the selected RR models, we applied the split-sample test procedure (Klemeš, 1986): the models can thus be tested in simulation mode, under meteorological conditions different from those of the calibration period. For each basin, the available time-series was split into several independent sub-periods. The models were successively calibrated on each sub-period and tested in validation mode on the remaining ones. The results are based on a total of 2093 validation tests, with 2.2 year long periods on average, for the 313 basins used here. Periods are quite short but this is not a problem since all models were strictly compared in the same conditions.

RESULTS AND DISCUSSION

In this section, we use model results to demonstrate: (i) that it is interesting to resort to large basins samples in order to compare RR models; and (ii) that the C_{2M} criterion is more valuable for such a comparison. We present here the results of the three selected RR models. They were tested on the whole sample of 313 basins and on randomly generated sub-samples as explained below.

To demonstrate the interest of large basin sets to compare the efficiencies of RR models, 500 random sub-samples of 5, 10 and 50 watersheds were drawn from the initial sample (313 basins). Fig. 2 clearly shows the high dispersion of mean NS criteria for the 500 random samples of 5, 10 and 50 watersheds, when compared to the mean NS efficiency over 313 basins (black cross). This figure shows that:

- when using a small number of basins, it is always possible to find samples of a limited number of basins, where model A is better than model B on one sample, and conversely model B is better than model A on the other;
- even with random sub-samples of 50 basins, it remains possible to find cases where M2 and M3 are better than M1 and M3 is better than M2, whereas the opposite conclusions are drawn on the whole sample.

The high dispersion of mean NS values averaged over 10 or even 50 basins is mainly caused by the basins where models fail dramatically. When the sample size increases, the weight of these basins is reduced.

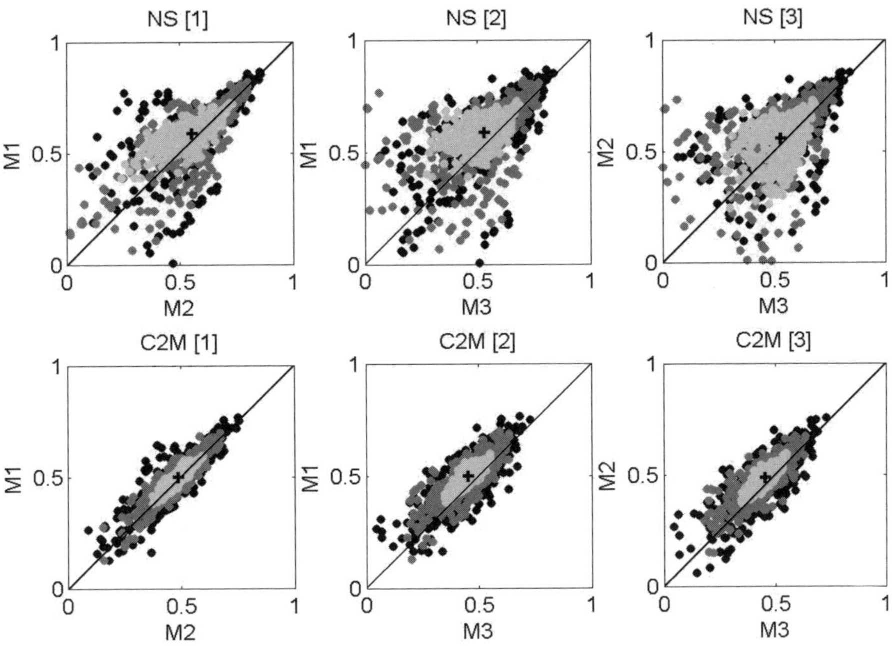

Fig. 2 Comparison of M1, M2 and M3 efficiencies of 500 random samples of 5, 10, 50 basins, with Nash-Sutcliffe and C_{2M} criteria (note that basins samples whose mean NS efficiency is lower than 0 are not shown). Black dots, 5 watersheds; dark grey dots, 10 watersheds ; Light grey dots, 50 watersheds.

The use of the C_{2M} criterion eases the comparison since the dispersion of efficiencies in Table 3 is much more limited with the C_{2M} than with the NS criterion. With random sub-samples of 10 or 50 basins, it is still possible to find samples of basins where any model is better than another. However, with the C_{2M} criterion, the dispersion of efficiencies' difference between two RR models decreases greatly

Table 3 Mean model efficiency using NS and C_{2M} criterion, over 313 watersheds, as a function of the chosen differentiation criterion.

	M1	M2	M3
Minimum NS	–815.2	–648.5	–1778.2
Mean NS	55.7	52.9	38.5
Maximum NS	89.5	87.0	85.4
NS Percentile			
0.1	10.3	0	–11.2
0.5	71.5	68.6	62.3
0.9	87.0	85.0	82.6
NS standard deviation	69.4	60.4	120.9
% of NS			
< –100 %	1.5	2.2	3.5
< 0 %	6	10.5	11.5
Mean C_{2M}	48.9	45.5	38.7
C_{2M} Percentile			
0.1	5.4	0	–5.0
0.5	55.6	52.2	45.2
0.9	77.0	74.0	70.3
C_{2M} standard deviation	28.0	28.2	31.2

when the sample size increases. In this case, 50 watersheds seem to be sufficient to discriminate the efficiencies of M1 and M3 models, or M2 and M3 models.

Fewer basins are needed to rank two models when using C_{2M} than with the NS criterion: when the difference of C_{2M} efficiency between two models over the whole sample is about 10 points (the case of M1 *vs* M3 and M2 *vs* M3), a sample of about 50 basins seems sufficient to differentiate these models, whereas 50 are clearly not enough in the case of the NS criterion. When the difference is smaller, the size of the required basin sample will increase, but there are less cases of obvious misinterpretation with the C_{2M} than with the NS criterion.

CONCLUSION

The objectives of this article were to advocate the use of large samples of basins to assess and compare RR models and to present the usefulness of a bounded version of the Nash-Sutcliffe criterion (called C_{2M}) in such an assessment.

The lack of a lower bound of the NS criterion is a real drawback and yields strongly skewed distribution of efficiencies over large test samples when, for any reason, very low performances are obtained on a few basins. As a consequence, it is hard to compute meaningful statistics to summarize the whole distribution. The C_{2M} criterion introduced here provides a bounded formulation of the NS criterion, varying within the interval [–1, +1]. The test of three model structures on a sample of 313 basins clearly shows that the C_{2M} criterion allows one to compute more meaningful mean model efficiencies over large test samples than the NS criterion, and therefore provides a more reliable comparison of model efficiencies. It is also shown that, given the variability of model efficiencies over an heterogeneous sample of basins, a large sample size (at least 50 basins) is clearly required to differentiate the efficiencies of

RR models. The smaller the difference between models, the larger the sample size required to warrant that the difference is significant.

Acknowledgements The authors would like to thank the institutions and researchers who provided data sets for model testing: Christian Scherer (French Ministry of Ecology and Sustainable Development); Bruno Rambaldelli (Météo France); Christian Zammit (Water and River Commission of Western Australia), Muguresu Sivapalan (Center for Water Research of the University of Western Australia); Mira Kobold (Environmental Agency of the Republic of Slovenia) Fransesc Gallart and Jerôme Latron (Jaume Almera Institute of Earth Sciences). Thanks are also due for the huge work of those who did the field data collection.

REFERENCES

Garrick, M., Cunnane, C. & Nash, J. E. (1978). A criterion of efficiency for rainfall–runoff models. *J. Hydrol.* **38**, 375–381.

Jakeman, A. J., Littlewood, I. G. & Whitehead, P. G. (1990) Computation of the instantaneous unit hydrograph and identifiable component flows with application to two small upland catchments. *J. Hydrol.* **117**, 275–300.

Klemeš, V. (1986) Operational testing of hydrological simulation models. *Hydrol. Sci. J.* **31**(1), 13–24.

Linsley, R. K. (1982) Rainfall–runoff models—an overview. In: *Proc. Int. Symp. on Rainfall–Runoff Modelling* (ed. by V. P. Singh), 3–22. Water Resources Publications, Littletown, Colorado, USA.

Makhlouf, Z. & Michel, C. (1994) A two-parameter monthly water balance model for French watersheds. *J. Hydrol.* **162**(3–4), 299–318.

Mathevet, T. (2005) Which rainfall–runoff model at the hourly time-step ? Empirical development and intercomparsison of rainfall–runoff models on a large sample of watersheds. PhD thesis, ENGREF University, Paris, France (in French).

Martinec, J. & Rango, A. (1989) Merits of statistical criteria for the performance of hydrological models. *Water Resour. Bull.* **25**(2), 421–432.

Nash, J. E. & Sutcliffe, J. V. (1970) River flow forecasting through conceptual models. Part I—A discussion of principles. *J. Hydrol.* **27**(3), 282–290.

Perrin, C., Michel, C. & Andréassian, V. (2001) Does a large number of parameters enhance model performance? Comparative assessment of common catchment model structures on 429 catchments. *J. Hydrol.* **242**(3–4), 275–301.

Vandewiele, G. L., Xu, C. Y. & Win, N. L. (1992) Methodology and comparative study of monthly models in Belgium, China and Burma. *J. Hydrol.* **134**, 315–347.

WMO (1975) Intercomparison of conceptual models used in operational hydrological forecasting. *Operational Hydrology Report no. 7, WMO no 429.* World Meteorological Organization, Geneva, Switzerland.

Xu, C. Y. & Vandewiele, G. L. (1995) Parsimonious monthly rainfall–runoff models for humid basins with different input requirements. *Adv. Water Resour.* **18**, 39–48.

Zhao, R J. & Liu, X. R. (1995) The Xinanjiang model. In: *Computer Models of Watershed Hydrology*, Chap. 7 (ed. by V. P. Singh), 215–232. Water Resources Publications, Littletown, Colorado, USA.

Large Sample Basin Experiments for Hydrological Model Parameterization: Results of the Model Parameter Experiment–MOPEX. IAHS Publ. 307, 2006.

Regionalization of dynamic watershed response behaviour

MAITREYA YADAV[1], THORSTEN WAGENER[1] & HOSHIN GUPTA[2]

1 *Department of Civil and Environmental Engineering, Pennsylvania State University, Sackett Building, University Park, Pennsylvania 16802, USA*
thorsten@engr.psu.edu

2 *Department of Hydrology and Water Resources, University of Arizona, Harshbarger Building, Tucson, Arizona 85721, USA*

Abstract Approaches to ungauged basin modelling typically use observable physical characteristics of watersheds (e.g. soil data) to directly infer hydrological model parameters, or they use regionalization methods based on parsimonious hydrological models. A different approach to streamflow prediction in ungauged basins is presented here where, instead of model parameters, the model independent hydrological response behaviour is estimated in the form of streamflow indices, and then regionalized with respect to physical characteristics of watersheds. Therefore, the approach uses a data driven regionalization method (under uncertainty) rather than the common hydrological model driven regionalization method. Ensemble predictions in ungauged basins can then be constrained by limits on acceptable hydrological model behaviour. This study utilizes data from 30 watersheds in the UK. Initial results show that the predictive uncertainty of the model can be reduced considerably through this new approach.

Key words ensemble predictions; hydrograph indices; prediction in ungauged basins; predictive uncertainty; regionalization; streamflow characteristics; watershed response

INTRODUCTION

Rainfall–runoff models are standard tools for hydrological analysis. One major limitation of currently available models is the need for adjustment of the model parameters using observed watershed response data to obtain reliable predictions (e.g. Sivapalan *et al.*, 2003; Wagener *et al.*, 2004). The problem is accentuated further when it comes to prediction in ungauged basins, where data for parameter estimation via calibration are not available. Two common approaches to overcome this problem in ungauged situations are: (a) the use of physically based models, and (b) the regionalization of model parameters using physical characteristics of watersheds. The introduction of physically based models was based on the hope that their parameters would be equivalent (or at least strongly related) to directly observable properties. However, differences in scale, over-parameterization and model structural error, have prevented this objective from (so far) being achieved (Beven, 1989). In the regionalization approach, a (typically parsimonious) hydrological model structure is selected, and calibrated to observable watershed response for a large number of gauged watersheds. Regression equations are then developed between the model parameters and physical characteristics of watersheds. This approach also suffers from model identification difficulties, model structure errors, and difficulties in finding an appropriate calibration strategy that appropriately considers the physical meaning of the model parameters (Wagener & Wheater, 2005).

The objective of the research reported here is to achieve a continuing reduction in predictive uncertainty, while maintaining reliable predictions, leading to an increased understanding of watershed function (Wagener *et al.*, 2004). In this study we introduce a new approach for improving predictions in ungauged basins that regionalizes model independent dynamic hydrological response characteristics (or indices) to physical characteristics of watersheds while considering uncertainty. The approach is applicable to any model (whether lumped or distributed) that can be run within a Monte Carlo framework, in contrast with other published approaches that can only be applied using relatively simple (identifiable) models (Wagener & Wheater, 2005). Initial results, using data from 30 watersheds in the UK, are presented here. Two of the watersheds were used for an independent evaluation of the approach.

REGIONALIZATION OF HYDROLOGICAL RESPONSE BEHAVIOUR

Dynamic response characteristics or response behaviour indices of a watershed can be derived from precipitation, evapotranspiration (or temperature) and streamflow time series of the watershed; examples include common descriptors of hydrograph shape such as runoff ratios and times to peak flow, etc. While indicators of this type are commonly used by the ecological community for the evaluation of flow regimes (e.g. Olden & Poff, 2003), they have only recently been (re)introduced in the context of hydrological model calibration (e.g. Yu & Yang, 2000; Shamir *et al.*, 2004). Other examples of such indices include runoff ratios, rising and falling limb densities, mean flow, exceedence of flow percentiles, etc. (e.g. Olden & Poff, 2003; Shamir *et al.*, 2004).

Our work extends these ideas by capitalizing on the information content inherent in such summary descriptors of watershed response, and by relating them to observable physical characteristics of the watersheds by means of regressive relationships. The idea of regionalizing such indices stems from the empirical observation that the amount of uncertainty involved in regionalizing hydrological model parameters can be large, particularly since it is difficult to account for the effects of model structural error during model calibration (Wagener & Wheater, 2005). Since the watershed response characteristics are not model-specific, uncertainties and confounding influences that might arise from the process of model identification are eliminated (or at least significantly reduced). Once regionalized, the behavioural information summarized by the response characteristics can be used as constraints on the model predictions, and facilitate, for example, a separation into behavioural and non-behavioural model sets using a binary classification approach. Therefore, regionalization in the context of this paper involves the development of regression relationships between watershed response characteristics and observable physical characteristics of watersheds.

REGIONALIZATION CASE STUDY

Watershed data

This study uses a set of 30 small to medium sized watersheds located throughout the UK (Fig. 1), covering a wide range of soil types, topography and land uses. Most of the watersheds have natural flow within 10% at their 95 flow percentile. Data for the selected watersheds was acquired from the UK National River Flow Archive (http://www.nwl.ac.uk/ih/nrfa). The precipitation and streamflow time series was taken from "Predictions in Ungauged Basins (PUB)—UK data downloads" at http://www.nwl.ac.uk/ih/nrfa. Temperature data was obtained from The British Atmospheric Data Centre (http://badc.nerc.ac.uk/home/index.html). Potential evapotranspiration was calculated from temperature data using Hargreaves equation (Maidment, 1993). Eleven consecutive years (1980–1990) of data were available for 29 watersheds.

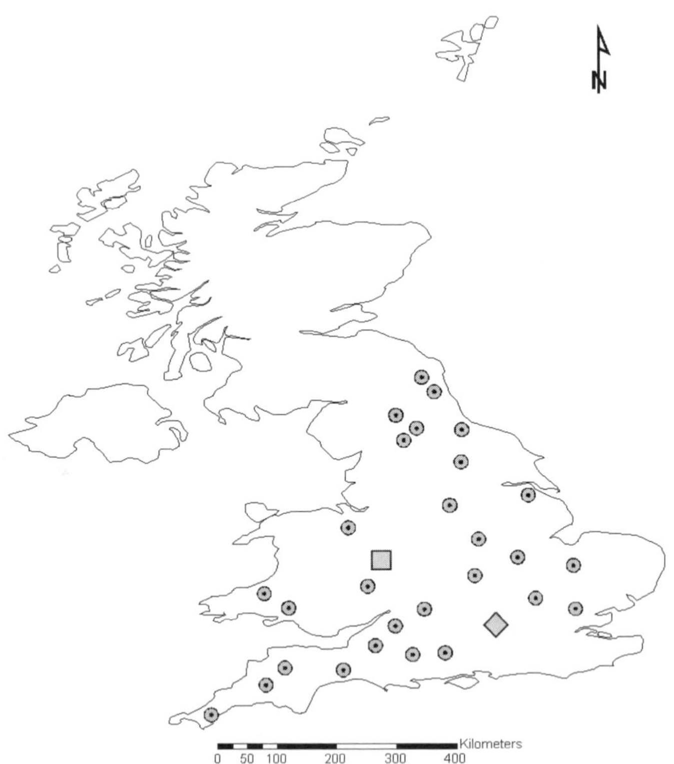

Fig. 1 Map of UK showing the location of watersheds used in this study. The square and diamond show the validation watersheds.

The time period used for the analysis was from 1 January 1983–31 December 1990; the average monthly values of rainfall, streamflow and potential evapotranspiration are plotted in Fig. 2(a–c). A normalized flow duration curve showing cumulative frequency of normalized flow values is also shown (Fig. 2(d)). The flows are normalized by the mean flow values to facilitate comparison. A steep slope in the flow duration curve indicates flashiness of the streamflow response to rainfall inputs whereas a flatter curve indicates a relatively damped response. It also represents the storage characteristics of the watersheds. Figure 2(d) shows the diversity in watersheds with respect to their hydrological response.

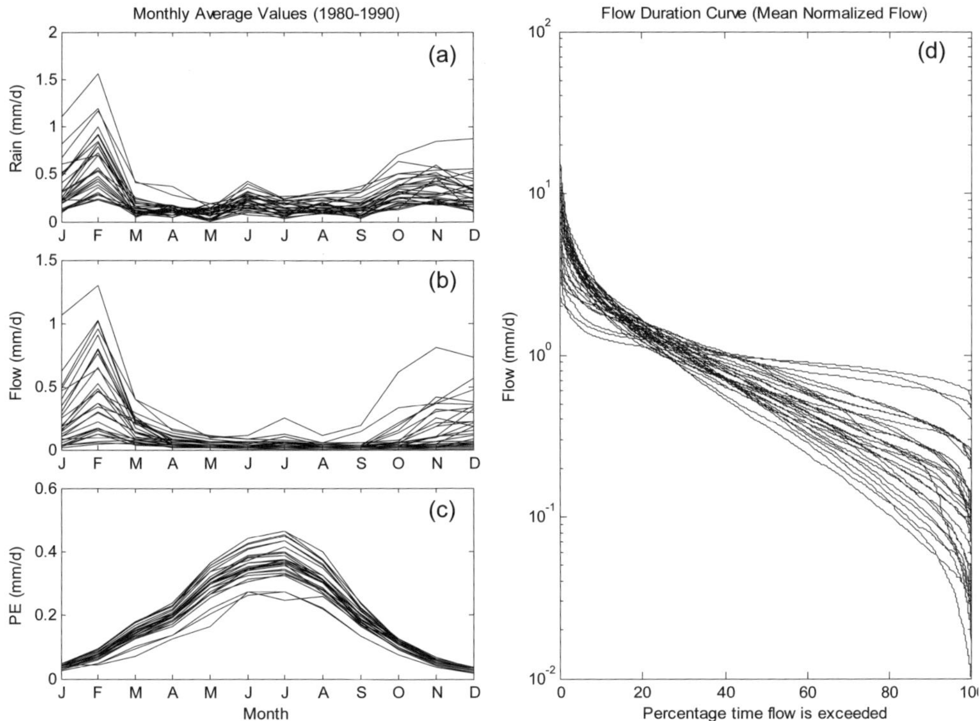

Fig. 2 Average monthly values of: (a) precipitation, (b) streamflow and (c) potential evapotranspiration for 30 UK watersheds. (d) Normalized flow duration curve.

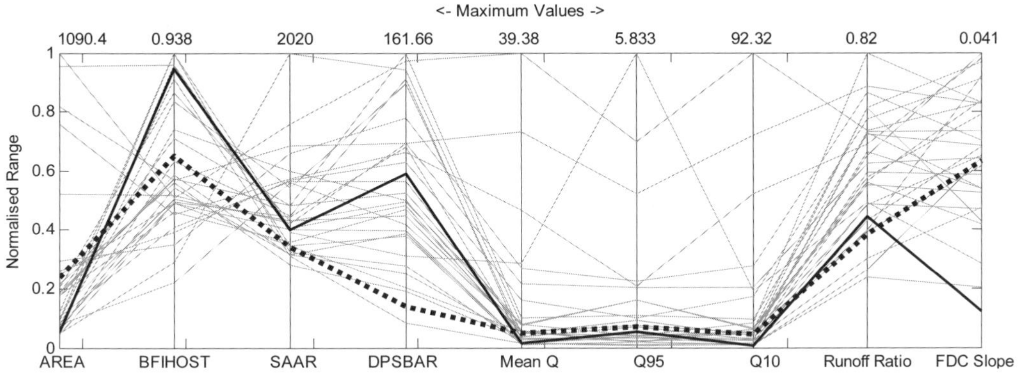

Fig. 3 Parallel coordinate plot showing the variation of response and physical characteristics among 30 UK watersheds. Maximum values for each characteristic are shown on the top axis. SAAR is the standard average annual runoff for the period 1961–1990. Validation watersheds are shown as black lines.

A comprehensive list of physical characteristics for all the watersheds was compiled from the National River Flow Archive (http://www.nwl.ac.uk/ih/nrfa) and the data CD accompanying the Flood Estimation Handbook (FEH, 1999). The main features of these watersheds are presented in a parallel coordinates plot in Fig. 3. The plot shows that most of the watersheds tend to have small area, small mean flow, small ten- and ninety-five percentile flow exceedence values. The two watersheds to be treated as "ungauged" in the study are shown by the continuous and dotted black lines and have very different characteristics from each other.

METHOD

The physical characteristics used in this study were BFIHOST and DPSBAR. BFIHOST (–) is the long-term average fraction of flow that occurs as baseflow—regionalized for the UK—and DPSBAR (m km^{-1}) is an index of watershed steepness (Boorman *et al.*, 1995). BFIHOST is estimated from a regression equation where BFIHOST is the independent variable and the HOST classifications (combining soils and geological information in the Hydrology Of Soil Types) are the dependent variables. The equation takes the form:

$$BFIHOST = a_1 * HOST_1 + a_2 * HOST_2 + ... + a_{29} * HOST_{29} \tag{1}$$

where $HOST_1 ... HOST_{29}$ are the proportions of each of the HOST classes, and $a_1 ... a_{29}$ are the regression coefficients (Boorman *et al.*, 1995).

Only two response characteristics were used in this study, runoff ratio and slope of the FDC. Runoff ratio is the ratio of mean annual streamflow, normalized by watershed area, to mean annual precipitation. The slope of the FDC was calculated by taking the part of curve between the 33% and 66% flow exceedence values of streamflow normalized by their means. Linear regression equations between individual response characteristics and physical characteristics for the 28 UK watersheds were developed based on an equation of the following form (Kottegoda & Rosso, 1997):

$$Y = \beta_0 + \beta_1 x_1 + \beta_2 x_2 + \cdots + \beta_{p-1} x_{p-1} + \varepsilon \tag{2}$$

where Y is the response characteristic of interest, $x_1, x_2, ..., x_{p-1}$ are p–1 physical characteristics with p regression coefficients ($\beta_0, \beta_1, \ldots, \beta_{p-1}$) and ε is an error term. Figure 4 shows the regression relationships between BFIHOST and FDC slope (Fig. 4(a)), and DPSBAR *vs* runoff ratio (Fig. 4(b)) under uncertainty. The coefficients of determination (R^2) were 0.69 and 0.58, respectively. The regression includes the estimation of prediction and confidence intervals; the confidence interval is a measure of the certainty (or uncertainty) of predicting the true (expected) value of the variable while the prediction interval is a measure of the certainty of predicting some future (possible) value of the variable. Since the uncertainty in prediction intervals includes the uncertainty in the regression parameters ($\beta_0, \beta_1, \ldots, \beta_{p-1}$) and any new measurement (Y), this interval is wider than the confidence interval, which considers uncertainty in regression parameters only, while the measurements are assumed to be random variables. Figure 4 also shows the watersheds sorted by drainage area (black corresponds to the smallest area and white corresponds to the largest area). The Roden River at Rodington, shown in square markers (dotted line in Fig. 3), and the Tillingbourne River at Shalford, shown in diamond markers (solid black in Fig. 3) are the two watersheds not used for developing the regression equations. These two watersheds were chosen intentionally to test the strength of the approach such that one

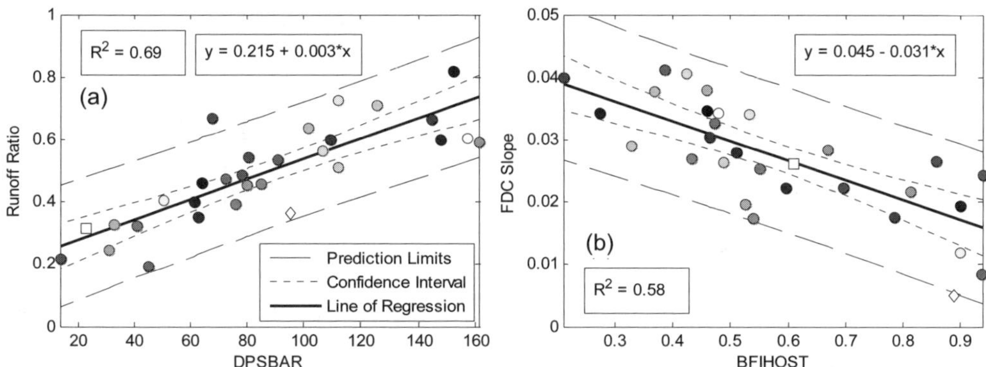

Fig. 4 Linear regression between response and physical characteristics for 28 UK watersheds. The square and diamond show the validation watersheds.

of them was close to the regression line and the other far away from it. The known values of DPSBAR and BFIHOST were used to calculate the confidence and prediction limits of runoff ratio and flow duration curve slopes from the regression equations for the two validation watersheds.

The simple lumped hydrological model (e.g. Wagener *et al.*, 2001) chosen for this study (Fig. 5) has five adjustable parameters, *Huz*, *b*, α, *Kq* and *Ks* (Table 1). It consists of a probability-distributed model as the soil moisture accounting model and a combination of a three-reservoir Nash Cascade for quick flow and a single reservoir slow flow routing model. The model was run within a Monte Carlo framework by randomly sampling 10 000 parameter sets so as to uniformly cover a predefined feasible space. For performance evaluation, the Nash-Sutcliffe Efficiency measure (NSE) (equation. (3)) and the Root Mean Squared Error measure (RMSE*) (equation (4)) were used, the latter computed using a Box Cox transformation applied to the observed and simulated flows:

$$NSE = 1 - \frac{\sum_{i=1}^{N}(q_i - \hat{q}_i)^2}{\sum_{i=1}^{N}(q_i - \frac{1}{N}\sum_{i=1}^{n} q_i)^2} \tag{3}$$

$$RMSE^* = \sqrt{\frac{1}{N}\sum_{i=1}^{N}(q_i^* - \hat{q}_i^*)^2} \tag{4}$$

$$q_i^* = (q_i^{\lambda} - 1)/\lambda \tag{5}$$

where q_i is mean daily streamflow (mm), $\hat{q}_i$ is the model simulated daily streamflow (mm), N is the length of time series of flow, q_i^* and $\hat{q}_i^*$ are Box Cox transformed values of daily observed and simulated streamflows, respectively (5), with a value of 0.3 for the transformation parameter λ. The two measures were selected because of their emphasis on fitting different parts of the time series of flow (i.e. peak flows and low flows, respectively).

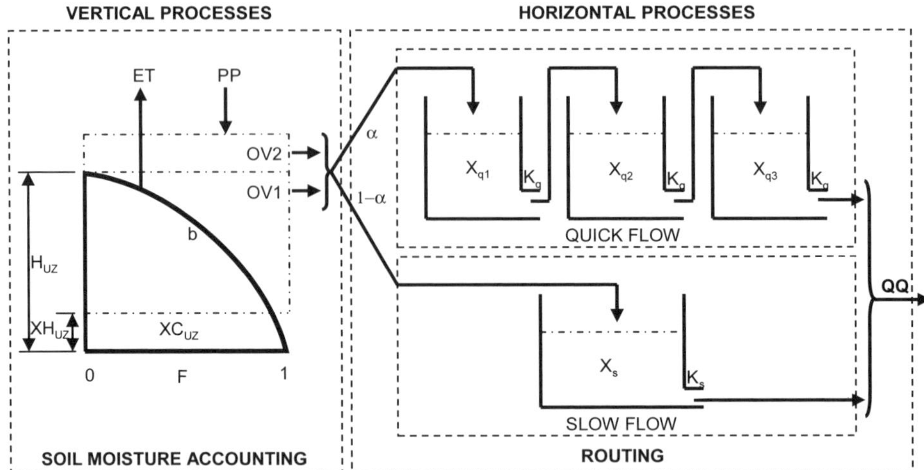

Fig. 5 Lumped 5-parameter model structure. ET and PP are potential evapotranspiration and precipitation respectively (mm). OV1 and OV2 are model simulated effective rainfall components (mm). X_i are states of individual buckets of the routing model. QQ is model simulated streamflow (mm). XH_{UZ} and XC_{UZ} are soil moisture accounting tank state contents (mm).

Table 1 Description of model parameters.

Parameter	Description	Unit	Min	Max
H_{UZ}	Maximum storage capacity of watershed	mm	1	300
b	Index describing spatial soil moisture distribution	–	0	2
α	Flow distribution coefficient	–	0	1
K_q	Residence time of quick flow reservoir	s^{-1}	0	1
K_s	Residence time of slow flow reservoir	s^{-1}	0	1

For selection of parameter sets giving acceptable simulations, those having response indices (runoff ratio or FDC slope) that lie within the confidence and prediction limits were considered behavioural, i.e. acceptable representation of the watershed. The method was first applied separately for each of the regression equations individually and then using the combination of both. The maximum and minimum simulated flows generated by the behavioural parameter sets (i.e. lying within the confidence and prediction intervals) were determined for each time step, and used to form the predictive ranges for the simulations.

RESULTS

The method described above was tested using the two "verification" watersheds as stated before. For reasons of brevity, only the results from the Roden River at Rodington (dotted line in Fig. 3, square in Fig. 4) are presented in detail here. The results obtained by using only one regression equation (runoff ratio *vs* DPSBAR) are shown in Fig. 6(a)–(c). The confidence and prediction intervals derived from the regression analysis have clearly constrained the parameter space in terms of the performance evaluation criterion used. Figure 6(c) shows the maximum and minimum simulated flows for these intervals and for the complete range of simulations. The 50-day period before the dashed vertical line was used as a model warm-up period. The observed streamflow is seen to lie fully inside the prediction intervals after the warm-up period. The number of behavioural simulations when flow is constrained by the prediction limits of the runoff ratio was 5764 (58%), and the corresponding number was 1857 (19%) for flow constrained by confidence limits of the runoff ratio.

The number of behavioural simulations decreased further when the constraints imposed by both regression equations were applied simultaneously (see Fig. 7(c)) resulting in a further narrowing of the confidence and prediction bands (compare with Fig. 6(c)). Again, the observed flow and best simulations lie within the predicted range. The confidence and prediction limits for this case are shown in Fig. 7(a) and (b). When the flow was constrained by the wider ranges (prediction limits) of both response characteristics the number of behavioural simulations was 1114 (11%), reducing to only 42 (0.4%) when the flow was constrained by the narrower confidence limits.

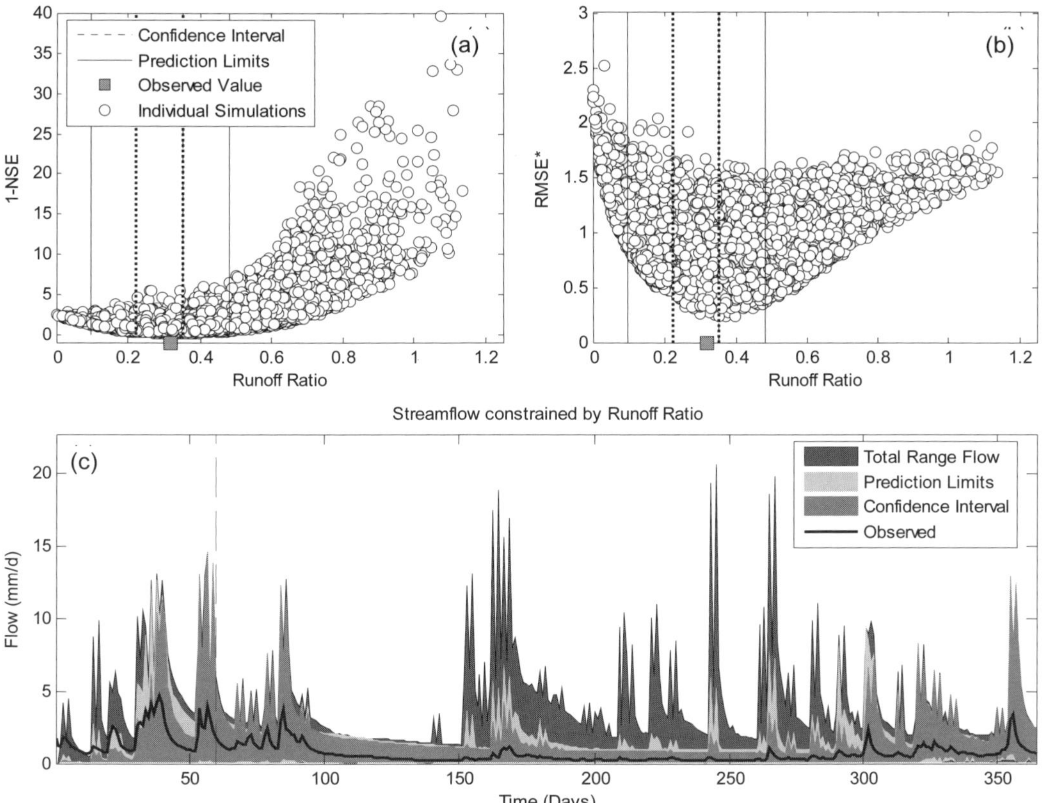

Fig. 6 Dotty plots of simulated runoff ratio *vs* (a) 1-NSE and (b) RMSE*. (c) Observed and (constraint) predicted streamflow ranges.

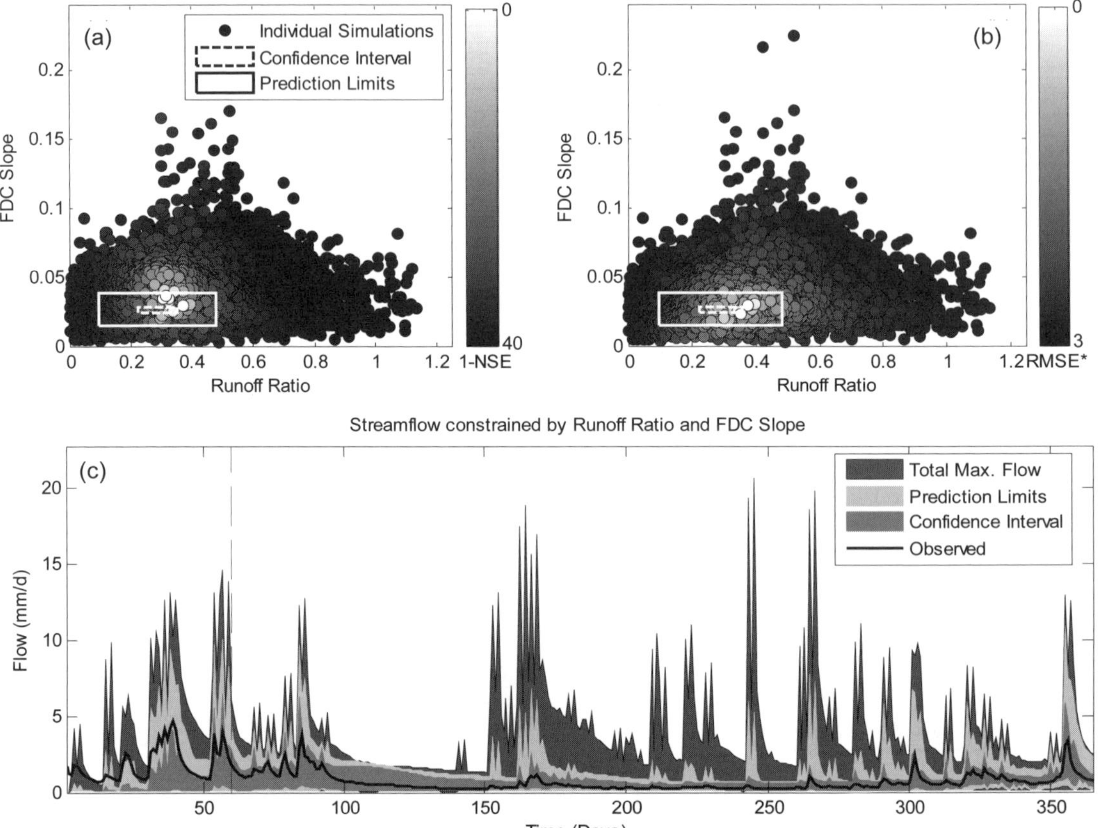

Fig. 7 Dotty Plots of runoff ratio *vs* slope of flow duration curve for: (a) 1-NSE and (b) RMSE*. (c) Observed and (constraint) predicted streamflow ranges.

DISCUSSION AND CONCLUSIONS

This paper presents an approach to reduce the uncertainty on predictions in ungauged basins through the regionalization of watershed characteristics using a framework that properly exploits the "uncertainty" information contained in the regionalization regression relationships. The initial results presented here illustrate the efficiency and ability of the approach in providing suitably constrained model-based hydrological predictions. More thorough testing will include the use of a larger set of informative indices and alternative hydrological models. Further, results with more statistical robustness can be obtained by a bootstrapping-type approach in which watersheds within the data set are treated in turn as ungauged. These results will be reported in due course. As always, we invite discussion and correspondence on this and related topics.

Acknowledgements Partial support for this work was provided by SAHRA under NSF-STC grant EAR-9876800, and the National Weather Service Office of Hydrology under grant numbers NOAA/NA04NWS4620012, UCAR/NOAA/COMET/ S0344674, NOAA/DG 133W-03-SE-0916. We thank The British Atmospheric Data Centre for providing the temperature data (http://badc.nerc.ac.uk/home/index.html). We thank the anonymous reviewer for helpful comments that improved the paper.

REFERENCES

Beven, K. J. (1989) Changing ideas in hydrology—the case of physically-based models. *J. Hydrol.* **105**, 157–172.

Boorman, D. B., Hollis, J. M. & Lilly, A. (1995) Hydrology of soil types: a hydrologically-based classification of the soils of the United Kingdom. *Institute of Hydrology Report No. 126*, Wallingford, UK.

FEH (1999) *Flood Estimation Handbook.* Institute of Hydrology, Wallingford, UK (with CD-Rom).

Kottegoda, N. T. & Rosso, R. (1997) *Statistics, Probability, and Reliability for Civil and Environmental Engineers.* McGraw-Hill, New York, USA.

Maidment, D. R. (ed.) (1993) *Handbook of Hydrology.* McGraw-Hill, New York, USA.

Olden, J. D. & Poff, N. L. (2003) Redundancy and the choice of hydrologic indices for characterizing streamflow regimes. *River Res. and Applic.* **19**, 101–121.

Shamir, E., Imam, B., Morin, E., Gupta, H. V. & Sorooshian, S. (2004) The role of hydrograph indices in parameter estimation of rainfall–runoff models. *Hydrol. Processes* **19**, 2187–2207

Sivapalan, M., Takeuchi, K., Franks, S. W., Gupta, V K., Karambiri, H., Lakshmi, V., Liang, X., McDonnell, J. J., Mendiondo, E. M., O'Connell, P. E., Oki, T., Pomeroy, J. W., Schertzer, D., Uhlenbrook, S. & Zehe, E. (2003) IAHS Decade on Predictions in Ungauged Basins (PUB), 2003–2012: Shaping an exciting future for the hydrological sciences. *Hydrol. Sci. J.* **48**(6), 857–880.

Yu, P. S. & Yang, T. C. (2000) Using synthetic flow duration curves for rainfall–runoff model calibration at ungauged sites. *Hydrol. Processes* **14**(1), 117–133.

Wagener, T., Boyle, D. P., Lees, M. J., Wheater, H. S., Gupta, H. V. & Sorooshian, S. (2001) A framework for the development and application of hydrological models. *Hydrol. Earth System Sci.* **5**(1), 13–26.

Wagener, T. & Wheater, H. S. (2005) Parameter estimation and regionalization for continuous rainfall–runoff models including uncertainty. *J. Hydrol.* **320**(1–2), 132–154.

Wagener, T., Sivapalan, M., McDonnell, J. J., Hooper, R., Lakshmi, V., Liang, X. & Kumar, P. (2004) Predictions in Ungauged Basins (PUB)—A catalyst for multi- disciplinary hydrology. *Eos, Trans. AGU* **85**(44), 451–452.

Wagener, T., Wheater, H. S. & Gupta, H. V. (2004) *Rainfall–Runoff Modelling in Gauged and Ungauged Catchments.* Imperial College Press, London, UK.

Large Sample Basin Experiments for Hydrological Model Parameterization: Results of the Model Parameter Experiment–MOPEX. IAHS Publ. 307, 2006.

Ungauged catchments: how to make the most of a few streamflow measurements?

CLAUDIA ROJAS-SERNA, CLAUDE MICHEL, CHARLES PERRIN & VAZKEN ANDREASSIAN

Cemagref, Hydrosystems and Bioprocesses Research Unit, Parc de Tourvoie BP 44, F-92163 Antony cedex, France

claude.michel@cemagref.fr

Abstract This paper deals with an alternative to the classical regionalization approach used to address the problem of ungauged basins with rainfall–runoff models. We chose a new, more practical orientation, which consists of making the most of a few discharge measurements. Indeed, it is not unrealistic to obtain such measurements by sending a gauging team to the spot where a hydrological prediction is needed. In this paper we aim to identify the parameters of a daily lumped rainfall–runoff model, GR4J, and we look for a strategy to calibrate its parameters using a few streamflow measurements, which are combined with *a priori* knowledge of the parameters. Results show this approach to be much more efficient than classical regionalization studies, as soon as about thirty measurements can be made, at random, during a period of three to five years.

Key words point measurements; rainfall–runoff modelling; regionalization; ungauged basin

INTRODUCTION

All problems posed to the hydrologist require the representation of the natural system by a model. Without a model, nothing can be said, computed or predicted. The type of model mostly needed by hydrologists is of the rainfall–runoff type. The most versatile type of rainfall–runoff model is the lumped model. Such a model has to be calibrated using rainfall–runoff data to obtain its parameters. When no data are available, i.e. when the studied watershed is ungauged, hydrologists are at pains to identify appropriate parameter values. The conventional approach consists in generalizing for a whole region regression relationships established for many gauged catchments monitored in that region. However, several studies have shown that regressions are not really effective when addressing the problem at the scale of a country (Lee *et al.*, 2005). Merz & Blöschl (2004) and Parajka *et al.* (2005) developed alternatives based on spatial proximity.

Working with a very parsimonious model (just four free parameters) we found that we were unable to understand how basic characteristics, such as the catchment area or the aridity index (ratio of the mean annual evapotranspiration to the mean annual precipitation) could influence the values taken by the model parameters. Thus, we chose a new avenue of research which aims at making the most of a few discharge measurements. We assume that we are looking for the parameters of a daily lumped rainfall–runoff model and that one day of measurements allows one to know the daily flow on that day.

The paper is organized as follows: (a) the GR4J model to be applied on an ungauged catchment is presented; (b) the data that have been used in this research to gain some statistical knowledge of GR4J parameters and to test the methods developed for ungauged catchments are described; (c) the method suitable when very few measurements are available is presented; and (d) the results and a conclusion are given.

THE GR4J MODEL

The GR4J model, which we will use throughout this paper, is described in detail by Edijatno *et al.* (1999) and Perrin *et al.* (2003) and we refer readers to the latter article for a complete description of the model. GR4J has four parameters to be calibrated. It is composed of a soil-moisture accounting procedure (parameter X_1 in mm), a transfer function combining a routing reservoir (parameter X_2 in mm) and a unit hydrograph (parameter X_3 in days), and a gain-loss function (parameter X_4 in mm). In order to make the calibration more efficient, it is carried out with transformed values of these parameters such that transformed parameters, noted x_i, belong to the interval $]-10,10[$.

The transformations applied to the four parameters of the GR4J model during the calibration process are:

$$x_1 = \log(X_1)$$

$$x_2 = \log(X_2)$$

$$x_3 = \frac{X_3 - 5}{0.45}$$

$$x_4 = \sinh^{-1}(X_4)$$

Based on the large sample of catchments on which GR4J has been calibrated, we have gained a sound knowledge of the prior distribution of the four parameters. Mean values and standard deviations of the transformed parameters are shown in Table 1.

Table 1 First two moments of the transformed GR4J parameters (from a sample of 1111 catchments).

Parameter	Mean value	Standard deviation
X_1	6.2	1.1
X_2	3.9	1.5
X_3	–6.1	3.7
X_4	–0.1	1.7

As a first approach towards using GR4J on ungauged catchments, we tried to explain the transformed parameters using three catchment features: the catchment area (A, km^2), the mean daily potential evapotranspiration (E, mm) and the probability of having a daily precipitation larger than 0.1 mm (W). The explanatory relationships found are:

$$x_1 = 6.3 + 0.001\log(A) + 0.05\log(E) + 0.1\log(W) \tag{1}$$

$$x_2 = 5.3 - 0.1\log(A) - 0.7\log(E) + 0.5\log(W) \tag{2}$$

$$x_3 = -8.7 + 0.4\log(S) + 0.4\log(E) + 0.5\log(W) \tag{3}$$

$$x_4 = 1.2 - 0.07\log(S) + 0.03\log(E) + 1.3\log(W) \tag{4}$$

Note that all these regression relationships have very low coefficients of determination, even though some explanatory variables (in bold type) have acceptable Student ratios (Table 2). Thus, these relations would not be sufficient to apply GR4J to ungauged basins.

Table 2 Coefficients of regression, Student ratios and coefficients of determination of the regression relationships of the GR4J parameters

Parameter	coefficients of regression	Student ratios	coefficient of determination (R^2)
x_1	$a_0 = 6.25$	49.66	0.0008
	$a_1 = 0.001$	0.05	
	$a_2 = 0.05$	0.42	
	$a_3 = 0.12$	1.12	
x_2	$a_0 = 5.31$	31.53	0.14
	$a_1 = -0.06$	–3.62	
	$a_2 = 0.67$	–4.68	
	$a_3 = 0.46$	3.35	
x_3	$a_0 = -8.68$	–21.19	0.06
	$a_1 = 0.41$	9.78	
	$a_2 = 0.44$	1.28	
	$a_3 = 0.50$	1.50	
x_4	$a_0 = 1.16$	5.92	0.04
	$a_1 = -0.07$	–3.27	
	$a_2 = 0.30$	1.83	
	$a_3 = 1.33$	8.31	

TEST DATA

In order to test our approach to determine GR4J parameters on an ungauged catchment, we used a large sample of catchments spanning five continents, assembled for this study: it is comprised of 1111 catchments, with areas from 0.1 to 50600 km^2, located in the United States (500 catchments), France (305 catchments), Mexico (260 catchments), Australia (32 catchments), the Ivory Coast (10 catchments) and Brazil (4 catchments). Note that the 428 basins of the MOPEX US database (Schaake *et al.*, this issue) as well as the 40 basins of the MOPEX French database (Chahinian *et al.*, this issue) are part of this international sample.

METHOD

Each of the 1111 catchments is successively considered as ungauged (with only a limited number of point measurements available) during the first half of the available flow series. It is treated with our approach and its efficiency in control (on the second half of the flow series) is assessed by comparing the results to those of a full calibration operation. To make the most of all data available, the previous operation is carried out a second time using the two periods in reverse order.

The proposed approach

Our approach blends prior knowledge and the information contributed by a few discharge measurements taken randomly during a given period of time. It is embodied into the calibration criterion C given as:

$$C = \alpha \frac{1}{4} \sum_{i=1}^{4} \left(\frac{x_i - x_i^0}{\sigma_i} \right)^2 + (1-\alpha) \frac{\sum_{j=1}^{N} \left(\sqrt{Q_j^{obs}} - \sqrt{Q_j^{calc}} \right)^2}{\sum_{j=1}^{N} Q_j^{obs}} \qquad (5)$$

where the parameters x_i are chosen to minimize C, σ_i is the standard deviation of parameter i as shown in the calibration process equations, x_i^0 is the set of prior parameters, Q^{obs} represents a measured daily flow at the basin outlet and Q^{calc} represents the flow calculated by the model using the set of parameters x_i. The parameter α is determined as a function of N, the number of daily measurements, in order to give the approach its maximum efficiency. The least square of the errors is calculated on the square roots of the outflow values in order to avoid giving to some large values an overriding role that could reduce the value of the other measurements. According to equation (5), when no measurements are available our best estimate is to use the prior set of parameters, x_i^0, which can be either the mean values displayed in calibration process equations or, alternatively, the values given by the four regression relationships given in equations (1) to (4).

Test of the proposed approach

We assume that the catchment of interest is now gauged and that a series of discharge data are available for a second period of time. Having determined the parameters x_i, by minimizing criterion C, we can assess the value of these parameter estimates against flow data available for the control period. We use the appraisal criterion F:

$$F = \frac{\sum_{k=1}^{K} \left[\left(Q_k^{obs} - \overline{Q}^{obs} \right)^2 - \left(Q_k^{obs} - Q_k^{calc} \right)^2 \right]}{\sum_{k=1}^{K} \left[\left(Q_k^{obs} - \overline{Q}^{obs} \right)^2 + \left(Q_k^{obs} - Q_k^{calc} \right)^2 \right]} \qquad (6)$$

Criterion F, proposed by Mathevet *et al.* (2005) is the bounded version of the Nash-Sutcliffe efficiency but has the new property of belonging to the interval]–1,1]. This calculation can be made for each half of the available flow record (when reversing their roles) and subsequently for each of the 1111 catchments. The mean value of the 2222 individual F values is denoted $\overline{F}$ and will be used to assess the efficiency of our approach. The standard deviation of the 2222 F values is denoted σ_F.

What is the difference between a gauged and an ungauged catchment?

We can tell the difference by just looking at the $\overline{F}$ value for both situations: for our 1111 catchments as for ungauged we have obtained $\overline{F}$ = 0.13 when using the mean values from Table 1, and $\overline{F}$ = 0.14 when using the values derived from the regression relationships (equations (1) to (4)). Clearly, very little can be gained from the most significant features of a catchment. Opposed to this, when considering all those catchments as gauged and calibrating GR4J in the conventional way, we obtain a much higher value of $\overline{F}$ = 0.36. Figure 1 shows the corresponding cumulative distribution of the F values instead of just their mean value. A shift of the distribution towards the right corresponds to an improvement. Our objective is to find a way to move the "ungauged" (left) distribution towards the "gauged" distribution with the help of a few daily measurements. Referring to the mean of the F values, the closer we will be to 0.36, the better our approach. Preliminary work on the same catchment sample (not reported here) showed that most models reported in the literature would yield, after full calibration, a $\overline{F}$ value in the range (0.19 to 0.36). Therefore, we could consider that a candidate method starts to be successful when $\overline{F}$ is greater than the threshold value of 0.19.

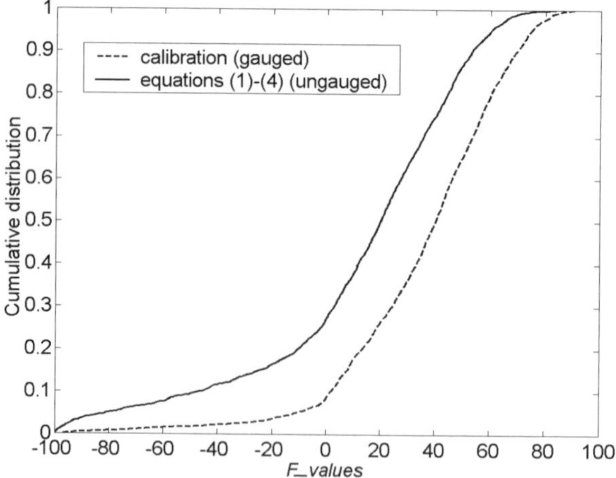

Fig. 1 Cumulative distribution of F values considering 1111 catchments either as gauged or as ungauged.

RESULTS

Using criterion *C*

Results corresponding to N successively equal to 5, 10, 20 and 50 are shown in Table 3.

Table 3 First results using criterion *C*.

N (number of available streamflow measurements)	α (respective weight of the prior knowledge)	$\overline{F}$ (efficiency criterion)
5	0.03	0.15
10	0.01	0.16
20	0.01	0.18
30	0.01	0.19
50	0.01	0.20

We see that with only 30 daily measurements, we can obtain the same statistical efficiency as with using the less efficient model reported in the literature but fully calibrated. These first results are encouraging.

Selection of the parameters to be calibrated

We had doubts concerning the σ_i used in equation (5). It is acknowledged that the standard deviations cannot properly perform their task of weighting the efforts with respect to each parameter. When the number of measurements is low, one is better advised to calibrate only one or two parameters.

The results bear out our concern with optimizing all four parameters of the GR4J model. Only when the number of measurements is greater than 50, does it become worthwhile to calibrate all four parameters.

Therefore, in our case of very few measurements, we dropped some parameters from criterion C. The corresponding results are shown in Table 4.

Table 4 Best selection of parameters to be calibrated as a function of the number of measurements available.

N (number of available streamflow measurements)	Parameters * optimized	α (respective weight of the prior knowledge, see equation (5))	$\overline{F}$ (efficiency criterion, see equation (6))
5	x_4, x_1	0.01	0.16
10	x_4, x_1	0.03	0.17
20	x_4, x_1, x_2	0.00	0.18
50	x_4, x_1, x_2	0.00	0.21

*Where: x_1, capacity of the soil-moisture accounting store; x_2, capacity of the routing reservoir; x_3, unit hydrograph time base; x_4, gain-loss function parameter.

CONCLUSION

To our knowledge, this is the first time that the problem posed by ungauged catchments has been broached by elaborating a strategy of direct discharge measurements at the point of interest. The first results presented in this paper show the value of this approach. Instead of measuring complex soil parameters, it seems more profitable to focus our efforts on acquiring a few direct measurements of the variable of interest, the discharge at the outlet of a catchment. The present article reports on the early stage of this strategy, when the first measurements are being acquired. We are continuing our work, aiming to increase the benefit that can be expected from a few direct streamflow measurements, and we will report further improvements later.

REFERENCES

Edijatno, Nascimento, N. O., Yang, X., Makhlouf, Z. & Michel, C. (1999) GR3J a daily watershed model with three free parameters. *Hydrol. Sci. J.* **44**(2), 263–277.

Chahinian, N., Mathevet, T., Habets, F. & Andréassian, V. (2005) The MOPEX 2004 French database: main hydrological and morphological characteristics. In: *Large Sample Basin Experiments for Hydrological Model Parametization: Results of the Model Parameter Experiment–MOPEX* (ed. by V. Andreassian, A. Hall, N. Chahinian & J. Schaake), 29–40. IAHS Publ. 307. IAHS Press, Wallingford, UK (this issue).

Lee, H., McIntyre, N., Wheater, H. & Young, A. (2005) Selection of conceptual models for regionalisation of the rainfall–runoff relationship. *J. Hydrol.* **312**, 125–147.

Mathevet, T., Andréassian, V., Perrin, C. & Michel, C. (2006) Large samples of watersheds: a necessary condition to compare the efficiency of rainfall–runoff models? *Water Resour. Res.* (in press).

Merz, R. & Blöschl, G. (2004) Regionalisation of catchment model parameters. *J. Hydrol.* **287**, 95–123.

Parajka, J., Merz, R. & Blöschl, G. (2005) A comparison of regionalisation methods for catchment model parameters. *Hydrol. Earth System Sci.* **9**(3), 157–171.

Perrin, C., Michel, C. & Andréassian, V. (2003) Improvement of a parsimonious model for streamflow simulation. *J. Hydrol.* **279**(1–4), 275–289.

Schaake, J. Cong, S. & Duan, Q. (2006) The MOPEX US database. In: *Large Sample Basin Experiments for Hydrological Model Parameterization: Results of the Model Parameter Experiment - MOPEX* (ed. by V. Andreassian A. Hall, N. Chahinian & J. Schaake), 9–28. IAHS Publ. 307. IAHS Press, Wallingford, UK (this issue).

Characterization of flow regimes for three French rivers using a 6-parameter continuous simulation rainfall–streamflow model: the importance of the parameter selection procedure

IAN G. LITTLEWOOD
Centre for Ecology and Hydrology, Wallingford, Oxfordshire OX10 8BB, UK
igl@ceh.ac.uk

Abstract An established 6-parameter rainfall–streamflow modelling package (PC-IHACRES) is applied to three French catchments using daily data sets made available to participants of a Model Parameter Estimation Experiment (MOPEX) workshop held in Paris, 2004. For one catchment, Le Loupe at Villeneuve-Loubet, the main aim is to characterize the flow regime over as wide a range of flows as possible, using a single set of model parameters selected by a multi-objective model parameter selection procedure. Using a more restricted, twin-objective, parameter selection procedure, a subsidiary aim for each of the three catchments (additionally Le Guillec at Trézilidé and Le Toulourenc at Malaucène), is to investigate how well a model calibrated on a sub-period of the available record performs on a different sub-period not used for model calibration. For Le Loupe at Villeneuve-Loubet, the 6-parameter model structure, with a single set of parameters calibrated over almost the entire available record, characterizes the flow regime between 5 percentile and 80 percentile flows. Split-sample analyses for Le Guillec at Trézilidé and Le Loupe at Villeneuve-Loubet, using two sub-periods of the available record in each case, demonstrate the difficulty of capturing a time-invariant characteristic response for those two catchments. A similar split-sample analysis suggests that Le Toulourenc at Malaucène has an essentially time-invariant characteristic response to rainfall. The results are discussed in the context of similar analyses reported in the literature, and with respect to the potential of the 6-parameter model for some aspects of regionalizing model parameters against physical catchment descriptors to enable flow simulation at ungauged sites.

Key words flow regime characterization; rainfall–streamflow modelling; regionalization; ungauged catchments

INTRODUCTION

In the late 1980s and early 1990s, computational advances in the application of unit hydrograph (UH) theory were marked by the development of the IHACRES approach to rainfall–streamflow modelling (e.g. Jakeman *et al*, 1990; Littlewood & Jakeman, 1994; Littlewood *et al.* 1997) (IHACRES: Identification of unit Hydrographs And Component flows from Rainfall, Evaporation and Streamflow data). For the first time, it became possible to systematically identify UHs for total streamflow (not a poorly-definable "direct flow" component of streamflow) from continuous records of rainfall and flow (not just "runoff event" data sets). Furthermore, the need was removed for prior separation of a baseflow component of streamflow, as is common in modelling approaches that identify a UH for "direct" flow only. Rather, the new methodology allows hydrograph separation as an integral part of the procedure, to give dominant quick- and slow-response components of streamflow.

The paper applies PC-IHACRES to characterize the rainfall–streamflow dynamics of three French catchments (43, 150 and 279 km^2), using data provided to participants of a Model Parameter Estimation Experiment (MOPEX) workshop held in Paris, 2004.

The model and modelling package

While a UH for total flow can be represented by one of a number of configurations of linear stores acting in parallel and/or series, the optimal configuration identifiable from information in commonly available rainfall and streamflow time series is usually two linear stores in parallel. These two linear stores generate dominant quick- and slow-response components of modelled streamflow (e.g. Littlewood & Jakeman, 1992). Typically, the IHACRES loss and UH modules each have three parameters (i.e. the whole model has just six parameters). Given its parametric parsimony, the potential for establishing statistical relationships between IHACRES model parameters and physical catchment descriptors has been recognized, investigated and commented upon by several researchers (e.g. Jakeman *et al.* 1992; Littlewood & Jakeman, 1994; Post & Jakeman, 1996, 1999; Sefton & Howarth, 1998; Littlewood, 2002, 2003; Kokkonen *et al.* 2003).

The PC-IHACRES modelling package (Littlewood *et al.*, 1997) applied in this paper incorporates the loss model structure introduced by Jakeman & Hornberger (1993). Equations (1) to (9) in Tables 1 and 2 define the model where: r_k, u_k, s_k, Q_k and t_k are respectively rainfall (mm), effective rainfall (mm), catchment wetness index (-), streamflow (mm) and air temperature (°C) at time-step k (in this paper $k = 1$ day); τ_w is a catchment drying time constant (days); f is a temperature modulation factor (°C^{-1}); R is a reference temperature (in this paper $R = 20$°C); C (mm^{-1}) is selected to give a water balance between the volumes of effective rain and observed streamflow over the model calibration period; $a_1^{(q)}$, $a_1^{(s)}$, $b_0^{(q)}$, $b_0^{(s)}$, a_1, a_2, b_0, and b_1 are coefficients of linear UH transfer functions; z^{-1} is the backward shift operator, i.e. $z^{-1}x_t = x_{t-1}$; and δ is a pure time delay (for all three catchments $\delta = 0$ days was found to give better models than $\delta = 1, 2, \ldots$ days). Equation (5) is the second-order transfer function representation of the two first-order transfer functions in equation (4).

Table 1 The IHACRES model.

Loss module		UH module	
$u_k = r_k \dfrac{(s_k + s_{k-1})}{2}$	(1)	$Q_k = \left[\left(\dfrac{b_0^{(q)}}{1+a_1^{(q)}z^{-1}}\right)+\left(\dfrac{b_0^{(s)}}{1+a_1^{(s)}z^{-1}}\right)\right]u_{k-\delta}$	(4)
$s_0 = 0$			
$s_k = Cr_k + \left(1-\dfrac{1}{\tau_w(t_k)}\right)s_{k-1}$	(2)	$Q_k = \left(\dfrac{b_0 + b_1 z^{-1}}{1+a_1 z^{-1} + a_2 z^{-2}}\right)u_{k-\delta}$	(5)
$\tau_w(t_k) > 1$			
$\tau_w(t_k) = \tau_w \exp(f(R - t_k))$	(3)		

Table 2 Unit hydrograph dynamic response characteristics (DRCs).

DRC	Quick flow		Slow flow	
Characteristic decay response times for data time step Δ, e.g. 1 day	$\tau^{(q)} = \dfrac{-\Delta}{\ln(-a_1^{(q)})}$	(6)	$\tau^{(s)} = \dfrac{-\Delta}{\ln(-a_1^{(s)})}$	(7)
Relative volumetric throughflow, where $V = \dfrac{b_0^{(q)}}{1+a_1^{(q)}} + \dfrac{b_0^{(s)}}{1+a_1^{(s)}}$	$v^{(q)} = \left(\dfrac{b_0^{(q)}}{1+a_1^{(q)}}\right)\left(\dfrac{1}{V}\right)$	(8)	$v^{(s)} = \left(\dfrac{b_0^{(s)}}{1+a_1^{(s)}}\right)\left(\dfrac{1}{V}\right)$	(9)

Preliminary selection of a set of parameters giving a best model fit can be made through a trade-off between a high coefficient of determination (D) and a small average relative parameter error ($ARPE$) for UH module parameters a_1, a_2, b_0, and b_1, as fully described elsewhere (e.g. Jakeman *et al.*, 1990; Littlewood *et al.*, 1997). The statistics D and $ARPE$ are defined by equations (10) and (11). In equations (10) and (11): Q_o is observed flow; Q_m is modelled flow; and the σ terms are standard deviations on the UH module parameters as indicated by subscripts (the UH parameter estimation algorithm employed in IHACRES yields a variance-covariance matrix for a_1, a_2, b_0, and b_1). For one catchment, as will be seen later, the twin-objective (D, $ARPE$) procedure was extended by subsequent trial-and-error adjustment of loss module parameter f in search of an improved match between the low-flow sections of flow duration curves for observed and modelled flows.

$$D = 1 - \frac{\Sigma(Q_o - Q_m)^2}{\Sigma(Q_o - \bar{Q}_o)^2} \tag{10}$$

$$ARPE = \left[\left(\frac{\sigma_{a_1}}{a_1}\right)^2 + \left(\frac{\sigma_{a_2}}{a_2}\right)^2 + \left(\frac{\sigma_{b_0}}{b_0}\right)^2 + \left(\frac{\sigma_{b_1}}{b_1}\right)^2\right] / 4 \tag{11}$$

The data

One option offered to MOPEX2004 workshop participants was to analyse daily data, 1 August 1995–31 July 2002, for the three catchments listed (with basic details) in Table 3.

Table 3 Basic catchments details.

Catchment	Code	Area (km^2)	Mean annual precip (mm)	Mean annual runoff (mm)	Runoff (%)	Mean air temp. (°C)	Gaps in record?
Le Guillec à Trézilidé	J3024010	43	1029 [a]	550 [a]	53 [a]	11.6	no
Le Toulourenc à Malaucène	V6035010	150	1115 [b]	354 [b]	32 [b]	9.2	yes
Le Loup à Villeneuve-Loubet	Y5615030	279	1043 [c]	511 [c]	49 [c]	11.8	yes

[a] 1996–2001; [b] 1999–2001; [c] 1997–2001.

Apart from the time series of daily rainfall, streamflow and air temperature, measurement units and basin areas, no other information about the catchments was known to the author at the time of analysis. Hereafter, the catchments will be referred to using the identification codes given in Table 3, rather than the names of the rivers and the sites at which they are gauged.

RESULTS

Catchment Y5615030

Model Y1 for Y5615030 was calibrated over the longest period of unbroken flow record (3 September 1996–31 July 2002), and has D = 0.836 and $ARPE$ = 0.01. The model-fit is shown in Fig. 1. For flows less than approximately the 20 percentile, model Y1 tends to overestimate flow increasingly with decreasing flow. As reported elsewhere (Littlewood, 2002, 2003), similar overestimation of low flows by *D-ARPE* trade-off PC-IHACRES models has been noted for several catchments in Wales and, in some cases, alleviated by subsequent manual fine-tuning of loss module parameter f as outlined above.

Figure 2 shows the effect of manual fine-tuning of parameter f for Y5615030, to give model Y4. (Models Y2 and Y3, not reported here, were intermediate; it is convenient to retain reference to models Y1 and Y4.) Table 4 gives the parameters of models Y1 and Y4, and their values of D and $ARPE$. Ideally, both f and τ_w should have been varied, but given the limited time available for analysis, parameter τ_w for model Y1 (9 days) was held constant while parameter f was varied to find model Y4. Model Y1 has the higher D, and supports the finding and comment by Wagener *et al.* (2004), made in the context of an investigation of several different, parametrically parsimonious, model structures, that "*A distinct difference in optimum parameter sets to fit high and low flows was found for all structures analysed ... This seems to be the*

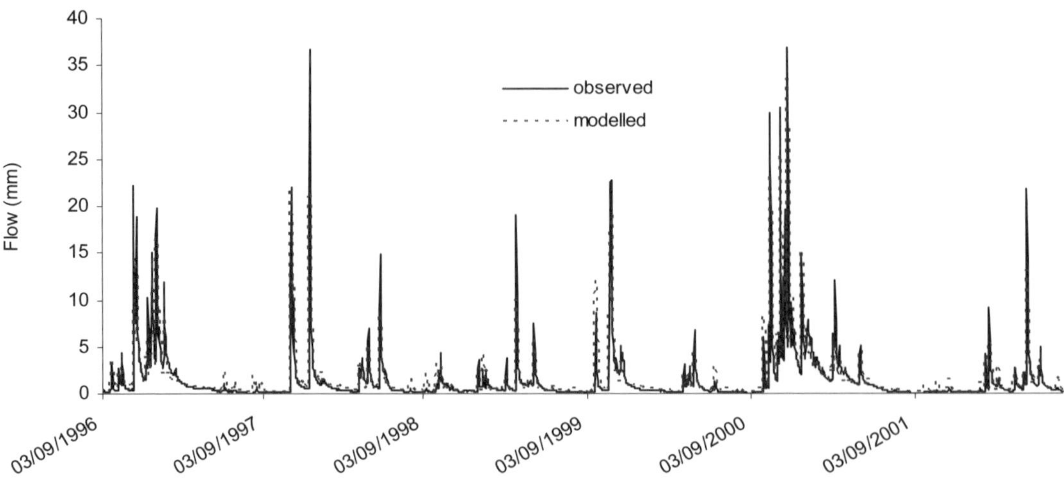

Fig. 1 Catchment Y561030 model Y1 3 September 1996 to 31 July 2002.

Table 4 Catchment Y5615030 models Y1 and Y4.

Model	*D*	*ARPE*	*f* ($^{o}C^{-1}$)	τ_w (days)	$1/C$ (mm)	$\tau^{(q)}$ (days)	$\tau^{(s)}$ (days)	$v^{(s)}$ (–)
Y1	0.836	0.01	0.0558	9	227	2.8	75	0.40
Y4	0.790	0.02	0.155	9	416	3.4	110	0.20

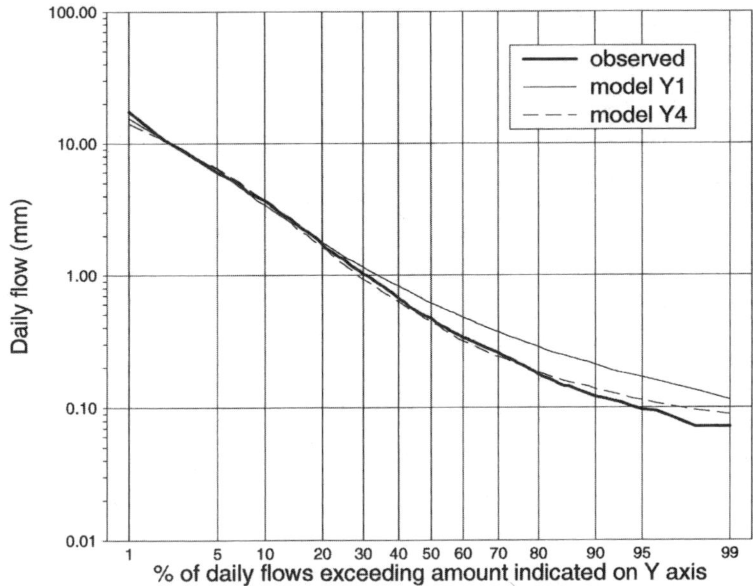

Fig. 2 Catchment Y5615030 flow duration plots 3 September 1996 to 31 July 2002.

main problem in currently available CRR model structures." (CRR: "catchment rainfall–runoff".) However, by accepting small reductions in: (a) *D* (recalling that maximizing *D* favours model-fit at high flows over model-fit at lower flows) and (b) the average precision on the UH parameters (i.e. an increase in *ARPE*), it can be seen in Fig. 2 that model Y4 extends the range of flows over which a given model structure performs well with a single set of six parameters. The region of overestimation is reduced from flows less than the 20 percentile (model Y1) to flows less than the 80 percentile (model Y4). In five out of seven catchments in Wales (129 km^2 to 1480 km^2), Littlewood (2003) fine-tuned parameter *f* in the same way, leading to good matches between observed and modelled flow duration curves (daily data) over even larger portions of the flow range, i.e. between the 5 percentile and 95 percentile (between the 1 percentile and 99 percentile in some cases).

All three catchments

Reverting to the restricted *D-ARPE* trade-off method of parameter selection for each catchment, models were calibrated over a sub-period of the available record (period

P1) and then applied to simulate flows over a sub-period not used for calibration (period P2). A model was then calibrated over period P2 and run in simulation mode over period P1.

Tables 5 and 6 show the results. Each of the six calibrated models accounts for 80%, or more, of the initial variance in observed streamflow. For V6035010, there is little difference between *D* for the calibration and simulation periods, suggesting that a good characterization of the rainfall–streamflow dynamics for that catchment would be obtained by calibrating a model over the whole available record.

Table 5 Split-sample analyses.

Catchment	Sub-period			*D*	*ARPE*
J3024010	20 Aug. 95–2 Sep. 1998	P1	C	0.796	0.16
	3 Sep. 1998–30 Jul. 2002	P2	S	0.805	–
	3 Sep. 1998–30 Jul. 2002	P2	C	0.865	0.07
	20 Aug. 1995–2 Sep. 1998	P1	S	0.670	–
V6035010	13 Nov. 1995–12 Oct. 1998	P1	C	0.803	0.04
	14 Nov. 1998–29 Jul. 2002	P2	S	0.802	–
	14 Nov. 1998–29 Jul. 2002	P2	C	0.804	0.04
	13 Nov. 1995–12 Oct. 1998	P1	S	0.794	–
Y5615030	3 Sep. 1996–13 Aug. 1999	P1	C	0.809	0.04
	14 Aug. 1999–31 Jul. 2002	P2	S	0.833	–
	14 Aug. 1999–31 Jul. 2002	P2	C	0.863	0.02
	3 Sep. 1996–13 Aug. 1999	P1	S	0.785	–

P1, sub-period 1; P2, sub-period 2; C, calibration period; S, simulation period.

Table 6 Split-sample analysis model parameters.

Catchment		τ_w (days)	f (°C^{-1})	$1/C$ (mm)	$\tau^{(q)}$ (days)	$\tau^{(s)}$ (days)	$v^{(s)}$ (–)
J3024010	P1	22	0.0620	241.0	1.68	50.4	0.802
	P2	10	0.0744	142.5	1.83	44.9	0.697
		(–54)	(20)	(–41)	(9)	(–11)	(0.105)
V6035010	P1	4	0.155	382.4	3.85	45.1	0.254
	P2	3	0.170	356.1	4.08	32.2	0.299
		(–25)	(10)	(–7)	(6)	(–29)	(0.045)
Y5615030	P1	13	0.0620	276.9	3.00	59.5	0.345
	P2	7	0.0558	201.4	2.62	80.9	0.445
		(–46)	(10)	(–27)	(–13)	(36)	(0.100)

P1, P2 – see Table 5.
Numbers in parentheses are percentage changes between P1 and P2 (absolute change for $v^{(s)}$).

However, for both J3024010 and Y561030, the model calibrated over the earlier of the two sub-periods performs better in simulation mode (in terms of *D*) over the later sub-period than it does over its calibration sub-period. Furthermore, the model calibrated over the later of the two sub-periods performs worse in simulation mode over the earlier sub-period. Correspondingly, some of the model parameters for each of J3024010 and Y561030 are quite different for the two sub-periods, suggesting that characterization of the flow regime has not been achieved as successfully (in terms of *D* and *ARPE*) for these two catchments as for V6035010.

DISCUSSION

IHACRES models calibrated using a multi-objective parameter selection procedure involving fine-tuning of parameter *f* for Y5615030, and for several Welsh catchments (Littlewood, 2002, 2003), give improved characterization of the low-flow portion of flow regimes, albeit with varying degrees of success. Further work is certainly required to assess the full potential of "whole flow regime" modelling using simple models like IHACRES. However, wide ranges of flow, including fairly high and low flows, can be modelled with a single set of six parameters, at least for some catchments and provided that a suitable, multi-objective, parameter selection procedure is followed. Where the objective is to establish statistical relationships between model parameters and physical catchment descriptors, to allow flow simulation at ungauged sites from inputs of rainfall and air temperature, it is important to ensure that the model performs well over as wide a range of flows as possible. Thus $\tau^{(q)}$ and $\tau^{(s)}$, for example, will be more likely to exhibit useful correlations with the physical catchment descriptors that may control the dominant quick- and slow-response components of streamflow.

It should always be remembered that the very lowest and highest flows for a given catchment, especially at and near the peaks of large floods, are typically more difficult to measure accurately than flows that occur more often. Partly for this reason, the very highest flows may require different modelling approaches. Furthermore, flows at and near hydrograph peaks are often not well defined by discrete daily time-step data; analysis of sub-daily data by the IHACRES method may help to model these high flows. However, as the data time-step decreases, or as catchment size increases, the spatial variance of rainfall within a cathment tends to increase. In either case a point will be reached when it is unreasonable to expect a spatially lumped model to perform well, and the catchment will need to be considered as several sub-catchments.

The split-sample analyses presented for all three catchments gave two types of result. For V6035010, models calibrated on either sub-period have about the same *D* and perform well in simulation mode on the other sub-period. The split-sample models for V6035010 have fairly similar model parameters, indicating that a good, time-invariant, characterization of the rainfall–streamflow dynamics exists for that catchment. For the other two catchments, although as good (or better) models, in terms of *D*, to those obtained for V6035010 were calibrated on one of the sub-periods, time-invariant characterization was not so successful. As shown in Table 6, for τ_w, *f*, $1/C$, $\tau^{(q)}$ and $\tau^{(s)}$ the percentage changes (absolute change for $v^{(s)}$) between periods P1 and P2 are less (or, in one case, the same) for V6035010 than for J3024010 and Y5615030. Catchments J3024010 and Y5615030 are perhaps more complex hydrologically than V6035010, such that time-invariant characteristic responses do not exist. If they do have time-invariant characteristic responses, it is possible that split-sample models calibrated using additional fine-tuning of parameter *f* (and τ_w), seeking a good match between flow duration curves for observed and modelled flows as an additional modelling objective, would give better models for J3024010 and Y5615030. Further analysis is required to investigate these points, using any additional time series that might exist for the catchments but were not available for the MOPEX workshop.

CONCLUDING REMARKS

This short paper has demonstrated the utility of PC-IHACRES for quickly analysing rainfall–streamflow dynamics at catchment-scale. Although the author was not familiar with the catchments, reasonably good model-fits were obtained in terms of D and, additionally in the case of Y5615030, a match between low-flow portions of flow duration curves for observed and modelled flows. Knowledge about the catchments and their hydrometry could have been useful; it is, of course, recommended that such information is used whenever possible.

While some modellers consider that different model structures, or different parameter sets for the same model structure, are required to model low and high flows at a given site, work presented here and elsewhere suggests that that is not always the case at daily time-step, provided that the very lowest or highest flows are not the main focus of interest. For a given model structure, the parameter calibration and selection procedure can be crucially important, especially when the modelling objective is to characterize catchment rainfall–streamflow behaviour, e.g. for model regionalization via relationships between model parameters and physical catchment descriptors to allow estimation of flow at ungauged sites from rainfall and air temperature (Littlewood *et al.*, 2002).

Rainfall–streamflow model parameters selected solely on the basis of D (the well-known Nash-Sutcliffe efficiency) are not good enough, especially for helping to devise regionalization schemes for predicting flow in ungauged catchments. This paper has demonstrated that a multi-objective modelling approach is essential for such work. Given the limitations of PC-IHACRES as a package, the additional use of flow duration curves, as described in the paper and elsewhere (Littlewood, 2002, 2003), requires considerable manual effort combined with assessment of flow duration curves "by eye". This hybrid, automatic-manual approach to parameter selection has the strong positive feature that it keeps the hydrologist closely involved in the modelling exercise, "*combining the strengths of manual and automatic methods*" (Boyle *et al.*, 2000). In order to facilitate the calibration of models for hundreds, or even thousands, of catchments, as required for systematic regionalization, it will be necessary to automate suitable multi-objective parameter selection procedures. Future rainfall–streamflow modelling software could incorporate the matching of flow duration curves for observed and modelled flows as an additional objective. While this is a useful goal, and probably will happen, manual aspects of model calibration should never be under-valued.

Acknowledgements The daily time-step catchment rainfall, streamflow and air temperature data used in this paper were prepared by Météo-France and Cemagref.

REFERENCES

Boyle, D. P., Gupta, H. V. & Sorooshian, S. (2000) Toward improved calibration of hydrologic models: combining the strengths of manual and automatic methods. *Water Resour. Res.* **36**, 3663–3674.

Jakeman, A. J, Littlewood, I. G. & Whitehead, P. G. (1990) Computation of the instantaneous unit hydrograph and identifiable component flows with application to two small upland catchments. *J. Hydrol.* **117**, 275–300.

Jakeman, A. J., Hornberger, G. M., Littlewood, I. G., Whitehead, P. G., Harvey, J. W. & Bencala, K. E. (1992) A systematic approach to modelling the dynamic linkage of climate, physical catchment descriptors and hydrologic response components. *Mathematics Computers in Simulation* **33**, 359–366.

Jakeman, A. J. & Hornberger, G. M. (1993) How much complexity is warranted in a rainfall-runoff model? *Water Resour. Res.* **29**(8), 2637–2649.

Kokkonen, T. S., Jakeman, A. J., Young, P. C. & Koivusalo, H. J. (2003) Predicting daily flows in ungauged catchments: model regionalization from catchment descriptors at the Coweeta Hydrologic Laboratory, North Carolina. *Hydrol. Processes* **17**, 2219–2238.

Littlewood, I. G. (2002) Improved unit hydrograph characterization of the daily flow regime (including low flows) of the River Teifi, Wales: towards better rainfall–streamflow models for regionalisation. *Hydrol. Earth System Sci.* **6**(5), 899–911.

Littlewood, I. G. (2003) Improved unit hydrograph identification for seven Welsh rivers: implications for estimating continuous streamflow at ungauged sites. *Hydrol. Sci. J.* **48**(5), 743–762.

Littlewood, I. G. & Jakeman, A. J. (1992) Characterisation of quick and slow streamflow components by unit hydrographs for single- and multi-basin studies. In: *Proc. Fourth General Assembly of the European Network of Experimental and Representative Basins* (ed. by M. Robinson) (Oxford, 29 September–2 October), published as Institute of Hydrology Report 120. Wallingford, UK.

Littlewood, I. G. & Jakeman A. J. (1994) A new method of rainfall–runoff modelling and its application in catchment hydrology. In: *Environmental Modelling,* Volume II (ed. by P Zanetti), 143–171. Computational Mechanics Publications, Southampton, UK.

Littlewood, I. G., Down, K., Parker, J. R. & Post, D. A. (1997) *The PC version of IHACRES for catchment-scale rainfall–streamflow modelling: User Guide*. Institute of Hydrology Software report. Wallingford, UK.

Littlewood, I. G., Jakeman, A. J., Croke, B. F. W., Kokkonen, T. S. & Post, D. A. (2002) Unit hydrograph characterisation of flow regimes leading to streamflow estimation in ungauged catchments (regionalisation). In: *PUB Communications* (ed. by P. Hubert, D. Schertzer, T. Takeuchi & S. Koide), (Brasilia, 20–22 November 2002). Also available at http://www.cig.ensmp.fr/~iahs/index.html.

Post, D. A. & Jakeman, A. J. (1996) Relationships between physical descriptors and hydrologic response characteristics in small Australian mountain ash catchments. *Hydrol. Processes* **10**, 877–892.

Post, D. A. & Jakeman, A. J. (1999) Predicting the daily streamflow of ungauged catchments in S. E. Australia by regionalising the parameters of a lumped conceptual rainfall–runoff model. *Ecol. Modelling* **123**, 91–104.

Sefton, C. E. M. & Howarth, S. M. (1998) Relationships between dynamic response characteristics and physical descriptors of catchments in England and Wales. *J. Hydrol.* **211**, 1–16.

Wagener, T., Wheater, H. S. & Gupta, H. V. (2004) *Rainfall-Runoff Modelling in Gauged and Ungauged Catchments.* Imperial College Press, London, UK.

Large Sample Basin Experiments for Hydrological Model Parameterization: Results of the Model Parameter Experiment–MOPEX. IAHS Publ. 307, 2006.

How informative is land-cover for the regionalization of the GR4J rainfall–runoff model? Lessons of a downward approach

LUDOVIC OUDIN[1], VAZKEN ANDRÉASSIAN[2], CÉCILE LOUMAGNE[2] & CLAUDE MICHEL[2]

1 *Université Pierre-et-Marie Curie, UMR 7619 Sisyphe, Case 105, 4 Place Jussieu, 75252 Paris cedex 5, France*
ludovic.oudin@ccr.jussieu.fr

2 *Cemagref, Hydrology and Water Quality Research Unit, PB 44, 92163 Antony cedex, France*

Abstract Prediction on ungauged basins is a big issue for operational hydrology and a challenge for scientists. For regionalization objectives, the first step of the development of *a priori* parameter estimation is to look for correlations between calibrated parameters and physical catchment descriptors. This article investigates the relationships between the parameters of the GR4J rainfall–runoff model and catchment vegetation characteristics over a large sample of 221 French catchments. First, the possible links between the calibrated parameters of the GR4J model and catchment vegetation types are investigated by linear regression. Then, we try to improve these relationships by introducing a more detailed description of the evapotranspiration process, explicitly taking into account vegetation types, following a downward approach. Results show that the GR4J model parameters cannot be determined directly from vegetation characteristics, and that the situation is not improved by a more detailed approach to evapotranspiration modelling.

Key words downward approach; evapotranspiration; land use; rainfall–runoff model; regionalization

INTRODUCTION

Vegetation type is one of the most often cited driving variables of catchment behaviour. This consensus has its roots in the numerous studies implemented by forest hydrologists on small catchments during the 20th century. Several reviews are available on this topic: see for example, Bosch & Hewlett (1982) or Andréassian (2004). Mainly based on paired catchment experiments, these studies consisted of deforestation and reforestation experiments, by which it was possible to demonstrate without doubt that forest cover could have an important role in the water balance at the catchment scale. However, the fact that vegetation has a role in a water cycle does not necessarily imply that vegetation is informative for regionalization objectives. Indeed, to use land cover for regionalization applications, we must be able: (i) to quantify its impact on the water cycle at the catchment scale; and (ii) to isolate its impact from other linked factors: soils, climate.

At first sight, the most rational approach to explicitly introduce vegetation characteristics into hydrological models is to use a mechanistic approach, with a physically-based model whose parameters are directly linked with vegetation types. Several large-scale experiments have supported the development of Soil-Vegetation-

Atmosphere Transfer (SVAT) schemes that describe the vertical fluxes of water and energy on the Earth's surface and try to explicitly take into account vegetation characteristics). However, the mechanistic approaches present two main drawbacks: they are often over parameterized (Franks & Beven, 1997; Avissar, 1998) and they suffer from process scaling problems (Grayson *et al.*, 1992).

The alternative approach proposed in this article is to start with a simple (i.e. parsimonious) rainfall–runoff model and try to introduce step-by-step more complexity to account for vegetation types: modifications would be accepted only if they improve model efficiency in terms of the output simulations (the streamflow simulations in our case). This approach is known as the downward approach (Klemeš, 1983; Sivapalan *et al.*, 2003), which attempts to predict the catchment behaviour by an interpretation of the observed response at the catchment scale. Specifically, we started with a simple representation of the catchment behaviour, and made the representation more complex only in response to improved results or improved ease of regionalization.

In this article, we tested this methodology by using data from 221 French catchments and the GR4J rainfall–runoff model, the possible links between GR4J calibrated parameters and catchment vegetation-type are investigated. Then, we try to improve these relations by introducing a more detailed description of the evaporation process, in order to explicitly take into account vegetation-types.

DATA AND METHODS

Catchment sample

The 221 French catchments are part of the sample used by Oudin *et al.* (2004) to discuss the use of potential evapotranspiration (PE) in rainfall–runoff modelling. Although France has a mainly temperate climate, its climate conditions are varied in this sample: Mediterranean conditions in the south of France, oceanic influences in the west and some continental features in the eastern part of the country. Basin sizes range from small (5.2 km^2) to medium (9387 km^2), with a median size of 88 km^2. Mean annual PE varies between 690 and 1340 mm, mean annual rainfall (P) between 620 and 1940 mm and mean annual streamflow between 23 and 1740 mm. The aridity index ($\overline{PE}/\overline{P}$) varies from 0.39 to 1.93 and the runoff coefficient from 0.03 to 1.05 (the value 1.05 corresponding to a karstic system).

Using a GIS, Plantier (2003) extracted the dominant cover types from the CORINE land cover classification (CEC, 1993). This classification relies on a 250 m grid resolution and was made by visual interpretation of high-resolution satellite images, e.g. Landsat-TM and SPOT-XS, at a 1:100 000 scale. Only two dominant classes were considered: forested and arable areas (others, like urban area or lake were very scarce over the catchments). The repartition of forested and arable catchments is quite symmetric: around 20% of the catchments are covered by more than 80% of forested land and symmetrically 20% of the catchments are covered by more than 80% of arable land. Hereafter, we will present the results in terms of percentage of forested area, since results for arable areas are symmetric.

The GR4J rainfall–runoff model

When trying to identify relationships between model parameters and physical catchment characteristics, it appears essential to use a parsimonious model. This avoids over parameterization, which is always detrimental to calibrated parameter precision. It seems obvious that a relationship between model parameters and catchment characteristics, if it exists, will be detectable using a parsimonious model. We used the GR4J model, a daily lumped rainfall–runoff model with only four parameters to calibrate, belonging to the family of soil moisture accounting models. A schematic diagram of the model and its parameters are shown in Fig. 1. For a detailed discussion of the model, see Perrin *et al.* (2003). The first two parameters regulate the water balance functions and the two others, the water transfer functions. These parameters are calibrated using a local search optimization algorithm described by Edijatno *et al.* (1999), with the Nash & Sutcliffe (1970) criterion (hereafter noted *NS*) used as an objective function.

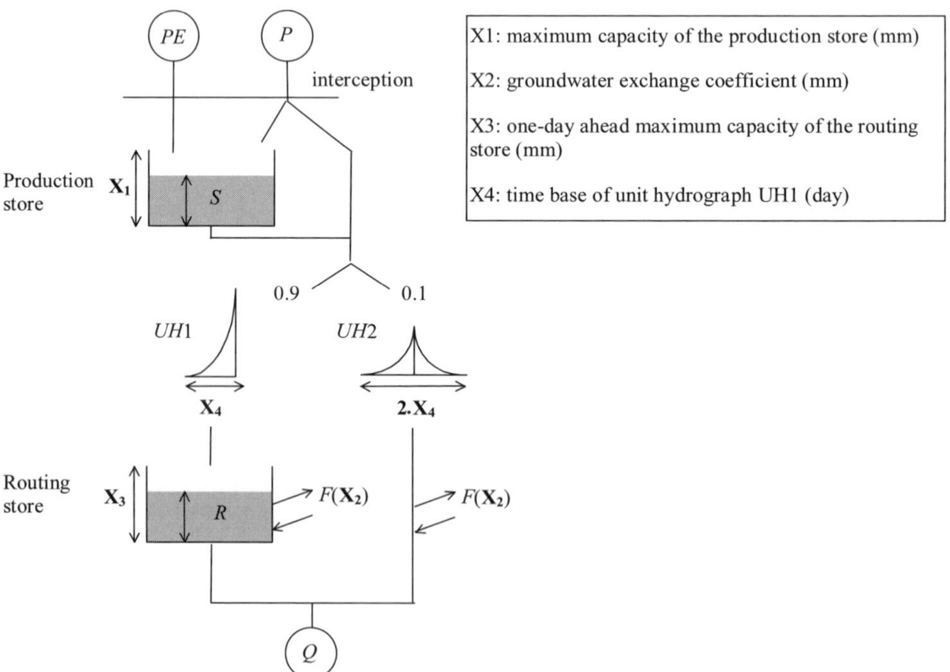

Fig. 1 Scheme of the GR4J rainfall–runoff model (*PE*, potential evapotranspiration; *P*, rainfall; *Q*, streamflow).

To assess the performance of the model, we used a split-sample test procedure (Klemeš, 1986): for each catchment, data time-series were split into two sub-periods. Then the model was calibrated on each sub-period and tested in validation mode on the other sub-periods. Two criteria were used to assess model efficiency on the validation periods. The first one is the standard Nash and Sutcliffe criterion:

$$NS = 100\left(1 - \frac{\sum_j \left(Q_{obs,j} - Q_{sim,j}\right)^2}{\sum_j \left(Q_{obs,j} - \overline{Q}\right)^2}\right) \tag{1}$$

where $Q_{obs,j}$ and $Q_{sim,j}$ are the observed and simulated streamflows on day j, and $\overline{Q}$ is the mean observed streamflow over the record period. The second criterion is based on the mean Cumulative Balance (CB) error of the model, written in relative terms (balance error) by:

$$CB(\%) = 100\left[1 - \left|1 - \frac{\sum_{i=1}^{n} Q_{sim,i}}{\sum_{i=1}^{n} Q_{obs,i}}\right|\right] \tag{2}$$

CB measures the ability of the model to correctly reproduce streamflow volumes over the studied period. Criterion CB is different from the first criterion in that it compensates for the errors at each time-step of the simulation.

A downward methodology to introduce vegetation descriptors into GR4J

The Penman-Monteith (Monteith, 1965) equation explicitly uses parameters linked with vegetation basin characteristics. Therefore, it is often considered as a first attempt to represent the soil–vegetation–atmosphere transfer and it remains the simplest SVAT scheme to implement. The Penman-Monteith formulation (with ET in m day^{-1}) can be written as follows:

$$ET = \frac{\Delta R_n + \gamma\left(e_a - e_s\right)\frac{\rho C_p}{r_a}}{\lambda\rho\left[\Delta + \gamma\left(1 + \frac{r_s}{r_a}\right)\right]} \tag{3}$$

where ET is the rate of evapotranspiration (in m day^{-1}), R_n is the net radiation (MJ m^{-2} day^{-1}), λ is the latent heat of vaporization (taken equal to 2.45 MJ kg^{-1}), ρ is the water density (1000 kg L^{-1}), e_a is the saturation vapour pressure (kPa) and e_s is the actual vapour pressure (kPa), C_p is the specific heat of the air, Δ is the slope of vapour pressure/temperature curve at equilibrium temperature (kPa °C^{-1}), γ is the psychrometric constant (taken equal to 6.6 10^{-2} kPa °C^{-1}), W is a wind speed function, r_s is the stomatal resistance and r_a is the aerodynamic resistance. Although the evapotranspiration processes may be too complex to be represented by these two resistances (Brutsaert, 1982), good correlations were obtained between modelled and measured evapotranspiration using this scheme (Allen *et al.*, 1998).

It is important to clarify a point that appears to be rather fuzzy in the literature: does the Penman-Monteith formulation refer to potential evapotranspiration, is it a formulation of reference crop evapotranspiration or a formulation of actual evapotranspiration? There are indeed multiple uses of this equation, depending on the formulation of the resistances (Wallace, 1995). For instance, the stomatal resistance may vary with the water content of the plant (Eagleson, 1978) and the vegetation-type, but in many cases, it is used as constant (equal to 69 s m^{-1}). Table 1 summarizes the rates of evapotranspiration computed with different formulations of r_s in the Penman-Monteith equation. At least, note that the original version of GR4J uses the stomatal resistance as a constant.

Table 1 Penman-Monteith equation and the corresponding rate of evapotranspiration.

Stomatal resistance	Data used for the Penman-Monteith equation			Computed rate of evapotranspiration
	Climatic data	Land use	Soil and vegetation water content	
$r_s^{RC}min$ = 69 sm^{-1}	✖			Reference crop potential evapotranspiration
$r_{s\,min}$	✖	✖		Potential evapotranspiration (surface dependent)
r_s	✖	✖	✖	Actual evapotranspiration

In order to get a more detailed representation of the evapotranspiration process, we used the Penman-Monteith equation as a formulation of potential evapotranspiration: the stomatal resistance will refer to the minimum of the stomatal resistance, which should depend on the land cover. Therefore, the term $r_{s\,min}$ is now determined by:

$$r_{s\ \min} = X_5 \tag{4}$$

where X_5 is an additional parameter to calibrate. As pointed out earlier, this formulation of r_s in the Penman-Monteith equation allows one to estimate a potential rate of evapotranspiration. Thus, the input PE (in m day^{-1}) is now computed by:

$$PE = \frac{\Delta R_n + \gamma(e_a - e_d)\dfrac{\rho C_p}{r_a}}{\lambda\rho\left[\Delta + \gamma\left(1 + \dfrac{X5}{r_a}\right)\right]} \tag{5}$$

where the aerodynamic resistance (s m^{-1}) is computed by as a function of the wind speed U (m s^{-1}):

$$r_a = \frac{208}{U} \tag{6}$$

This scheme was chosen because it is easy to implement in a soil moisture accounting rainfall–runoff model. Note that this implementation is fairly simple compared to other existing SVAT models. But, since GR4J's structure is initially parsimonious, we wished to propose a SVAT module of assorted parsimony. Other implementations were tested (Oudin, 2004), including more complex SVAT schemes such as the GRHUM rainfall–runoff model (Loumagne *et al.*, 1996). These investigations, not reported here, yield similar results to those presented hereafter.

Following the recommendations of the downward approach (Klemeš, 1983), two conditions are to be fulfilled to accept the modified structure:

(a) The first condition concerns the model efficiency in validation mode. Since the first purpose of an empirical rainfall–runoff model such as GR4J is to simulate streamflow, it is essential that the modifications do not degrade model performance, especially if additional parameters are added (Nash & Sutcliffe, 1970).
(b) If the two models yield similar performance, a second condition concerns the relationship between model parameters and observed vegetation types: a more detailed approach can only be justified if its additional parameters enable a physical interpretation.

RESULTS

Links between GR4J parameters and vegetation catchment characteristics

As a first step, we wanted to investigate the possible links between GR4J calibrated parameters and vegetation catchment characteristics. Fig. 2 compares the calibrated parameters and the percentage of forest cover: each point represents the parameter value averaged over the calibration periods and the error bars indicate the range of the calibrated parameters, large error bars meaning uncertain parameters values. There is no apparent relationship between catchment vegetation attributes and model parameters, even if "uncertain" parameter values are not considered. This is quite disconcerting since one would hope to find a relationship between vegetation attributes and, at least, the maximum capacity of the soil moisture accounting (SMA) store.

The absence of relationships between a catchment's vegetation descriptors and model parameters corroborates previous studies related to regionalization for a large number of catchments (e.g. Merz & Bloschl, 2004). There may be two reasons why finding relationships between vegetation characteristics and GR4J model parameters proves to be difficult:

(1) the lumped GR4J model may have a too crude representation of the evapotranspiration process to benefit from land cover information;
(2) vegetation may have only a marginal impact on catchment hydrological behaviour, and it may have served as an index for a second driving variable (such as soil) in the studies that have established a significant link.

As modellers, we focused on the first hypothesis. It is indeed possible that model parameter values hold some information from vegetation, but that the formulation of the model structure is inadequate and does not allow it to be revealed.

Would a more physically-oriented structure be more adequate?

To address this issue, we decided to compare the performance of the original GR4J model with those of a modified structure of GR4J, which presents a more detailed description of the evapotranspiration processes, involving explicit vegetation-related

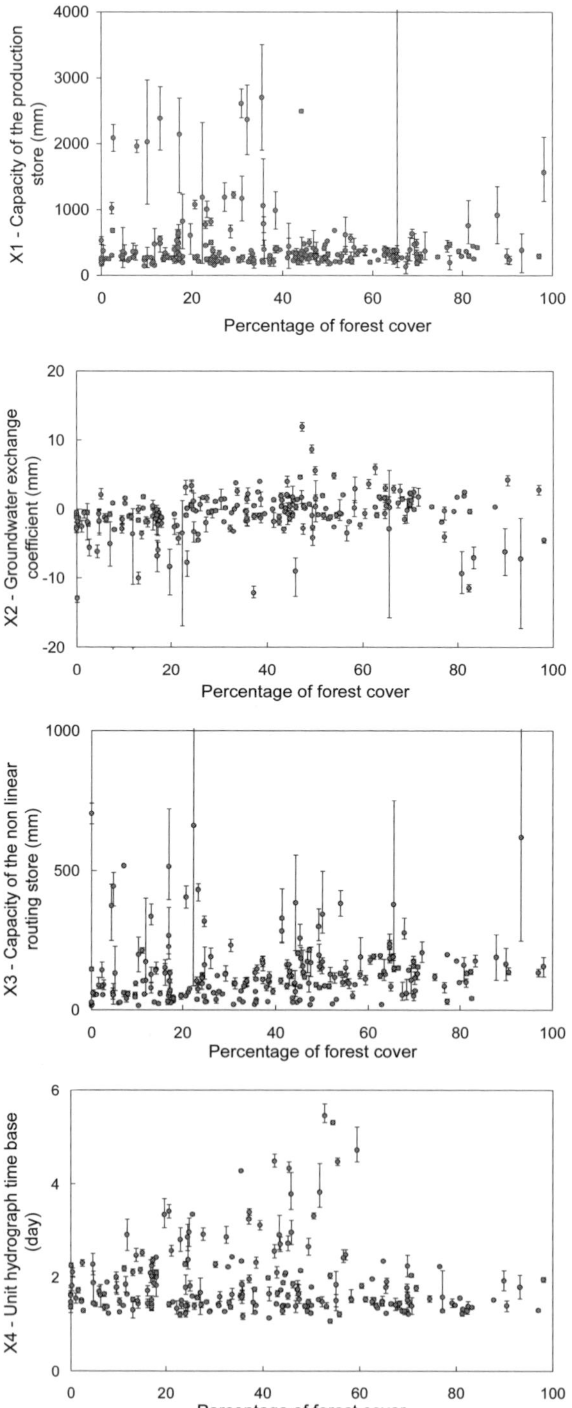

Fig. 2 Calibrated values of the GR4J model plotted against the percentage of forest cover. Error bars show the range of the calibrated parameters over one catchment.

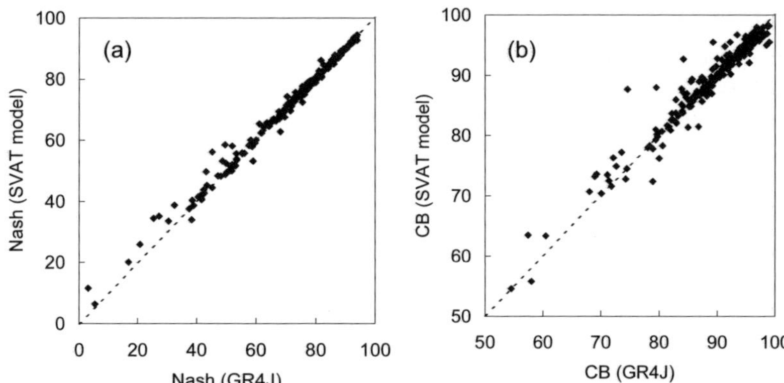

Fig. 3 Performance of the SVAT-oriented structure against the initial GR4J model, in terms of: (a) Nash and Sutcliffe criteria and (b) the water balance criterion.

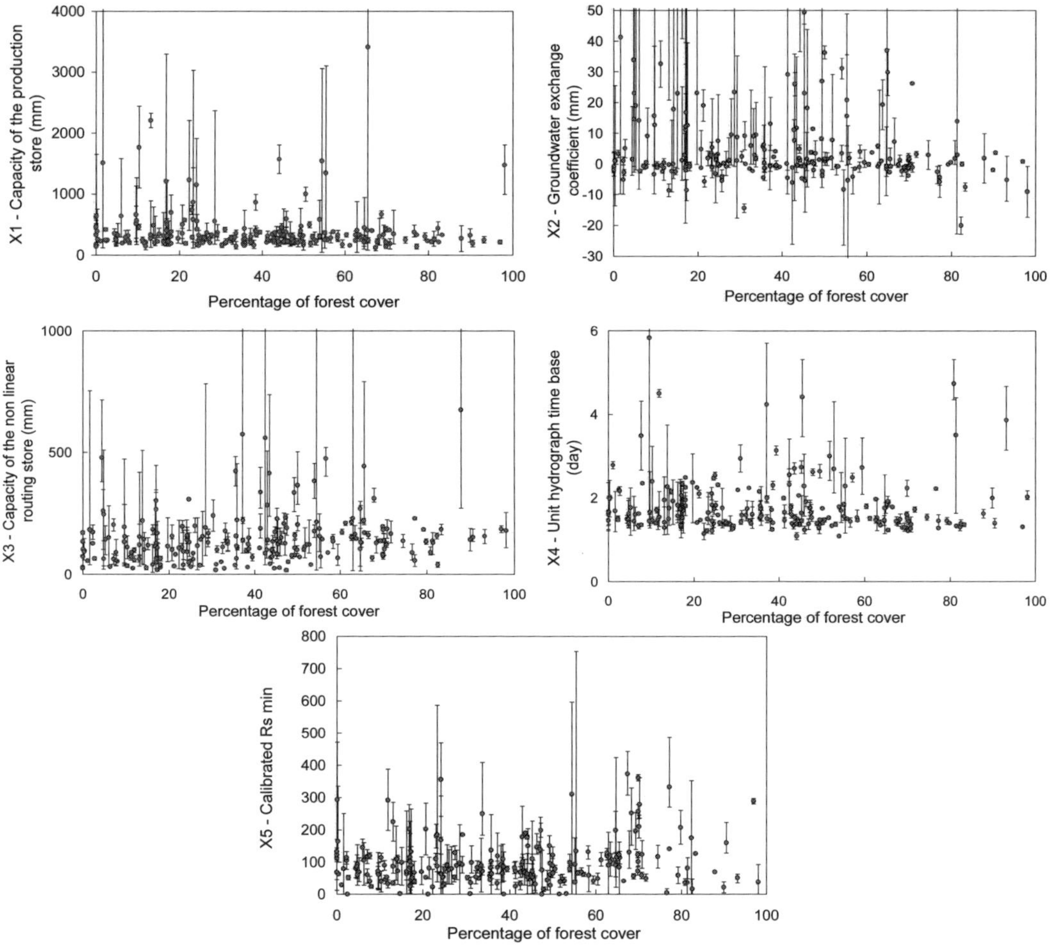

Fig. 4 Calibrated values of the SVAT-oriented model plotted against the percentage of forest cover. Error bars show the range of the calibrated parameters over one catchment.

parameters. Fig. 3 presents the Nash-Sutcliffe and water balance criteria obtained with the original rainfall–runoff model, and the modified structure over the 221 catchments in validation mode. Strikingly, the two structures perform very similarly and no systematic gain is obtained with the refined structure. These findings are supported by previous research, see e.g. Perrin *et al.* (2001) and Schulz & Beven (2003): increasing model complexity does not necessarily increase its performance.

Fig. 4 compares the calibrated parameters of the SVAT structure and the catchment attributes. In comparison with the relationships plotted in Fig. 2, the situation has not improved:

– The four initial GR4J parameters remain impossible to predict from the catchment vegetation attributes.
– The additional parameter does not appear to be related to vegetation type, while it is in theory related to the minimum value of the stomatal resistance. Besides, large error bars observed suggest that this parameter is not determined precisely, and increases uncertainty on the calibration of other parameters, particularly X2, which is related to groundwater exchanges modelling.

CONCLUSION

In this paper, we tried to establish relationships between the GR4J rainfall–runoff model parameters and catchment vegetation characteristics over a large and varied sample of 221 French catchments. We found no satisfying relationships, strengthening previous findings of similar large-scale experiments (Merz & Bloschl, 2004). Given the lack of such relationships, we modified the part of the model structure handling evapotranspiration. This new structure, based on the Penman-Monteith scheme, is more physically based and was expected to lead to more significant correlations between model parameters and catchment vegetation characteristics. However, this modification was unsuccessful since there was no improvement of the GR4J model performance in validation mode over the 221-catchment sample and the regionalization relationships were not improved.

One could argue that the model structure is still not satisfactory from a physical point of view. This may be a cause of our failure to find relationships between model parameters and catchment vegetation attributes. However, if such relationships exist, it should have been easy to detect them using a parsimonious model.

Acknowledgments Streamflow data on French catchments were provided by the HYDRO database of the French Ministry for the Environment. The authors are also grateful to Jean-Paul Goutorbe and Bruno Rambaldelli of Météo France for the French climatic data. Funding from the Programme National de Recherche en Hydrologie and Région Ile-de-France is also acknowledged.

REFERENCES

Allen, R. G., Smith, M., Perrier, A. & Pereira, L. S (1998) Crop evapotranspiration—Guidelines for computing crop water requirements. *Irrigation and Drainage Paper no. 56.* FAO, Rome, Italy.

Andreassian, V. (2004) Waters and forests: from historical controversy to scientific debate. *J. Hydrol.* **291**, 1–27.

Avissar, R. (1998) Which type of soil-vegetation-atmosphere transfer scheme is needed for general circulation models: A proposal for a higher-order scheme. *J. Hydrol.* **212–213**(1–4), 136–154.

Bosch, J. M. & Hewlett, J. D. (1982) A review of catchment experiments to determine the effect of vegetation changes on water yield and evapotranspiration. *J. Hydrol.* **55**, 3–23.

Brutsaert, W. (1982) *Evaporation into the Atmosphere*. Kluwer, Dordrecht, Germany.

CEC (1993) Corine Land Cover. Technical Guide. Office des publications officielles de communautés européennes.

Eagleson, P. S. (1978) Climate, soil and vegetation. 3. A simplified model of soil moisture movement in the liquid phase. *Water Resour. Res.* **14**, 722–730.

Edijatno, Nascimento, N., Yang, X., Makhlouf, Z. & Michel, C. (1999) GR3J: a daily watershed model with three free parameters. *Hydrol. Sci. J.* **44**(2), 263–278.

Franks, S. W. & Beven, K. J. (1997) Estimation of evapotranspiration at the landscape scale: a fuzzy disaggregation approach. *Water Resour. Res.* **33**(12), 2929–2938.

Grayson, R. B., Moore, I. D. & McMahon, T. A. (1992) Physically based hydrologic modeling. 2. Is the concept realistic? *Water Resour. Res.* **28**(10), 2659–2666.

Henderson-Sellers, A., McGuffie, K. & Pitman, A. J. (1996) The Project for Intercomparison of Land-surface Parametrization Schemes (PILPS); 1992 to 1995. *Climate Dyn.* **12**(12), 849–859.

Klemeš, V. (1983) Conceptualization and scale in hydrology. *J. Hydrol.* **65**(1–3), 1–23.

Klemeš, V. (1986) Operational testing of hydrologic simulation models. *Hydrol. Sci. J.* **31**(1), 13–24.

Loumagne, C., Chkir, N., Normand, M., Ottlé, C. & Vidal-Madjar, D. (1996) Introduction of soil/vegetation/atmosphere continuum in a conceptual rainfall–runoff model. *Hydrol. Sci. J.* **41**(6), 889–902.

Merz, R. & Bloschl, G. (2004) Regionalization of catchment model parameters. *J. Hydrol.* **287**(1–4), 95–123.

Monteith, J. L. (1965) Evaporation and the environment. In: *The State and Movement of Water in Living Organisms* (ed. by G. E. Fogg) (XIXth Symp.), 205–234. Cambridge University Press, Swansea, UK.

Nash, J. E. & Sutcliffe, J. V. (1970) River flow forecasting through conceptual models. Part I—a discussion of principles, *J. Hydrol.* **10**, 282–290.

Oudin, L. (2004) Looking for a relevant potential evapotranspiration model for lumped rainfall–runoff modelling. PhD Thesis, ENGREF, Paris, France (in French).

Oudin, L., Andréassian, V., Perrin, C. & Anctil, F. (2004) Locating the sources of low-pass behavior within rainfall–runoff models. *Water Resour. Res.* **40**(11), 101–114 doi:10.1029/2004WR003291.

Perrin, C., Michel, C. & Andréassian, V. (2001) Does a large number of parameters enhance model performance? Comparative assessment of common catchment model structures on 429 catchments. *J. Hydrol.* **242**, 275–301.

Perrin, C., Michel, C. & Andréassian, V. (2003) Improvement of a parsimonious model for streamflow simulation. *J. Hydrol.* **279** (1–4), 275–289.

Plantier, M. (2003) Prise en compte de caractéristiques physiques du bassin versant pour la comparaison des approches globale et semi-distribuée en modélisation pluie-débit, Rapport de DEA thesis, Université Louis Pasteur–ENGEES / Cemagref (Antony), Strasbourg, Switzerland.

Schulz, K. & Beven, K. (2003) Data-supported robust parameterisations in land surface-atmosphere flux predictions: towards a top-down approach. *Hydrol. Processes* **17**(11), 2259–2277.

Sivapalan, M., Blöschl, G. Zhang, L. & Vertessy, R. (2003) Downward approach to hydrological prediction. *Hydrol. Processes* **17**(11), 2101–2111.

Wallace, J. S. (1995) Calculating evaporation—resistance to factors. *Agric. Forest Meteorol.* **73**(3–4), 353–366.

Zhang, L., Dawes, W. R. & Walker, G. R. (2001) Response of mean annual evapotranspiration to vegetation changes at catchment scale. *Water Resour. Res.* **37**(3), 701–708.

Large Sample Basin Experiments for Hydrological Model Parameterization: Results of the Model Parameter Experiment–MOPEX. IAHS Publ. 307, 2006.

Simulation of streamflow by a regionalized lumped rainfall–runoff model over Luxembourg

B. HINGRAY[1], F. GUEX[1], D. GUEX[1], S. PUGIN[1], A. MUSY[1], L. PFISTER[2], A. IDRISSI[2], J. F. IFFLY[2] & L. HOFFMANN[2]

1 *Hydrology and Land Improvement Laboratory, Swiss Federal Institute of Technology (EPFL) CH 1015 Lausanne, Switzerland*
benoit.hingray@epfl.ch

2 *CREBS-Cellule de Recherche en Environnement et Biotechnologies, Centre de Recherche Public – Gabriel Lippmann, 162a Avenue de la Faïencerie, L-1511 Luxembourg, Luxembourg*

Abstract A parsimonious hydrological model was applied to simulate continuous discharge series on 17 gauged sub-catchments of the Alzette Basin in Luxembourg. The model has only three calibration parameters: the two first Model Parameters (MPs) are involved in the flow production process and in the baseflow reconstitution, while the third governs the quick flow component. The calibrated MP values obtained for nine calibration catchments were regionalized for the Alzette physiographic and hydrological area and regional equations were derived from a stepwise regression analysis linking these MP values to different Catchment Physical Characteristics (CPCs). To allow for a robust regionalized model, a manual two-step and bi-signal calibration procedure was applied to determine the optimal MP for each calibration catchment. The regionalization was next improved using the observed correlation between two parameters. The regional equations obtained produce both hydrological meaning and statistical significance. The regionalized model was evaluated on the eight gauged catchments that were not used for the development of the relationships between MPs and CPCs.

Key words flow simulation; parameter regionalization; rainfall–runoff; reservoir based model

INTRODUCTION

Long-term discharge series allow the derivation of relevant hydrological information such as flow duration curves or discharges quantiles. For ungauged catchments, such series can be reconstructed via continuous simulation thanks to regionalized hydrological models. The possibility of estimating the model parameters on the basis of external information obtained, for example, from gauged catchments is a significant hydrological issue (e.g. Abdulla & Lettenmaier (1997), Sefton & Howarth (1998) and Perrin (2000)). Regionalization works are mostly based on regression relationships developed between model parameters (MPs) and catchment physical characteristics (CPCs). These approaches may be severely limited, particularly as a result of too high a number of parameters or of the strong correlation that may exist between some of them.

The regionalization work presented here was carried out with a parsimonious conceptual hydrological model for 17 catchments of different sizes (18 km^2 to 1176 km^2) and geological substrates. These catchments are part of the Alzette basin, located in Luxembourg, and thus belong to a small and fairly homogenous region from a climatic, hydrological and physiographical point of view (Pfister *et al.*, 2000). The work was undertaken as part as the Pan-European FRHYMAP (Flood Risk scenario and Hydrological Mapping) project (Hoffmann, 2001).

Data

The Alzette basin is mainly situated in the Grand-Duchy of Luxembourg (Fig. 1). Data (time series as well as physiographic data) come from the hydro-climatological database built and validated by the Gabriel Lippmann Public Research Center in Luxembourg.

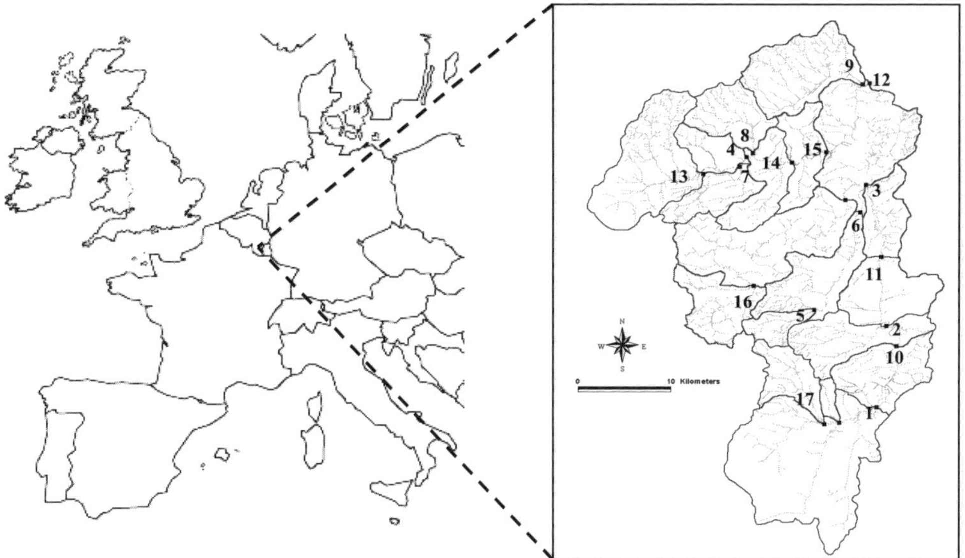

Fig. 1 Experimental catchments and gauging stations of the Alzette basin (Alzette outlet is 12.)

The area of the catchments ranges from 18 to 1175 km^2; the largest catchment includes all other catchments. The elevation of the catchments is quite homogenous as the average elevation ranges from 295 to 390 m a.m.s.l. The development of regional equations that link the optimal MP values for each experimental catchment to different CPCs is undertaken for a set of nine "calibration" sub-catchments of the Alzette basin. They are representative of the different physiographic and geological types observed on the whole Alzette basin. The other 8 sub-catchments of the Alzette basin are used for the validation of these relationships.

Twenty-four daily raingauges and five hourly raingauges located inside or in the proximity of the Alzette basin were used to compute an average hourly rainfall series for each catchment. Series of daily potential evapotranspiration (*PET*) were calculated using the *Penman-Monteith* relation (Monteith *et al.*, 1990) from climatic data series available at the Luxembourg-City Airport station for different types of land use regrouped in permanent grassland, cropland, forest and urban areas. For each catchment, an average daily *PET* series was next determined according to the surface area proportion of each land use in that catchment.

Hydrological model description

For the present application we use a simplified version of the SOCONT hydrological model developed for small alpine catchments (Schaefli, 2005; Schaefli *et al.*, 2005). This version is adapted for catchments with regimes not influenced by snow accumulation and melt processes. In this case, the model simulates a continuous hourly discharge series from only hourly rainfall and PET time series.

SOCONT is based on a deterministic, conceptual and "storage oriented" representation of the catchment hydrological behaviour. The model is expected to describe both the quick and slow components of the streamflow observed in the river. The total rainfall is thus divided after interception in infiltrated rainfall and net rainfall that supply the "soil reservoir" and the "quick flow reservoir", respectively. The model finally connects infiltrated and effective rainfall, as well as actual and potential evapotranspiration, through the filling rate of the soil reservoir, S/A, where A is the maximum storage capacity of the soil reservoir and S is the storage level, at time t. The baseflow discharge is linearly related to S. The net rainfall is routed through two non-linear reservoirs producing the fast component of flow.

The model has three parameters to be calibrated: A, the maximum storage capacity of the slow reservoir, and K and β the recession constants of the slow and rapid reservoirs, respectively.

Model calibration

The parameters optimization is a two-step and multi-signal manual procedure (Niggli *et al.*, 2001). This method, which is only achievable because of the limited number of parameters, guarantees, in the opinion of the authors, a robust calibration, contrary to automatic calibration algorithms which are subject to many numerical uncertainties. The robustness of the calibration procedure used here is a necessary condition for regionalization works (Abdulla & Lettenmaier, 1997).

The parameters that condition the simulated baseflow component (provided by the model's soil reservoir) are calibrated so that the simulated baseflow reproduces at best a reference baseflow. The reference baseflow discharge series is obtained from the observed discharges series using a numerical hydrograph separation algorithm, the Baseflow Index method described by Kaden (1993).

The classical Nash criterion (Nash & Suttclife, 1970) was used to evaluate the model's ability to reproduce this reference baseflow. As already highlighted by Niggli *et al.*, 2001, the only two parameters governing the baseflow, A and K, are correlated. A power function ($K = a.A^b$) provides a fairly good approximation of the relation between the different potentially optimal (A,K) sets (Fig. 2). Only a limited number of these couples provide an acceptable reconstitution of the simulated baseflow series. This acceptable range depends on the catchment. The exponent b of the A–K relationships is however more or less the same for all catchments (same slopes in the logarithmic plot in Fig. 2). Note that assuming a steady state of the slow reservoir behaviour (obtained for a constant rainfall intensity) and no losses via evapotranspiration, it is possible to analytically solve the ordinary differential equation governing the slow reservoir behaviour and to find the analytical expression of the baseflow discharge (equation (1)). This expression shows that the baseflow discharge only depends on a unique and global parameter $u = A.K$.

$$Q_{base}^* = u\left[\sqrt{1+\left(\frac{u}{2.P_{eff}}\right)^2} - \left(\frac{u}{2.P_{eff}}\right)\right].S_{bv} \qquad (1)$$

where Q_{base}^* is the baseflow discharge obtained for a steady state behaviour of the slow reservoir when filled with a constant effective recharge rate P_{eff} and S_{bv} is the catchment surface area. As the real baseflow discharge $Q_{base}(t)$ is rather constant in

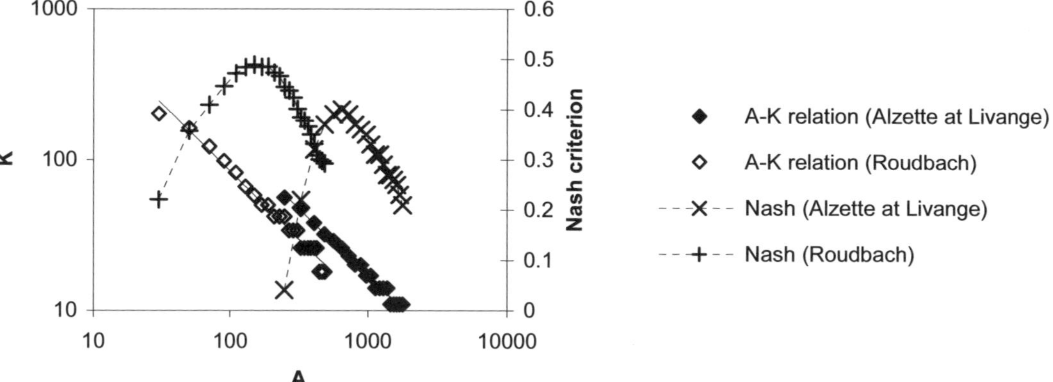

Fig. 2 log–log relation between *A* and K for two catchments and Nash criteria corresponding to each *A-K* couple.

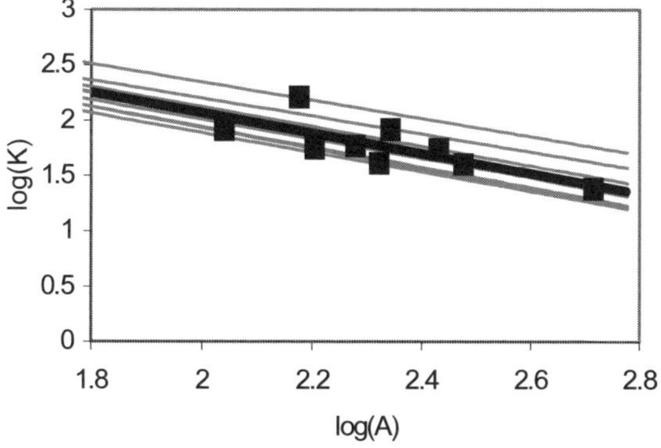

Fig. 3 the ten log-log linear relations obtained between *A* and *K* for the different calibration catchments. Bold line is the regional (*K,A*) relation, bold squares are the calibrated *A–K* couples.

time (when compared to the total discharge variations), the variable Q_{base}* corresponds in a first approximation to the mean baseflow value obtained with the mean effective rainfall intensity P_{eff} estimated for a N years observation period. Using the slow reservoir to reproduce the baseflow series or, more roughly, a given *BFI* value thus constrains the value of the *u* parameter and next the *A.K* value.

The second step of the calibration procedure is to determine the best parameters *A*, *K* and β for the reconstitution of the total flow. The *A* and *K* values are here forced to follow the at-site *K*(*A*) relationship previously calibrated. The optimal parameters set (*A*, *K*(*A*), β) is identified for each studied catchment via the optimization of the objective function *C*4 estimated on the total discharges:

$$C4 = 1 - \frac{\sum_i \left(Qobs_i + \overline{Qobs}\right)\left(Qsim_i - Qobs_i\right)^2}{\sum_i \left(Qobs_i + \overline{Qobs}\right)\left(\overline{Qobs} - Qobs_i\right)^2} \qquad (2)$$

Criterion *C*4 was chosen because it gives more weight to high flows than the classical Nash-Sutcliffe criterion. Its optimal value is 1. For the nine calibration catchments, the *C*4 criterion value varies for the calibration period (1997–1998) from 0.71 to 0.86 and for the validation period (1999–2000) from 0.70 to 0.84.

Regionalization

The catchments are described by a number of physical descriptors relative to their morphology (such as the Gravelius coefficient, the elongation and relief factors, the maximal river network length and the maximal drainage density), to their geology (and especially the percentage of impervious substrate (marl, schist, clay or silt)) and to the land use (such as proportion of urban areas, cropland, forest, grassland and areas dedicated to mining extraction). No climatic characteristics were used because the hydro-climatologic environment is homogenous over the whole study area (Pfister, 2000). A preliminary analysis investigated the correlations between CPCs in order to identify possible interdependencies and to produce subsets of independent CPCs. For each of the three model parameters *A*, *K* and β, an automatic stepwise regression procedure (Draper & Smith, 1981) was next used to determine the most significant explanatory variables.

For parameter β (recession coefficient of the rapid reservoir), only variables relative to the catchment size (such as catchment area *S*, catchment perimeter *P*, river network length *LRES*) were significant at 5% level in the regression models. For the best regression, using the catchment surface area (*S*); the coefficient of determination value (R^2) calculated between calibrated and regionalized β values reaches 0.96 (equation (3)).

For the parameter *A* (maximum storage capacity of the slow reservoir), different regression models with similar performance were obtained. The two explanatory variables retained are the global slope index *IG* and the ratio of impervious substrates %*IMP*. The selected variables were significant at 5% level and the R^2 value calculated between calibrated and regionalized *A* values is 0.67 (equation (4)). Note that both variables can have a physical relationship with *A*. With a higher proportion of impervious substrates the catchment water storage capacity is lower, and the higher the slope the lower is the substrate depth and thus the storage capacity.

No explanatory variable was found to be sufficiently relevant for the *K* parameter (recession coefficient, slow reservoir). An estimate of *K* was obtained by the regionalization of the relations $K = a.A^b$ previously highlighted between *A* and *K* for each catchment. A least square adjustment was therefore made in a log-log diagram for *A*–*K* couples previously calibrated for each catchment (Fig. 2). The regional model for *K* estimation is then derived from this regional power function (equation (5)) and from the regional regression expression previously identified for *A*. The resulting regional model is rather efficient: the R^2 value reaches 0.60. In order to improve the *K* regional model, an attempt was made to regionalize the residuals of this relation. The improvement was not significant and the explanatory variables of the residuals were not relevant.

The three regional equations above were used to calculate the parameters for the nine calibration catchments and for the eight validation catchments. Figure 4 compares the regionalized (using the regional equations) and the calibrated parameters, for both calibration and validation catchments. The R^2 values for the validation catchments are 0.97, 0.58 and 0.35 for β, *A* and *K*, respectively. This confirms that the regionalized equation for β is the most successful whilst the equations determined for *A* and *K* are weaker, but nevertheless acceptable. These results correspond to those obtained from similar previous studies (Edijatno, 1991; Makhlouf, 1994): the most satisfactory regional equations also concerned the streamflow routing parameter, relating this parameter to the catchment area; it was however more difficult to develop significant equations for the water balance parameters.

Model Parameter	Regional equation	R^2	Equation
Recession coefficient, rapid reservoir	$\beta = 960.S^{0.73}$	$R^2 = 0.96$	(3)
Maximum storage capacity, slow reservoir	$A = 71600 \,.\, IG^{-0.38} \,.\, \%IMP^{-1.25}$	$R^2 = 0.67$	(4)
Recession coefficient, slow reservoir	$K = 8090 \,.\, A^{-0.92}$	$R^2 = 0.60$	(5)

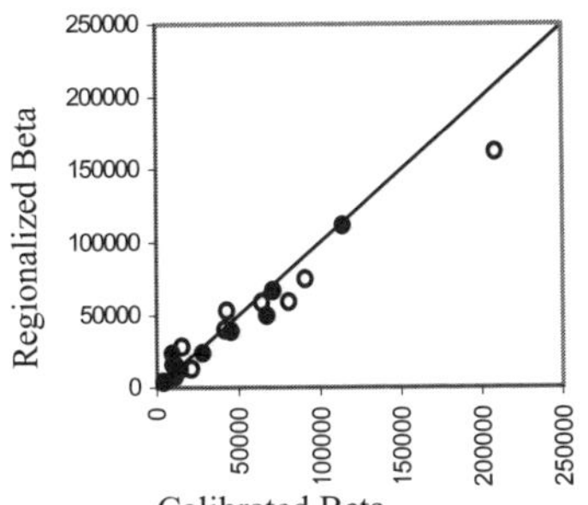

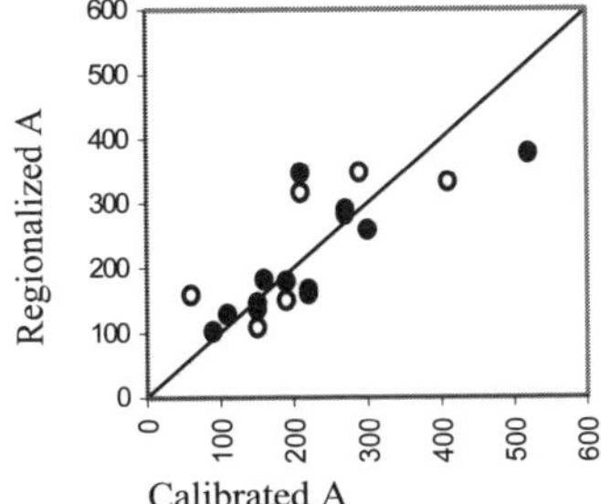

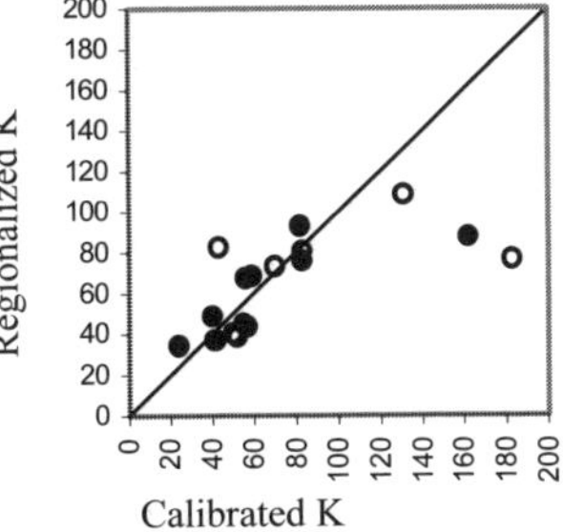

Fig. 4 Scatterplots showing the relationships between calibrated and regionalized parameters for the calibration catchments (full circles) and the validation catchments (empty circles).

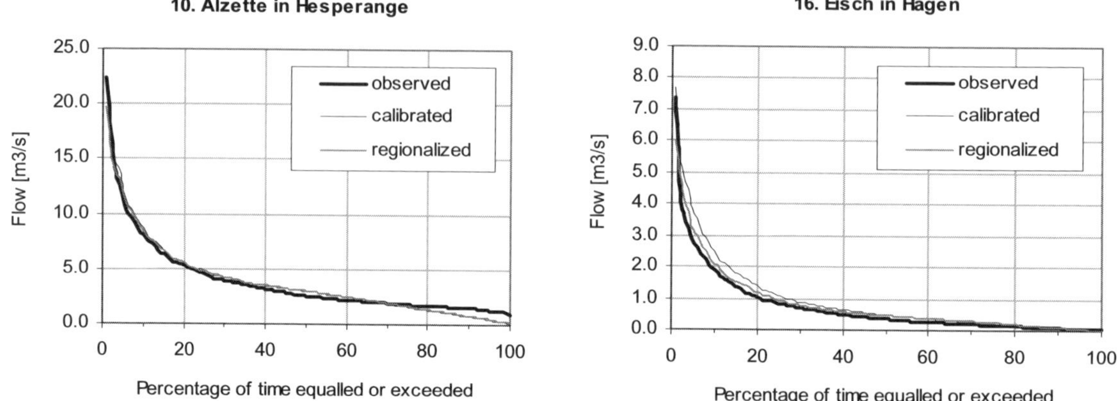

Fig. 5 Flow duration curves of observed, calibrated and regionalized flows for two of the validation catchments.

For four of the eight validation basins the *C*4 criterion values obtained for the total discharge with the regionalized model are equivalent to or higher than values obtained when the model is calibrated on the catchment. The *C*4 average value is 0.74 for the regionalized model and 0.81 when the model is calibrated, respectively. The flow duration curves of the observed, calibrated and regionalized flows (Fig. 5) for the validation catchments are very well reproduced for almost all catchments. The values of Nash efficiency coefficient calculated between simulated and observed daily flows from these curves are all above 0.95.

CONCLUSION

A parsimonious hydrological model was applied to simulate continuous hourly discharge series on several catchments of the Alzette Basin-Luxembourg characterized by various land use and lithological characteristics. The calibrated Model Parameters obtained for nine calibration sub-catchments were regionalized for the Alzette context and different regional equations were derived from a stepwise regression analysis linking these model parameters values to different Catchment Physical Characteristics.

The model, in its calibrated form as well as in its regionalized form, provides quite good results for the reproduction of the hourly discharges. The mean interannual discharges, the annual variability of hourly discharges (not presented) as well as the flow duration curves computed on daily discharges are very well reproduced.

The model requires only few data. It is thus possible to perform a very simple and robust manual calibration of its three parameters. Moreover, the regionalization work, if needed, may also produce, thanks to the relation highlighted between two of the three parameters, robust regional relationships between model parameters and catchment physical characteristics with both hydrological meaning and statistical significance. Such a simple rainfall–runoff model can thus provide an interesting simulation tool for practitioners that have to determine some important characteristics of flows for gauged and ungauged catchments.

Acknowledgments This work was carried out as part of the Pan-European research project FRHYMAP belonging to the IRMA-SPONGE European Program. It was founded both by the European Community and the Swiss National Fund for Research.

REFERENCES

Abdulla, F. A. & Lettenmaier, D. P. (1997) Development of regional parameter estimation equations for a macroscale hydrologic model. *J. Hydrol.* **197**, 230–257.

Draper, N.R. & Smith, H. (1981) *Applied Regression Analysis*, second edn. John Wiley & Sons, New York, USA.

Edijatno (1999) Mise au point d'un modèle élémentaire pluie-débit au pas de temps journalier. Thèse de doctorat, Université Louis Pasteur/ENGEES, Strasbourg, France.

Hoffmann, L. (2001) FRHYMAP: Flood Risk and Hydrological Mapping. Umbrella Program IRMA-SPONGE, NCR Publication (04/2001), 23–25.

Kaden, U. (1993) *Etude de la séparation des écoulements sur différents bassins de l'Alsace. Application et automatisation de la technique du Baseflow Index "BFI"*, Mémoire de maîtrise, Université Louis Pasteur-UFR de Géographie, Strasbourg, France.

Makhlouf, Z. (1994) Compléments sur le modèle pluie-débit GR4J et essai d'estimation de ses parameters, Thèse de doctorat, Université Paris XI Orsay, France.

Monteith, J. L. & Unsworth, M. (1990) *Principles of Environmental Physics*, second edn. Arnold, UK.

Nash, J. E. & Sutcliffe, J. V. (1970) River flow forecasting through conceptual models. Part I, a discussion of principles. *J. Hydrol.* **10**(3), 282–290.

Niggli, M., Hingray, B. & Musy, A. (2001). *WRINCLE—Water Resource: INfluence of CLimate change in Europe. Methodology for producing runoff maps and assessing the influence of climate change in Europe.* EC framework IV Project ENV 4970452.

Perrin, C. (2000) Vers une amélioration d'un modèle global pluie-débit au travers d'une approche comparative. Thèse de doctorat, Institut National Polytechnique de Grenoble, CEMAGREF, France.

Pfister, L. (2000) Analyse spatio-temporelle du fonctionnement hydro-climatologique du bassin versant de l'Alzette (Grand-Duché du Luxembourg). Détection de facteurs climatiques, anthropiques et physiogéographiques générateurs de crues et d'inondations. PhD thesis, Université Louis Pasteur, Strasbourg, Centre de Recherche Public—Gabriel Lippmann, Luxembourg.

Pfister, L., Iffly, J. -F., El Idrissi, A. & Hoffmann, L. (2000) Vérification de l'homogénéité physiogéographique, hydrologique et climatique du basin de l'Alzette (Grand-Duché de Luxembourg) en vue d'opérations de régionalisation, Archives de l'Institut grand-ducal du Luxembourg. *Sect. Sci. Nat. Phys. Math.* NS **43**, 239–253.

Schaefli, B. (2005) Quantification of modeling uncertainties in climate change impact studies on water resources: application to a glacier-fed hydropower production system in the Swiss Alps. PhD Thesis N°3225. EPFL, Lausanne. (http://library.epfl.ch/theses).

Schaefli, B., Hingray, B., Niggli, M. & Musy, A. (2005) A conceptual glacio-hydrological model for high mountainous catchments. *Hydrol. Earth Syst. Sci.* **9**, 95–109.

Sefton, C. E. M. & Howarth, S. M. (1998) Relationships between dynamic response characteristics and physical descriptors of catchments in England and Wales. *J. Hydrol.* **211**, 1–16.

Large Sample Basin Experiments for Hydrological Model Parameterization: Results of the Model Parameter Experiment–MOPEX. IAHS Publ. 307, 2006.

Regionalization of a monthly rainfall–runoff model for the southern half of France based on a sample of 880 gauged catchments

NATHALIE FOLTON & JACQUES LAVABRE
Cemagref, BP 31, 13612 Aix-en-Provence, France
nathalie.folton@cemagref.fr

Abstract The aim of this study was to provide a means of determining reference low flows at any point in the hydrographic network in the southern half of France, using a continuous monthly rainfall–runoff modelling approach. The model adopted here has two parameters that were calibrated on 880 catchments. The accuracy of the calibrated model was excellent for assessing the mean annual discharge, and limited errors were observed in predicting low flows. A regionalization procedure was performed for the two parameters, using an upward approach and adopting a grid of set values based on environmental data. Based on comparisons with optimum parameters, a regional map of corrected parameters was drawn up. This model satisfactorily simulates mean annual discharge, and the level of uncertainty observed on low flow values is acceptable, given the difficulty in estimating extremely low flows.

Key words low flows; rainfall–runoff model; regionalization

INTRODUCTION AND SCOPE

The French Water Law of 1992, enforced by decree no. 93-742 in March 1993, sets the 5-year return period mean monthly flow (QMNA5) as the standard legal basis to deliver authorizations for wastewater discharge and freshwater withdrawal in rivers. The minimum discharge of rivers downstream of dams is defined by the article L 232-5 of the French "Code Rural" (Law no. 84-512 of 2 June 1984) as a percentage of the mean annual discharge.

However, these legal provisions raise problems as to how to determine these reference flows at any point of the national hydrographic network. Of course, the discharge of some rivers has been recorded at gauging stations. However, in most cases flows are still unknown. The French Environment Agencies (DIREN) and State water services (MISE) responsible for implementation of the water policy are confronted by this lack of knowledge on flows. In the absence of any overall studies, agencies have to rely on their own empirical field knowledge to extrapolate data from the information collected by the available recording networks.

The development of streamflow models based on rainfall data may help to solve these problems. One of the advantages of this approach is its broad scope, since it is not restricted to a single hydrological variable. The regionalization must be made here on the parameters of the rainfall–runoff model. It provides a means to simulate historical flow time series at any point of the drainage network. From these simulated flow series, it is possible to deduce the hydrological variables of interest. The GR2M model (Makhlouf & Michel, 1994) was used here. It simulates the transformation of rainfall into runoff at a monthly time step. This model was tested on catchments ranging in size from a few km^2 to more than 1000 km^2. Its main advantage is its simplicity and the limited number of parameters to optimize (only two free parameters). An improved version of the model was proposed by Folton & Lavabre (2006) to cope with the uncertainties found on some mountainous catchments. A snowmelt module was introduced in the model, without adding new free parameters, to ease the regionalization procedure. All the runoff variability is therefore accounted for by only two parameters.

To apply the model to an ungauged catchment, it is necessary to first relate model parameters to the physical and/or climatic characteristics of the catchment. The method classically used to regionalize the model parameters consists of establishing correlations of this kind. The aim here is to identify the physical variables that affect the rainfall–runoff transformation, and then to look for a regression relationship between model parameters and physical characteristics. This regionalization approach has been widely used by hydrologists. Many authors have adopted it using various models under various climatic conditions: Servat & Dezeter (1992), Sefton & Howarth (1998), Seibert (1999), Post & Jakeman (1999), Perrin (2000) and Mwakalila (2003), to name but a few. Other authors such as Vandewiele & Atlabachew (1995) applied a spatial interpolation technique to the parameters. Fernandez & Vogel (2000) used simultaneous parametric calibration methods and expressions relating the parameters to physical descriptors. The success rates obtained with all these methods depend on the model used and the regional context, but the outcome was often not very satisfactory. The best results were often obtained in specific contexts. More recently, solutions based on neighbouring regions were proposed, especially for flood quantile estimation. To determine the similarity between two stations, physiographic and hydrological data are used in methods known as the “regions of influence” method developed by Burn (1990), and the canonical correlation analysis method (Ouarda *et al.*, 2001). Hydrological homogeneous regions are often identified for one or two flood quantiles, and the analysis has to be repeated to deal with other variables.

In the present study, a classical parametric regionalization approach was adopted. However, the method tested here also involved the use of an original upward procedure. A physical grid was set to assess model parameters. Then a regional grid based on the comparison with optimum calibration parameters was used to correct parameter values.

DATA AND MODEL

Study area

The study area includes 43 French administrative departments and is approximately 256 000 km^2. The contrasts of the climate in this region are extreme, ranging from the Mediterranean coastal climate to the high Alpine mountainous climate. The mean annual rainfall ranges from 500 to 2500 mm. This yields a strong hydrological variability:

- the mean annual discharge ranges from only a few L s^{-1} km^2 near the coastline to 70 L s^{-1} km^2 in some mountain catchments;
- the 5-year return period mean monthly flow (QMNA5) value ranges from 0 to more than 15 L s^{-1} km^2.

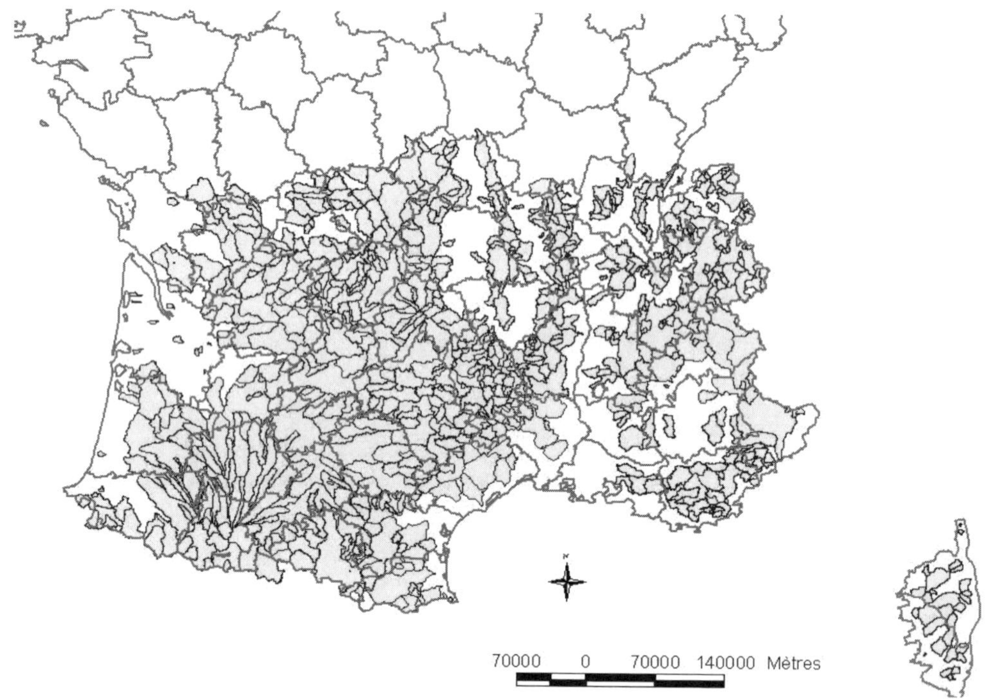

Fig. 1 Map of the study catchments.

All the available hydrological information was used in this study. Data on 880 catchments were used. 36% are smaller than 100 km^2, and 88% smaller than 1 000 km^2 (Fig. 1). Data collected at approximately 3200 rainfall recording stations and 68 potential evapotranspiration (PET) measuring stations were used.

We also used 1 km^2 pixel maps of mean monthly temperatures and mean monthly rainfall, which were drawn up using the AURHELY's method (Benichou & Le Breton, 1987). Temperature data were used to draw up a PET map (Folton & Lavabre, 2001, 2004). Topographic information was used to calculate catchment contours and derive monthly catchment rainfall maps (Folton & Lavabre, 2001).

A large variability of the mean annual discharge was observed over the study area. For more than 68% of the catchments, the mean annual runoff was below 750 mm, i.e. approximately 25 L s^{-1} km^2, but for a limited number of catchments (less than 10%), it was larger than 1250 mm, i.e., approximately 40 L s^{-1} km^2. Table 1 shows that mean annual discharge logically increases with mean annual rainfall.

Table 1 Distribution of catchments in classes of mean annual runoff and mean annual discharge. (scale: 1 mm = 0.0317 L s^{-1} km^2).

		Mean annual rainfall classes (mm)								
		0–250	250–500	500–750	750–1000	1000–1250	1250–1500	1500–1750	1750–2500	
Mean annual runoff classes (mm)	0–250	–	–	2.6%	12.9%	0.6%	0.1%	–	–	→ 68%
	250–500	–	–	0%	19.8%	12.6%	0.5%	–	–	
	500–750	–	–	–	1.2%	13.3%	4.7%	0.1%	0.2%	
	750–1000	–	–	–	0.4%	3.1%	5.6%	2.5%	0.1%	
	1000–1250	–	–	–	0.1%	1.7%	3.8%	4.7%	0.2%	
	1250–1500	–	–	–	0.1%	0.4%	0.7%	2.1%	1.0%	
	1500–1750	–	–	–	–	0.2%	0.4%	1.1%	1.5%	→ 9%
	1750–2500	–	–	–	–	–	–	0.4%	1.2%	

The GR2M rainfall–runoff model

The GR2M model used here is a monthly lumped rainfall–runoff model. Its structure includes two reservoirs. Figure 2 shows a diagram of the model. To account for snow or ice influences in mountainous catchments, a monthly snowmelt module was added. Each of the two reservoirs included in the model depends on a parameter.

XV_1 is the parameter of the production function. This parameter governs the catchment water balance via a multiplication factor applied to the monthly discharges. In the model structure, this parameter corrects only the output flow. XV_1 is positive without upper bound. Values approaching 0 indicate a low yielding catchment, whereas values greater than 2 correspond to mountainous catchments, where the rainfall and snowfall inputs are generally underestimated by the gauging network.

XV_2 is the parameter of the transfer function. It determines the temporal distribution of output discharge and accounts for the time lag between rainfall and runoff values. It actually determines the part of direct runoff that does not enter the transfer store. The value of XV_2 ranges between 1 and 2. A catchment responds more quickly as XV_2 approaches 1. In fact, if XV_2 is equal to 1, all the net rainfall for a given month is directly transferred to the outlet during this month. In this case, it will not be stored in the transfer store. Such values are characteristic of highly responsive catchments. If XV_2 is equal to 2, all the net rainfall enters the transfer reservoirs. This occurs in the case of catchments showing a high level of inertia.

MODEL RESULTS

Model performance

The analysis of mean annual discharge and QMNA5 values simulated by the model indicates its ability to produce accurate reference low flow estimates (when calibrated).

The mean annual discharge values were predicted extremely accurately by the model (Fig. 3(a)). The coefficient of determination R^2 (coefficient of correlation measure), has a value of 99.8%. The simulation of the QMNA5 variable were less accurate, but still acceptable (Fig. 3(b)). The overall R^2 was 89.7%.

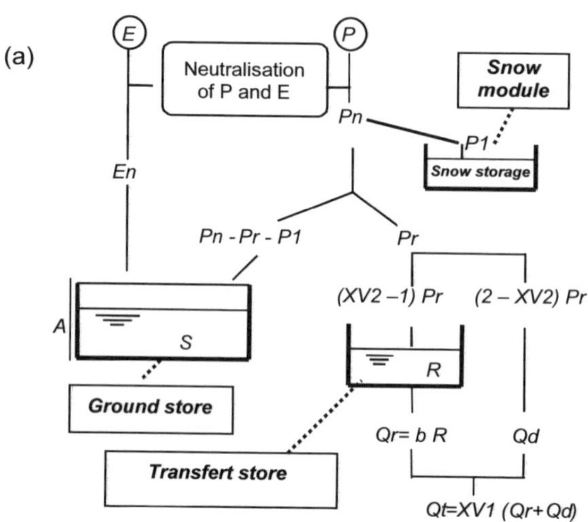

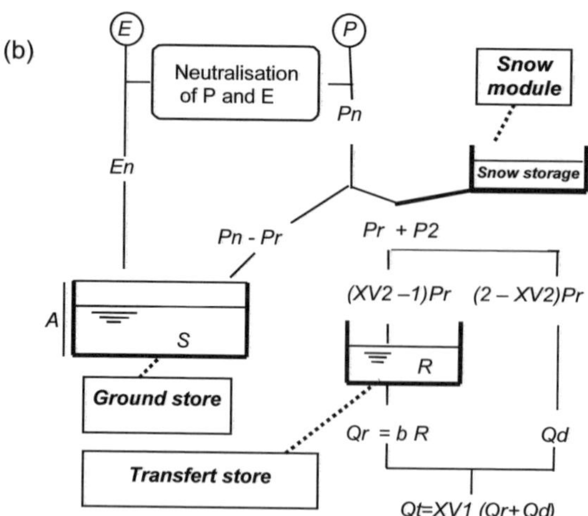

Fig. 2 Model structure with (a): storage of a part of net rainfall (*Pn*) in "snow module"; and (b): snowmelt. *P*, rainfall; *E*, evapotranspiration; *Pn*, net rainfall after evaporation; *Pr*, runoff-effective rainfall; *En*, net evapotranspiration; *A*, maximum capacity of the ground store; *S*, level of this store; Qd, direct flow; Qr, discharge from the transfert store with $b = 0.4$; *R*, level of this store; *Qt*, total flow; *XV1*, *XV2*: parameters.

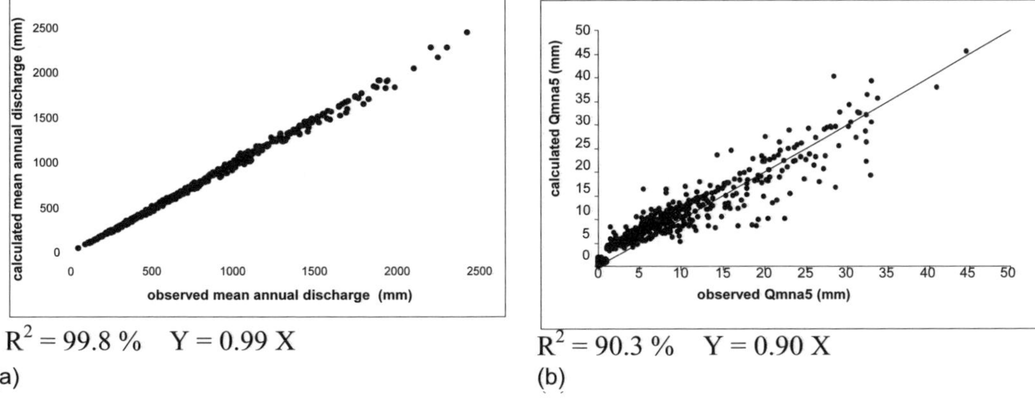

Fig. 3 Comparison of the observed and calculated values (using parameters calibrates for the GR2M model).

Parameter variability

No link could be established between the two model parameters and catchment area (the coefficients of determination were found to be not significant). There is also no significant correlation between model parameters and mean annual rainfall. The lack of significant correlations observed in both cases shows that model parameters mainly reflect the hydrological behaviour of the catchments. It is also important to note that the production and transfer parameters are independent, which guarantees better efficiency.

XV_1 is the production parameter. For similar rainfall, runoff is greater as the value of XV_1 increases. As shown in Fig. 4, for 75 % of the catchments, the value of XV_1 ranges between 0.75 and 1.25. For about 1% of the catchments, the value of this parameter was very low, since it was below 0.5. Conversely, for 8% of the catchments XV_1 showed high values.

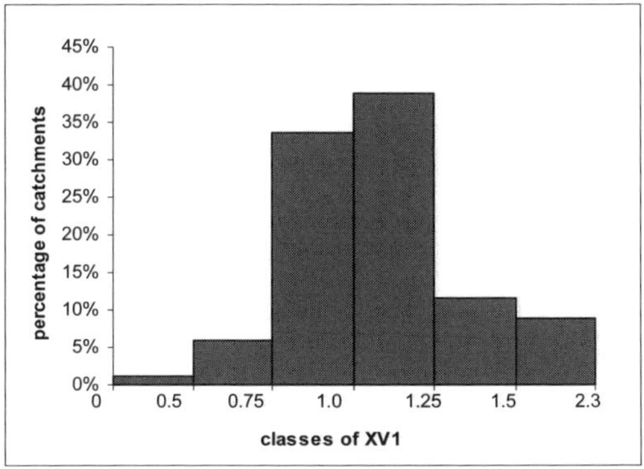

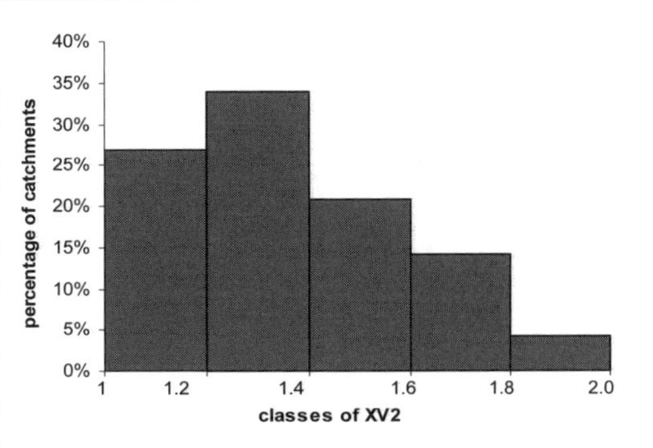

Fig. 4 Distribution of parameters XV_1 and XV_2 for the 880 catchments.

By its construction, the transfer parameter XV_2 ranges between 1 and 2. A value of around 1 reflects a highly reactive hydrological behaviour; whereas a value of around 2 means that all the rainfall is transferred by the routing reservoir and that the runoff will therefore be delayed. 55% of the catchments showed XV_2 values between 1.2 and 1.6, which corresponds to classical hydrological behaviour. Of the catchments (mountainous catchments), 4% had a greatly delayed runoff and 27%, an extremely fast runoff.

A parametric sensitivity analysis was performed to determine the influence of each parameter on the mean annual discharge and on the QMNA5. Figures 5 and 6 show the variation of these parameters with mean annual discharge and the QMNA5. As shown in Fig. 5, XV_1 was quite proportional to the mean annual discharge, whereas the effects of XV_2 on the mean annual discharge were negligible. Conversely, Fig. 6 shows the regular increase of XV_1 and XV_2 parameters when the QMNA5 values increase. On average, catchments with extremely low flows were associated with low XV_1 and XV_2 values, whereas catchments with consistently low flows were associated with high XV_1 and XV_2 values.

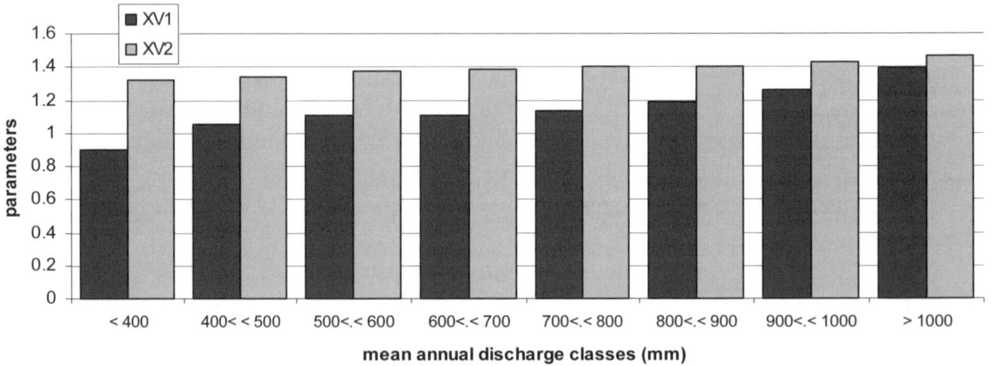

Fig. 5 Variation of XV_1 and XV_2 parameters with mean annual discharge.

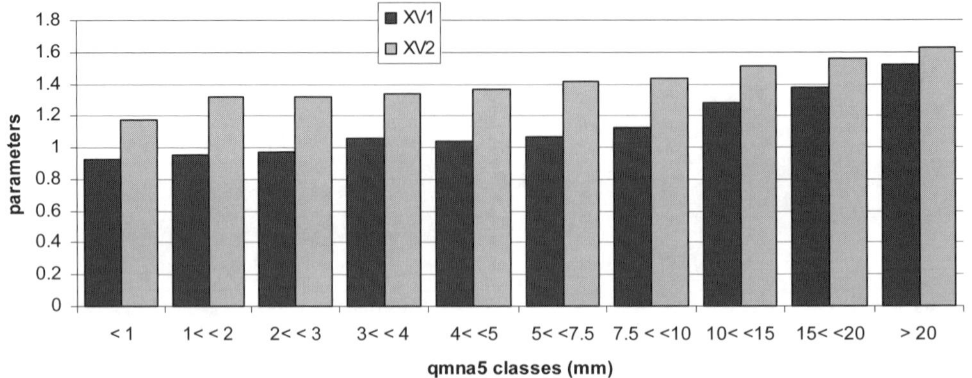

Fig. 6 Variation of XV_1 and XV_2 parameters with QMNA5.

THE REGIONALIZATION METHOD

Spatial information used in the regionalization procedure

The calibrated value of the two parameters model being available for the 880 catchments, our aim was to draw up a map over the whole study area for these parameters. It was therefore proposed to link the values of the two model parameters with some physical descriptors of the catchments: geology, land cover and altitude.

The following spatial information digitised on a 1 km^2 grid was used:

- The INRA Europan Soil Geographical Database on the 1/1 000 000 scale was used. The parent material class was used to determine the material from which the soil was originally formed. The system of classification used is based on the types of rock rather than on the geological age. This system of classification was developed at the European level using the chemical composition, the grain size and the origin of materials. There are 10 classes of materials in France.
- Based on the CORINE land cover database all over France, several simplified typological patterns were identified. The CORINE data system is based on a heirarchicalized 3-level list of 44 items classified in terms of the main land cover patterns. The following five categories were selected here to characterize the impact of land cover on runoff: artificial landscapes, cultivated agricultural landscapes, natural landscapes, semi-natural landscapes and forests.
- A Digital Elevation Model was used to determine the mean altitude of the catchments.

The attempts made to link model parameters with these physical features turned out to be most disappointing. Due to the low correlation obtained in the multiple regression analyses, it was not possible to link model parameters with the selected catchment descriptors. This means that the differences in the catchments' hydrological behaviour reflected by the two model parameters involved uncertainties of several kinds: rainfall amounts, runoff values and the modelling approach. These cumulated errors have a level of impact similar to the variability of the runoff due to the specificity of the catchments. It was therefore very difficult to interpret the values of XV_1 and XV_2 directly derived from catchment descriptors. In addition, the explanatory variables used here are lumped variables that did not reflect the heterogeneity of the catchments that may also vary in time (Perrin, 2000). The only significant trend was observed between parameter XV_1 and the mean altitude of the Alpine catchments. But this correlation was due to the fact that parameter XV_1 accounts for the uncertainty introduced by the underestimation of rainfall. Many authors, such as Servat & Dezetter (1992), Sefton & Howarth (1998), Seibert (1999), Post & Jakeman (1999), Perrin (2000) and Merz & Blöschl (2004), have faced similar problems. In studies made at a regional level, more satisfactory correlations have often been obtained. For example Makhlouf (1994) obtained encouraging results in regionalizing parameters of a previous version of the GR2M model in Brittany (western France). Correlations are less difficult to obtain for a specific zone than for a whole country, as noticed by Johansson (1994) and Post (1999). These authors concluded that the significant correlations obtained in their study were apparently due to the more homogeneous behaviour of the catchments studied.

Given these problems, it was proposed to use an upward regionalization procedure involving the use of physical descriptors. The idea here was to use a physical assessment grid to evaluate the two model parameters, and then to compare the parameters calculated from these grids with those given by the model at each gauged catchment. This comparison yielded a map of corrective coefficients defined as the ratio between the parameters evaluated using the physical grid and those obtained by calibration. A map of corrective coefficients was then drawn up for each parameter. The final grid was obtained by combining the physical grid and the map of corrective coefficients.

Regionalization methods

To draw up the physical parameter evaluation grid, the previous spatial data were used. These included:

- 5 classes of vegetation cover
- 10 geological classes

These data were interpreted as substratum retention data. A multiplication coefficient corresponding to the median spatial value of each parameter was attributed to each of the classes. The coefficient to be applied to parameter XV_1 was such that the parameter increased with decreasing substrate retention values, and *vice versa* in the case of the coefficient applicable to XV_2. The significant correlation previously observed with the altitude was taken into account by including an altitude correction coefficient ($Ccor_{Altitude}$), to improve the estimation of model parameters in the Alpine catchments.

The final grids for the two parameters were eventually drawn up pixel by pixel, by multiplying the grids of physical corrective coefficients by the median value of the parameter.

$$XV_{physique} = Ccor_{Vegetation} * Ccor_{Geology} * Ccor_{Altitude} * \text{Median value of the parameter}$$

The Alpine region showed the highest XV_1 values, whereas the sandy Aquitaine catchments showed the lowest values.

These parameter values were integrated for each of the 880 catchments. It was thus possible to compare the XV_1 and XV_2 values obtained by model calibration with those obtained using the physical estimation grid.

A corrective coefficient defined as the ratio between the "physical" value and the calibrated value was therefore calculated for each catchment. The next step consisted of drawing up a grid (with pixels corresponding to 1 km^2) giving the corrective coefficient applicable to each of these parameters. The regional approach is based here on 2/3 of the catchments. The catchments with the smallest areas were selected to avoid giving too much weight to the large catchments. These larger catchments will be used for control purposes to check the results of the regionalization procedure. Figure 8 gives the corrective ratio applicable to XV_1 in the case of the calibrated sample. This ratio was obtained by spatially drawing ratio values, starting from the larger catchments and finishing with the smaller catchments. It was then necessary to fill the empty spaces in the grid, corresponding to the zones where there was no catchment data. We used an automatic procedure, in which the study area was automatically segmented successively and the pixels completed by calculating the mean corrective coefficient between the patterns of segmentation (i) and (i – 1). Eight successive segmentation steps were carried out in this way on our study area.

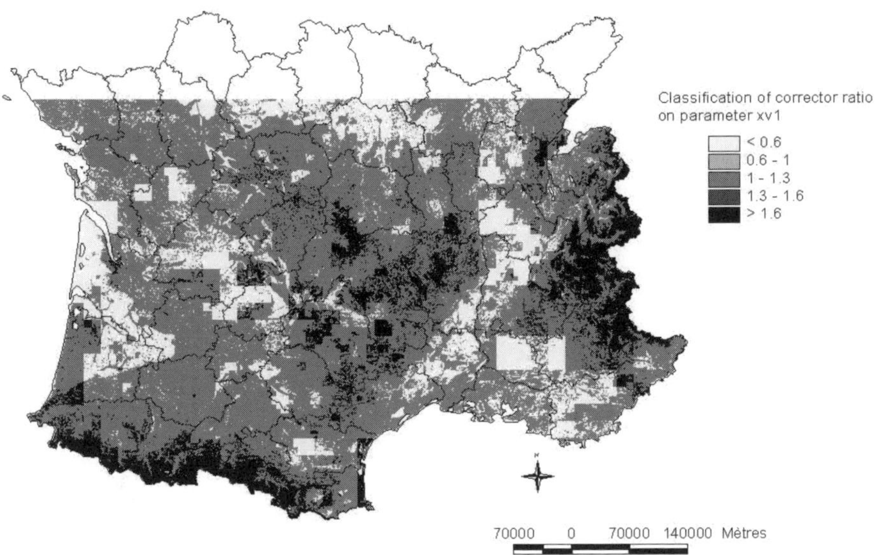

Fig. 7 Map of corrective ratios applicable to parameter XV_1 after the regionalization procedure.

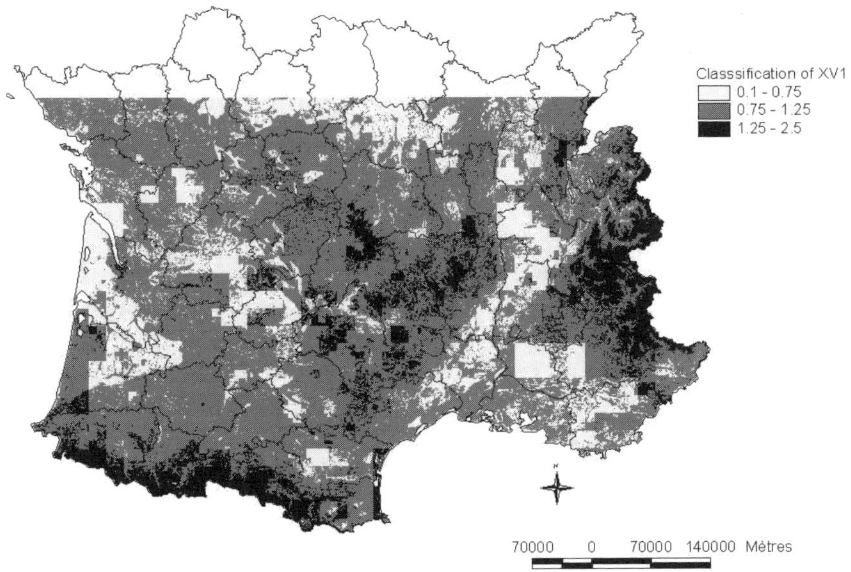

Fig. 8 Final map of regional parameter XV_1.

Let us take the following notations:

- D, the domain under investigation, which has an area of 1280 × 1280 km^2;
- M_D, the mean corrective coefficient calculated on domain D. This mean is weighted by the surface area of the studied catchments;
- *n*, the number of successive segmentation steps, $n = 8$.

At each segmentation step, the value of the corrective coefficient (*Ccor*) for a pixel was obtained by applying the following formula:

$$Ccor = \frac{M_d}{2_n} + \sum_{i-1}^{n} \frac{M_{D/4^i}}{2^{n-i+1}} \quad \text{if } M_{D/4^i} = \text{no data, then } M_{D/4^i} = M_{D/4^{i-1}}$$

The final grid of the corrective ratios applicable to XV_1 was obtained after performing eight segmentation steps. Performing further segmentation steps did not significantly improve the results.

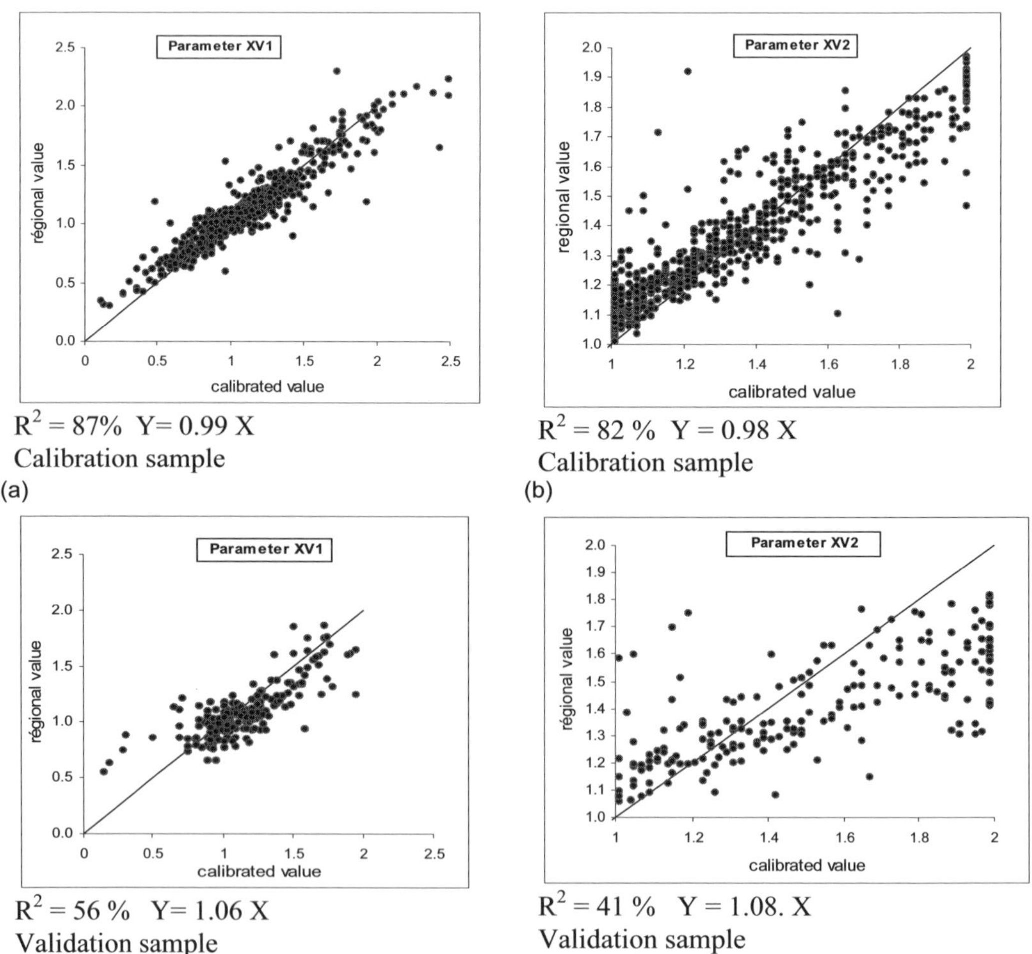

Fig. 9 Comparisons of calibrated parameters and regionalized.

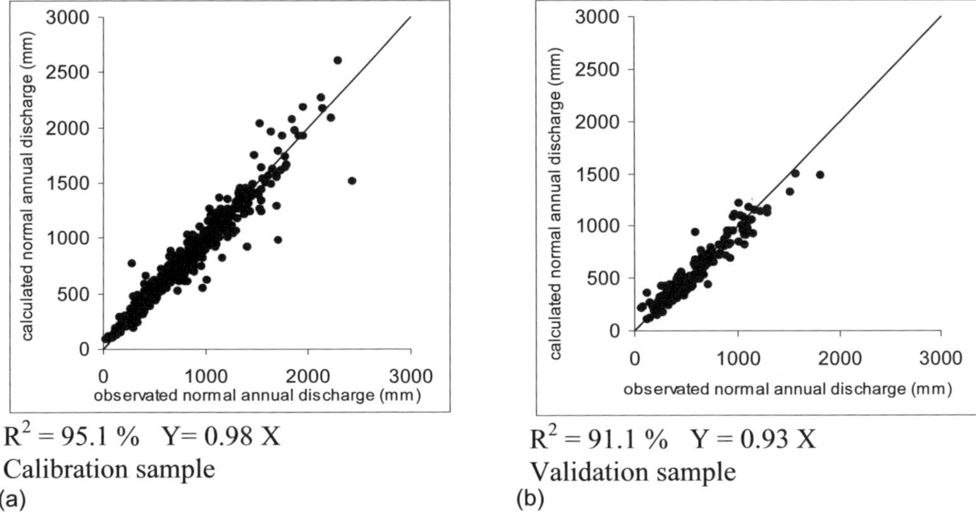

Fig. 10 Mean annual discharge obtained with the regional model.

The final stage in the estimation of parameter XV_1 consisted of simply multiplying the map of physical parameters XV_1 by the corresponding corrective map. These two maps are given in Figs 7 and 8.

The procedure was repeated with parameter XV_2.

RESULTS

The regional parameter grids can be used to obtain mean values by integration over the study catchments. Fig. 9(a)–(d) compare the calibrated and regionalized parameters, for the calibration and validation catchments samples.

Using the regionalized parameters XV_1 and XV_2, the rainfall–runoff model can now be used to simulate the monthly flow time series.

Figure 10(a),(b) shows the simulated mean annual discharges. The values are accurately simulated during the calibration model, in spite of a slight underestimation. The results obtained after the parameter regionalization were slightly less accurate than in calibration, but they generally remain very satisfactory. The R^2 coefficients were greater than 90% in both samples.

In terms of QMNA5 variable, the accuracy of the regional model was acceptable, although the points showed greater scatter (Fig. 11).

CONCLUSION

The rainfall–runoff modelling approach used in the present study was found to very accurately simulate the monthly discharge series. Using these simulated series, satisfactory estimates of the reference low flow discharge variables (the mean annual discharge and the QMNA5) were obtained.

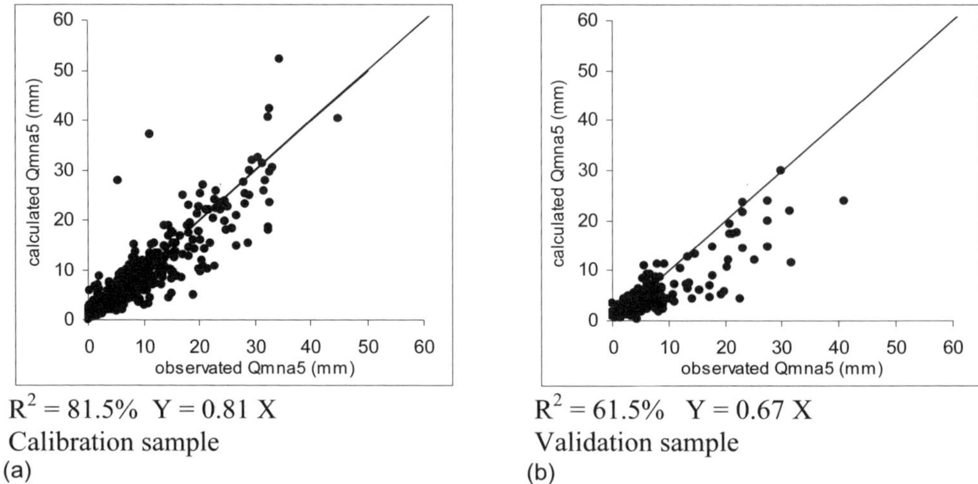

Fig. 11 Results obtained for the QMNA5 using the regional model.

The model was calibrated on 880 catchments located in the southern half of France. Maps of the two parameters used in the model were obtained. Using this information, a regionalization approach was used to evaluate model parameters at any point of the whole study region.

The regionalization procedure involved several phases:

- drawing up a grid for evaluating model parameters from physical characteristics,
- establishing a corrective coefficient, defined for each catchment as the ratio between the parameter value obtained from physical characteristics and the parameters obtained by calibration.
- automatically interpolating the corrective coefficients,
- overlaying the grid of corrective coefficients with the grid based on physical criteria.

We therefore ended up with maps of the two model parameters.

Using this information to estimate parameter values, the rainfall–runoff model could be applied to simulate monthly discharge time series, from which reference low flow variables could be deduced.

The comparisons between the values obtained by this procedure and the observed values show that the mean annual discharges were accurately simulated by the model. Despite the scatter observed, the accuracy of the 5-year return period monthly low flow (QMNA5) given by the model was satisfactory.

REFERENCES

Benichou, P.& Le Breton, O. (1987) Prise en compte de la topographie pour la cartographie des champs pluviométriques statistiques. *La Météorologie*, 7e série **19**, 23–34.

Burn, H. D. (1990) Evaluation of Regional Flood Frequency Analysis with a region of influence approach. *Water Resour. Res*. **26**(10), 2257–2265.

Fernandez, W., Vogel, R. M., Sankarasubramamian A. (2000) Regional calibration of a watershed model. *Hydrol. Sci. J.* **45**(5), 689–707.

Folton, N & Lavabre, J (2006) Amélioration d'un modèle pluie débit au pas de temps mensuel. Cemagref, France (Personal communication).

Folton, N. & Lavabre, J. (2004) Estimation des débits de référence d'étiage. Nouvelle estimation de l'ETP. Nouvelle optimisation des paramètres du modèle pluie-débit. Nouvelle approche régionale de la modélisation de la pluie en débit aux régions Languedoc-Roussillon, Provence-Alpes-Côte d'Azur, Corse, Auvergne, Limousin, Rhône-Alpes, Rapport d'étude Cemagref.

Folton, N. & Lavabre, J. (2001) Estimation des débits de référence d'étiage. Cartographie régionale de l'évapotranspiration potentielle. Calcul des pluies de bassin. Rapport d'étude Cemagref, France.

Johansson, B. (1994) The relationship between catchment characteristics and the parameters of a conceptual runoff model: a study in the south of Sweden. In: *FRIEND: Flow Regimes from International Experimental and Network* Data (ed. by P. Seuna, A. Gustard, N. W. Arnell & G. A. Cole), 475–482. IAHS Publ. 221. IAHS Press, Wallingford, UK.

Makhlouf, Z. & Michel, C. (1994) A two-parameter monthly water balance model for French watersheds. *J. Hydrol.* **162**(3–4), 299–318.

Makhlouf, Z. (1994) Compléments sur le modèle pluie-débit GR4J et essai d'estimation de ses paramètres. Thèse de Doctorat, Université Paris XI Orsay, France.

Merz, R. & Blöschl, G. (2004) Regionalization of catchment model parameters model. *J. Hydrol.* **170**, 277–291.

Mwakalila, S. (2003) Estimation of stream flows of ungauged catchments for river basin management. *Physics Chem. Earth* **28**, 935–942.

Ouarda, T. B. M. J., Girard, C., Cavadias, G. S. & Bobée, B. (2001) Regional flood frequency estimation with canonical correlation analysis. *J. Hydrol.* **254**, 157–173.

Paturel, J. E., Servat, E. & Vassiliadis, A. (1995) Sensitivity of conceptual rainfall- runoff algorithms to errors in input data—case of the GR2M model. *J. Hydrol.* **168**, 111–125.

Perrin, C. (2000) Vers une amélioration d'un modèle global pluie-débit au travers d'une approche comparative. Thèse de Doctorat, Institut National Polytechnique de Grenoble, Grenoble, France.

Post, D. A. & Jakeman, A. J. (1999) Predicting the daily streamflow of ungauged catchments in SE Australia by regionalizing the parameters of a lumped conceptual rainfall–runoff model. *Ecol. Modelling* **123**, 91–104.

Sefton, C. E. M. & Howarth, S. M. (1998) Relationships between dynamic response characteristics and physical descriptors of catchments in England and Wales. *J. Hydrol.* **211**, 1–16.

Seibert, J. (1999) Regionalization of parameters for a conceptual rainfall–runoff model. *Agric. For. Met.* **98–99**, 279–293.

Servat, E. & Dezetter, A. (1992) Modélisation de la relation pluie-débit et estimation des apports en eau dans le nord-ouest de la Coôte d'Ivoire. *Hydrologie Continentale* **7**(2), 129–142.

Vandewiele, G. L. & Atlabachew, E. (1995) Monthly water balance of ungauged catchments obtained by geographical regionalization. *J. Hydrol.* **170**, 277–291.

Modélisation hydrologique et regionalisation en Afrique de l'Ouest et centrale

J. E. PATUREL[1], E. SERVAT[1], A. DEZETTER[1], J. F. BOYER[1], C. LAROCHE[2], L. MOUNIROU[2], H. LUBES-NIEL[1] & G. MAHE[1]

1 *UMII, HydroSciences Montpellier, case MSE, Place E. Bataillon, 34095 Montpellier cedex 5, France*
paturel@msem.univ-montp2.fr

2 Groupe *EIER-ETSHER, BP 594, Ouagadougou 01, Burkina Faso*

Résumé L'économie des pays en développement dépend fortement de la ressource en eau. Pour mieux la planifier et mieux la gérer, il faut mieux la connaître dans sa dimension spatiale et dans sa dimension temporelle. Depuis près de 20 ans, l'évaluation des ressources en eau en Afrique de l'Ouest et Centrale constitue la finalité des travaux menés par HydroSciences Montpellier. Certains facteurs conditionnent les objets de nos études et les outils employés: le manque de données disponibles impose d'utiliser des modèles pluie—débit peu gourmands en données et des outils d'analyse robustes; les plans de développement de ces régions s'appuient sur des analyses à grand pas de temps et d'espace; le contexte de changement climatique observé depuis 1970 n'a pas touché de la même façon toutes les régions. Ce papier dresse quelques résultats de travaux, utilisant des outils ou des méthodes développés à HydroSciences, et des perspectives liés à la régionalisation en Hydrologie.

Mots clefs Afrique de l'Ouest; bassins non jaugés; changement climatique; grands échantillons de données; modélisation Pluie – Débit; régionalisation

Hydrological modelling and regionalization in West and Central Africa

Abstract The economy of developing countries strongly depends on water resources. For a better management of water resources, we need to know better its spatio-temporal variability. For the last 20 years water resources assessment in Western and Central Africa has been the main research field of the HydroSciences Montpellier (HSM) laboratory. There are three main external constraints to our studies: the lack of available data which drives our choice for rainfall–runoff models using only few input data and robust analysis tools; the large time and space scales which are those used by development plans; and the climate changes since 1970 which have had impacts of various magnitude and spatial scale across the region. This paper summarizes some results using tools and methods developed at the HSM Laboratory, and draws some perspectives of research in the field of regionalization in hydrology.

Key words West Africa; ungauged basins; climatic change; large sample data; rainfall – runoff model; regionalization

INTRODUCTION

Depuis près de 20 ans, l'évaluation des ressources en eau en Afrique de l'Ouest et Centrale constitue la finalité des travaux menés par l'équipe VAHYNE d'Hydro-

Sciences Montpellier. Pour se placer du point de vue de l'aménageur et du gestionnaire, on travaille à partir de données disponibles auprès des services nationaux, gestionnaires des réseaux de mesure. Les modèles pluie-débit alors utilisés doivent être peu "gourmands" en données et très robustes, pour tenir compte de la qualité et de la densité des points de mesures. Notre choix s'est donc porté sur des modèles conceptuels globaux. Les pas d'espace balayés vont de quelques dizaines de kilomètres carrés à quelques dizaines de milliers de kilomètres carrés. Quant au pas de temps, il a été journalier et est devenu, depuis quelques années, mensuel.

Après avoir décrit le type de modèles utilisés, la banque de données mise en place et le contexte hydoclimatique de la région étudiée, ce papier présente des résultats de travaux liés à la régionalisation (essayer d'établir des lois à une échelle régionale permettant de déterminer les valeurs des paramètres pour des bassins non jaugés en exploitant les informations sur des sites jaugés voisins sur lesquels on peut caler le modèle) et utilisant des outils ou des méthodes développés à HydroSciences:

- une analyse de l'évolution de la relation Pluie-Débit au pas de temps mensuel (GR2) sur ces régions en relation avec le changement climatique observé,
- les résultats d'une première tentative de prédétermination des paramètres d'un modèle journalier (GR3) à partir de données liées aux données pluviométriques de la station considérée et à la géographie et à l'occupation du sol du bassin versant,
- une analyse de l'information nécessaire minimale pour caler correctement un modèle mensuel (GR2).

Quelques perspectives envisagées de recherche closent cet article.

CHOIX DES MODELES PLUIE-DEBIT

Comme l'ont déjà écrit Edijatno *et al.* (1999), "notre ambition est d'arriver à une simulation pas trop décevante en utilisant une représentation la plus simple possible du processus pluie-débit, dépendante d'un très faible nombre de paramètres, qu'il restera à caler pour s'adapter à la rivière que l'on veut étudier". *A priori*, on se place du point de vue de l'aménageur et du gestionnaire; on travaille à partir de données disponibles auprès des services nationaux gestionnaires des réseaux de mesure. Les modèles pluie-débit alors utilisés doivent être peu "gourmands" en données et très robustes pour tenir compte de la qualité et de la densité des points de mesures. Notre choix s'est donc porté sur des modèles conceptuels. Les pas d'espace balayés vont de quelques dizaines de kilomètres carrés à quelques dizaines de milliers de kilomètres carrés. Quant au pas de temps, il est journalier ou mensuel. Nous avons travaillé à partir de deux versions des modèles Génie Rural (GR):

- une version journalière (GR3J) développée par Edijatno & Michel (1989),
- une version mensuelle (GR2M) développée par Makhlouf (1994). que nous avons adaptée à notre contexte climatique et scientifique (Ouédraogo, 2001; Ardoin, 2004).

Nous renvoyons à ces auteurs pour une description complète des modèles utilisés.

PRESENTATION DE LA BASE DE DONNEES ET DU CONTEXTE CLIMATIQUE

Un "Système d'Informations Environnementales sur les Ressources en Eau et leur Modélisation" (SIEREM, Boyer, 2003; Boyer *et al.*, 2004) a été mis en place (www.hydrosciences.fr/SIEREM) pour les besoins de recherche de l'équipe HSM et regroupe des données chronologiques (séries hydroclimatiques rattachées à une station de mesure) et des données spatiales (caractérisant le milieu physique).

Les données hydro-climatiques proviennent de plusieurs sources: programmes de recherche où des complètements de données ont pu être faits, gestionnaires nationaux ou institutionnels de bases de données hydrologiques. Profitant de notre connaissance du milieu africain, nous avons extrait ce que nous avons considéré comme la source la plus fiable et comblé ses lacunes par les autres sources potentielles. Nous avons pu ainsi rassembler plus de:

- 27 500 stations-années de données météorologiques (autres que données pluviométriques),
- 70 000 stations-années de données pluviométriques,
- 20 000 stations-années de données hydrométriques,
- à des pas de temps généralement journaliers ou mensuels.

Les informations sur le milieu physique sont issues de cartes thématiques (géologie, pédologie, couverture végétale, ...) qui ont été numérisées par nos soins. A partir de développements avec des SIG et MNT, il est possible de disposer pour chacun des bassins versants de la base (plus de 350 bassins) d'une "donnée Sol" issue du CD-Rom, Digital Soil Map of the World (FAO-UNESCO, 1974–1981). La FAO a proposé une classification des sols à laquelle sont associées des classes de capacité de rétention en eau. Cette donnée nous est utile dans le cadre de la modélisation.

Le traitement d'une partie de ces informations nous a permis de mettre en évidence le phénomène climatique, aujourd'hui, bien connu: autour de 1970, un changement climatique a été observé qui se traduit par des modifications au sein des chroniques de pluies et de débits (et donc de coefficients d'écoulement) (Servat *et al.*, 1998).

ANALYSE DE L'EVOLUTION DE LA RELATION PLUIE-DEBIT ET CONSEQUENCES SUR LES TRAVAUX LIES A LA MODELISATION ET A LA REGIONALISATION

Le contexte hydroclimatique précédemment décrit nous a permis de distinguer, en général, 2 périodes, de part et d'autre de 1970. Sur chacune de ces périodes, les modèles ont été calés et les modifications éventuelles de leurs paramètres obtenus par la construction de régions de confiance autour du jeu optimum de paramètres ajusté ont été étudiées. Nous renvoyons à Lubès-Niel *et al.* (2003) pour le détail des méthodes de construction des régions de confiance et d'analyse.

La Fig. 1 montre les résultats obtenus pour un des paramètres du modèle GR2M. Il apparaît bien que les couples de paramètres optimisés sont différents de part et d'autre des années 1970, même si, en fonction des chevauchements possibles des régions de confiance, divers degrés de signification des changements sont possibles. La relation

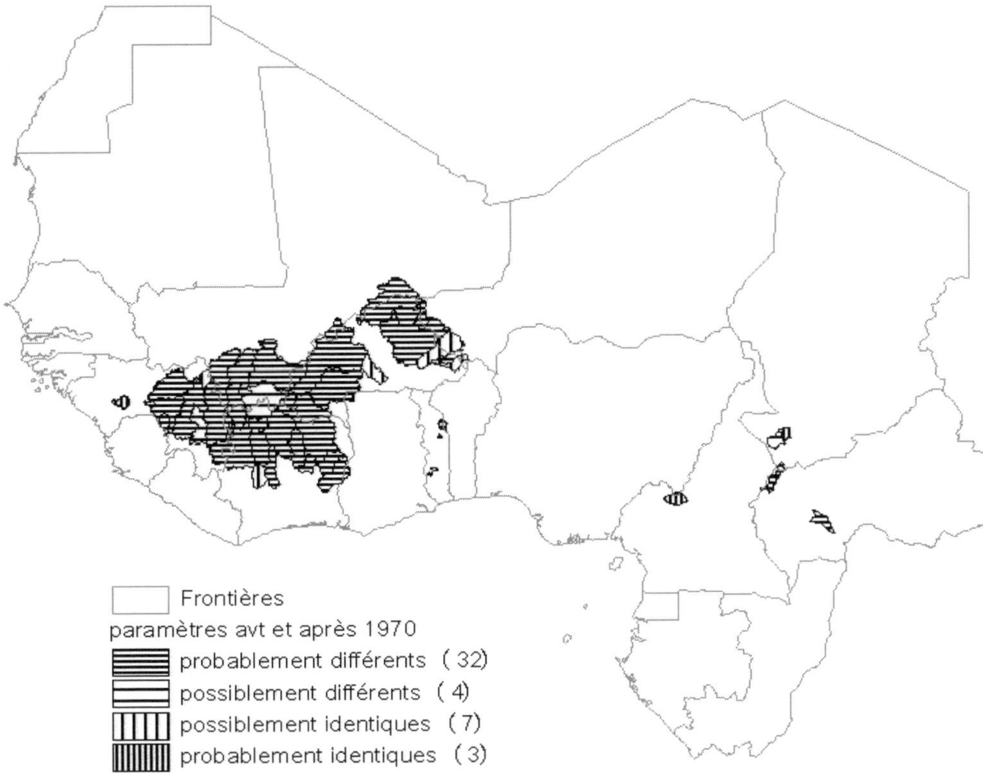

Fig. 1 Comparaison des valeurs d'un paramètre de GR2M obtenues par calage avant et après 1970 (d'après Dufresne, 2004).

Tableau 1 évolution des coefficients d'écoulement sur quelques bassins du Burkina Faso et du Niger (d'après Mahé *et al.*, 2005).

Rivière	Station	Système	Surface km²	Coeff. d'écoul. avant 1972 %	Coeff. d'écoul. après.1972 %	Ratio après./avant 1972 %
Nakambe	Wayen	Volta	20800	1.4	2.9	+ 108
Gorouol	Alcongui	Niger	44850	1.6	2.2	+ 40
Gorouol	Koriziena	Niger	2500	4.5	8.8	+ 95
Gorouol	Dolbel	Niger	7500	6.6	8.6	+ 32
Dargol	Tera	Niger	2750	6.2	8.4	+ 37
Dargol	Kakassi	Niger	6940	3.9	6.2	+ 57
Sirba	Garbe Kourou	Niger	38750	2.0	3.2	+ 61
Goroubi	Diongore	Niger	15350	2.3	2.3	0
Diamangou	Tamou	Niger	4030	3.5	2.3	– 35
Tapoa	Campement W	Niger	5330	0.9	1.0	+ 10
Mékrou	Barou	Niger	10500	(11)	(7)	– 33
Bani	Douna	Niger	101600	15.8	6.0	– 62

Pluie-Débit a donc été modifiée de part et d'autre de 1970. Pour compléter ce constat, le tableau 1 révèle que les coefficients d'écoulement peuvent augmenter sur certains bassins sahéliens du Burkina Faso et du Niger, entraînant même des écoulements plus importants alors que le contexte pluviométrique est déficitaire (Mahé *et al.*, 2005). Les raisons évoquées sont liées aux activités humaines et à l'impact de la variabilité climatique.

En termes de données d'observation, la conséquence est qu'un bassin pour lequel on ne dispose de données qu'avant 1970 peut être considéré comme non jaugé après 1970. A cela s'ajoutent le manque de soutien financier des états africains pour leurs services nationaux gestionnaire des réseaux de mesure de données hydroclimatiques et le désengagement de l'ORSTOM (ex IRD) de ces services à partir des années 1970.

Le constat est donc simple: les données se font de plus en plus rares et sont de bien moindre qualité; la région Afrique de l'Ouest et Centrale tend à devenir un vaste espace non jaugé (!); la nécessité d'avoir à sa disposition des données en quantité suffisante limite l'estimation des ressources en eau sur des bassins versants non jaugés.

Aussi, la transposition des observations a déjà été le souci de nombreux hydrologues à l'ORSTOM, aujourd'hui IRD. Certains, tel Rodier (1964), y ont répondu en proposant des relations empiriques. D'autres ont développé des méthodes originales (Le Barbé *et al.*, 1993). Pour sa part, l'équipe HydroSciences a proposé l'utilisation de modèles globaux conceptuels Pluie Débit.

ESTIMATION DE RESSOURCES EN EAU ET TENTATIVE D'EXTENSION AUX BASSINS NON JAUGES

A partir de données journalières, des travaux de modélisation ont été menés pour permettre une estimation des écoulements à l'exutoire d'un bassin versant en milieu de savane dans le nord-ouest de la Côte d'Ivoire (Servat, 1993; Servat & Dezetter, 1993). Afin de pouvoir appliquer les résultats obtenus à des bassins non jaugés, il a été envisagé des prédéterminations des paramètres du modèle journalier (GR3J) à l'aide d'équations mettant en jeu différents types de variables:

– variables propres au bassin versant (surface et coefficient de Gravelius);
– variables pluviométriques (cumuls sur certaines périodes et ratios pluviométriques);
– variables liées à l'occupation du sol du bassin versant (proportions de savane, de forêt arbustive, de zones cultivées, et de zones d'habitat).

La qualité des ajustements obtenus, mesurée par le coefficient de détermination, ne varie qu'entre 0.323 et 0.870. Pour toute poursuite de ces travaux, plus que le résultat en tant que tel (assez décevant), ce qu'il faut retenir est:

– l'importance des totaux pluviométriques de début de la saison des pluies,
– l'importance de la couverture végétale (portion de forêt arbustive et d'espaces cultivés),

pour prédéterminer les paramètres du modèle GR3J.

Des essais ont été menés sur d'autres zones géographiques de Côte d'Ivoire: zone de forêt et zone de "transition" (entre forêt et savane). Ils se sont révélés négatifs.

MISE AU POINT D'UNE MÉTHODOLOGIE D'EXPLOITATION DE DONNÉES PONCTUELLES DE DÉBITS ET DE DONNÉES RÉGIONALES POUR L'ESTIMATION DES PARAMÈTRES D'UN MODÈLE PLUIE-DÉBIT

Devant les échecs rencontrés depuis plus de trente ans par les méthodes classiques qui n'utilisent aucune information débimétrique sur les bassins, on se propose d'abandonner ici l'idée de trouver une estimation satisfaisante des paramètres du modèle sans aucune estimation de débit et d'explorer une voie différente où l'on exploiterait à la fois une information régionale et une information plus ponctuelle sur les débits. L'objectif est de déterminer le niveau minimal d'information nécessaire pour obtenir une estimation fiable des paramètres des modèles pluie-débit. Nous avons défini une méthodologie d'approche qui puisse nous permettre de remplir cet objectif.
Remarque: avant de débuter ce travail, des essais à partir d'une version journalière de GR ont révélé que même en supprimant aléatoirement 70% des données disponibles d'une chronique de débit, on pouvait caler le modèle de la même façon que si l'on disposait de toutes les données disponibles, et ce pour un bassin en régime tropical pur. Ce résultat montre donc que si l'information est très rare, il peut être quand même être possible de caler correctement un modèle hydrologique; il suffit qu'il y ait l'information nécessaire et suffisante (Vernay, 2004).

Les résultats présentés se basent sur des données d'une soixantaine de bassins versants en Afrique de l'Ouest (Laroche *et al.*, 2005)

Détermination d'une information issue d'une approche régionale

L'information régionale retenue est la nature du régime d'écoulement du bassin versant tel qu'il a été défini par Rodier (1964). Cette classification (Tableau 2) se base sur des observations, sur des études et sur une très bonne connaissance du milieu. Même si elle date d'avant les changements climatiques observés en Afrique de l'Ouest et en Afrique Centrale, les résultats obtenus semblent confirmer l'option prise.

Tableau 2 Quelques caractéristiques de régimes hydrologiques rencontrés en Afrique de l'Ouest.

Régime hydrologique	Limites des isohyètes (mm)	Principaux bassins hydrologiques	Principales caractéristiques
Sahelien	300 à 750	Sirba, Dargol, Gorouol, Nakambé, Tapoa et une partie du bassin du fleuve Niger	Longue saison sèche, crue sporadique, importantes pertes par évaporation
Tropical pur	750 à 1200	Mouhoun, Nakambé, Comoé supérieure, et une partie du bassin du fleuve Niger	Présente une saison de hautes eaux en juillet à début octobre et une saison de basses eaux de début décembre à début juin.
Tropical de transition	Pan > 1200	Bandama, Sassandra et Niger supérieur	Il se distingue du régime tropical pur par une saison de hautes eaux beaucoup plus longue et une saison de basses eaux bien moins sévères.
Tropical de transition variante dahomeenne	1200 à 1400	Bassin moyen de la Comoé, du Couffo, du Mono, de l'Ouémé, et de tous les petits cours d'eau du centre du Bénin et du Togo.	Les étiages sont beaucoup plus rigoureux que dans le régime tropical de transition. Les modules sont plus faibles. A cet égard, ce régime ressemble beaucoup au régime tropical pur. La différence essentielle est une période de hautes eaux plus longue, mais pas plus abondante au total.

Utilisation d'une information locale

L'information locale n'est autre qu'une partie de la chronique elle-même. L'information nécessaire dépendra de l'architecture du modèle et des caractéristiques de ce dernier, GR2M dans ce cas.

Pour les hydrogrammes simples d'Afrique de l'Ouest et Centrale (hydrogramme unimodal avec une période plus ou moins longue où les écoulements sont très faibles, voire nuls), comme pour les bassins en régime hydrologique sahélien ou tropical pur, l'information à fournir est constituée:

– des écoulements de hautes-eaux qui constituent une bonne approximation des volumes annuels écoulés,
– et d'une partie des écoulements de basses–eaux qui avec les débits de pointe de l'hydrogramme peuvent permettre de caler la décrue.

Les travaux se continuent pour les bassins en régime équatorial où les hydrogrammes sont généralement bimodaux et les écoulements de basses–eaux sont bien souvent loin d'être négligeables. Nous anticipons sur les résultats à venir mais il est probable que pour ces bassins, l'information nécessaire sera certainement plus importante pour caler correctement le modèle.

PERSPECTIVES

Depuis quelques années, les travaux d'HydroSciences Montpellier se focalisent en particulier sur le pas de temps mensuel et sur de grands pas d'espace (# 2500 km^2), échelles de temps et d'espace auxquelles il est raisonnable de planifier des actions en matière de ressources en eau à l'échelle d'un grand bassin versant ou d'une région.

Les résultats décevants que nous avons présentés sur un essai de prédétermination de paramètres d'un modèle ne prouvent cependant pas qu'il ne faille pas continuer dans cette voie: c'est sans doute que les données utilisées jusque là (se limitant bien souvent aux seules données climatiques) n'expliquent pas à elles seules les écoulements; c'est que d'autres caractéristiques inhérentes aux bassins de réception des précipitations expliquent bien plus fortement les écoulements (les résultats présentés le montrent). A l'heure des images satellites et des systèmes d'information géographique, une quantité d'information (liée à la physiographie des bassins, liée au paysage ...) est dorénavant disponible au sein de laquelle il faut chercher. Il ne semblerait pas inintéressant de refaire "tourner" des ACP ou tout autre outil d'analyses statistiques exploratoires en combinant toutes ces nouvelles données.

La Fig. 2 montre que géographiquement un certain nombre de bassins se comportent de la même façon en termes d'évolution d'un des paramètres du modèle GR2M de part et d'autre de 1970. Ce même comportement de certains bassins qui est propre au modèle (mais peut-être à d'autres modèles?) pourrait traduire un même comportement hydrologique de ceux-ci (sans pour autant affirmer que le modèle choisi reproduit exactement le comportement hydrologique du bassin versant considéré) et il pourrait être intéressant de composer ces techniques d'analyse avec ces comportements en modélisation.

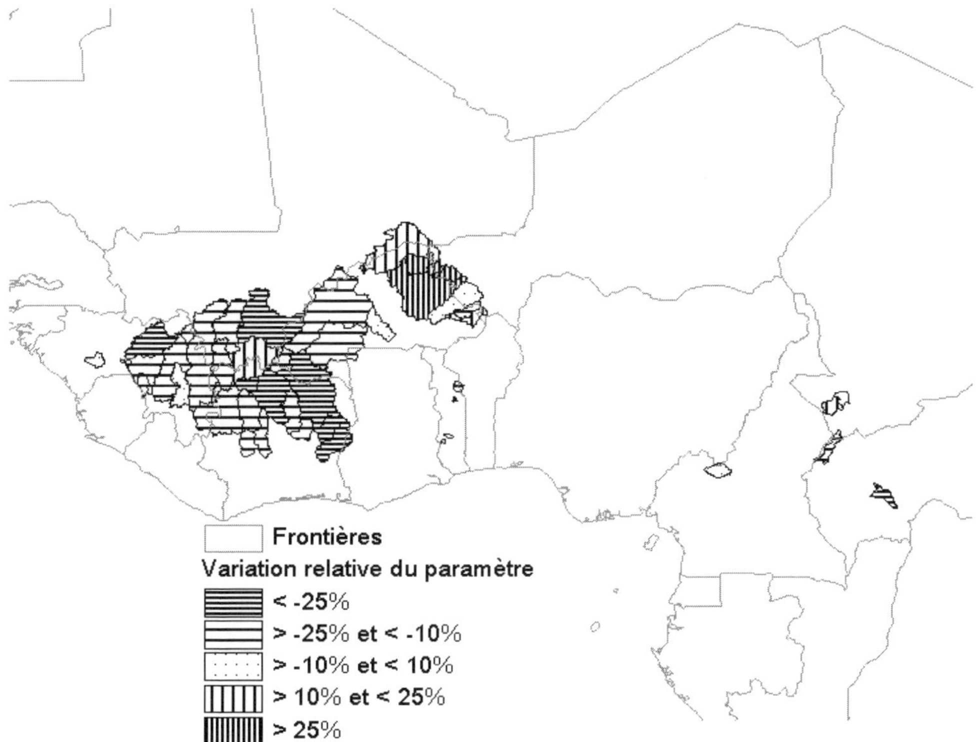

Fig. 2 Écarts relatifs entre les valeurs d'un paramètre de GR2M obtenues par calage avant et après 1970 (d'après Dufresne, 2004).

Hormis les modèles GR, l'équipe a utilisé d'autres modèles tels que le modèle VUB, le modèle WBM et le modèle de Yates. D'autres existent et mériteraient d'être appliqués à la zone d'étude sur laquelle nous travaillons. Ces modèles sont simples d'utilisation et les concepts sur lesquels ils s'appuient le sont aussi. Seuls changent la prise en compte des données d'entrée, l'agencement des réservoirs, les équations de vidange des réservoirs, la prise en compte ou non d'un éventuel apport souterrain, …

Comme il n'existe pas de modèle conceptuel global unique, il apparaît comme vain de ne vouloir utiliser qu'un même modèle sur l'Afrique de l'Ouest. La Fig. 3 montre que le modèle GR2M donne de bons résultats (valeurs de Nash élevées) sur une partie de l'Afrique de l'Ouest mais que ses performances sont bien moindre ailleurs. L'idée serait de tester un ensemble de modèles sur les données existantes et d'analyser les différents résultats obtenus et, en particulier, quel est le modèle qui se cale le mieux et pourquoi? Cela est-il lié aux données disponibles ou à des facteurs explicatifs évidents? … L'objectif serait de proposer ainsi une cartographie des "meilleurs" modèles en fonctions de critères qu'il faut définir.

Parmi les critères de choix, on peut citer ceux liés au calcul d'une fonction coût ou d'une fonction objectif. Mais ils ne me semblent pas suffisants. Le critère primordial pourrait être la quantification des incertitudes qui accompagnent les résultats de la modélisation.

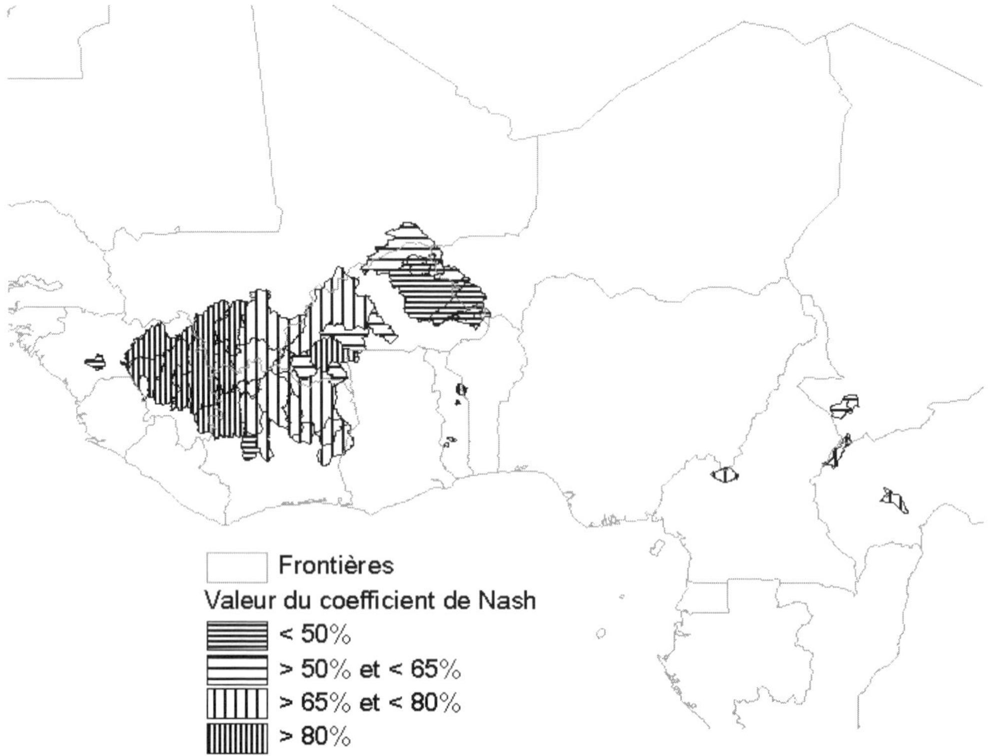

Fig. 3 valeur du coefficient de Nash pour quelques bassins ouest-africains (d'après Dufresne, 2004).

CONCLUSIONS

Les différents contextes, scientifiques, hydroclimatiques et environnementaux, de l'Afrique de l'Ouest et Centrale ont conditionné la nature des travaux et les outils développés par HydroSciences Montpellier, depuis près de 20 ans:

- l'Afrique de l'Ouest et Centrale où les données hydrométriques se font de plus en plus rares et sont de qualité de plus en plus douteuses en dépit des efforts accomplis par les services nationaux gestionnaires des réseaux. Même sur des bassins jaugés, l'information est très parcellaire et on peut se considérer parfois dans une situation proche de celle des bassins non jaugés.
- Le contexte hydroclimatique qui a beaucoup changé depuis 35 ans. A la variabilité climatique, s'est superposé un développement effréné des activités humaines qui ont entraîné une modification profonde de la réponse des bassins versants aux événements pluvieux.
- Le modèle GR2M, modèle mensuel à 2 paramètres, se prête bien à ce milieu puisqu'il est peu "gourmand" en données et est robuste.

Parmi ces travaux, de nombreuses études ont été menées sur la thématique de la régionalisation. Différentes voies de recherche ont été entreprises qui nécessitent d'autres développement, en particulier en raison des nouvelles données qui peuvent être acquises par les images satellitaires.

Pour plus de détail sur les méthodes ou les résultats exposés, le lecteur est invité à se référer aux références des auteurs de ce papier.

RÉFÉRENCES

Ardoin, S. (2004) Variabilité hydroclimatique et impacts sur les ressources en eau de grands bassins hydrographiques en zone soudano-sahélienne. Thèse de Doctorat, Université de Montpellier II, Montpellier, France.

Boyer, J. F. (2003) Développements et technologies mises en œuvre à l'IRD pour la construction du Système d'Information Environnemental sur les Ressources en Eau et leur Modélisation – SIEREM. Séminaire Gestion et mise à disposition des données environnementales, CEMAGREF, décembre 2003, Lyon, France.

Boyer J. F., Servat E. & Dieulin C. (2004) A specific method of modeling and designing environmental information systems - the example of SIEREM, an environmental information system for water resources. BALWOIS, Mai 2004, Ohrid, Macédoine.

Dufresne, M. (2004) Modélisation pluie-débit GR2M en Afrique de l'Ouest—Régions de confiance des paramètres du modèle. Stage de recherche, ENGEES.

Edijatno & Michel C. (1989) Un modèle Pluie-Débit journalier à 3 paramètres. *La Houille Blanche* **2**, 113–121.

Edijatno, Nascimento, N. O., Yang, X., Makhlouf, Z. & Michel C. (1999) A daily watershed model with three free parameters. *Hydrol. Sci. J.* **44**(2), 263–277.

FAO (1974–1981) *Digitized Soil Map of the World and Derived Soil Properties*. FAO, Rome, Italy.

Laroche, C., Mounirou, L. A., Mar, A. L., Paturel, J. E. (2005). Necessary hydrometric information for a good estimation of hydrological functionning of watersheds in western Africa. Poster VIIème Assemblée Scientifique de l'AISH, 3–9 Avril 2005, Foz do Iguaçu, Brésil.

Le Barbé, L., Ale, G., Millet, B., Texier, H., Borel, Y. & Gualde R. (1993) *Les ressources en eaux superficielles de la République du Bénin*. ORSTOM, Paris, France.

Lubes-Niel, H., Paturel, J. E. & Servat, E. (2003) Study of parameter stability of a lumped hydrologic model in a context of climatic variability. *J. Hydrol.* **278**, 213–230.

Mahé, G., Paturel, J. E., Servat, E., Conway, D. & Dezetter, A. (2005) The impact of land use change on soil water holding capacity and river modelling in the Nakambe River, Burkina Faso. *J. Hydrol.* **300**(1–4), 33–43.

Makhlouf, Z. (1994) Compléments sur le modèle pluie - débit GR4J et essai d'estimation de ses paramètres. Thèse de Doctorat, Cemagref Antony, Université Paris XI, Orsay, France.

Ouédraogo, M. (2001) Contribution à l'étude de l'impact de la variabilité climatique sur les ressources en eau en Afrique de l'Ouest—Analyse des conséquences d'une sécheresse persistante: normes hydrologiques et modélisation régionale. Thèse de Doctorat, Université de Montpellier II, Montpellier, France.

Rodier, J. (1964) *Régimes hydrologiques de l'Afrique Noire à l'ouest du Congo*. ORSTOM, Paris, France.

Servat, E. (1993) Evaluation régionale des ressources en eau: application à la Cote d'Ivoire—Rapport de synthèse du programme ERREAU, ORSTOM, Abidjan.

Servat, E. & Dezetter, A. (1993) Rainfall–runoff modelling and water resources assessment in northwestern Ivory Coast—tentative extension to ungauged basins. *J. Hydrol.* **148**, 231–248.

Servat, E., Paturel, J. E., Kouamé, B., Travaglio, M., Ouedraogo, M., Boyer, J. F., Lubès-Niel, H., Fritsch, J. M., Masson, J. M. & Marieu B. (1998) Identification, caractérisation et conséquences d'une variabilité hydrologique en Afrique de l'Ouest et Centrale. In: *Water Resources Variability in Africa during the XXth Century* (ed. by E. Servat, D. Hughes, J. M. Fritsch & M. Hulme) (Abidjan'98, Abidjan, Côte d'Ivoire, Novembre 1998), 323–338. IAHS Publ. 252, IAHS Press, Wallingford, UK.

Vernay, M. (2004). Changement climatique et impacts sur les normes hydrologiques: caractérisation et apport de la modélisation. DEA, ENPC.

Large Sample Basin Experiments for Hydrological Model Parameterization: Results of the Model Parameter Experiment–MOPEX. IAHS Publ. 307, 2006.

Predicting river flow statistics at ungauged locations—a hydrostochastic approach

ERIC SAUQUET[1], LARS GOTTSCHALK[2], IRINA KRASOVSKAIA[2] & ETIENNE LEBLOIS[1]

1 *Cemagref, Hydrology - Hydraulics Research Unit, 3 bis quai Chauveau CP220, F-69336 Lyon cedex 09, France*
eric.sauquet@cemagref.fr

2 *Department of Geosciences, University of Oslo, PO Box 1022 Blindern, N-0315 Oslo, Norway*

Abstract The specific objective of this paper is to provide an overview of the collaborative projects between the University of Oslo (Norway) and Cemagref (France). The research undertaken concentrated on the development of methods for estimating statistical properties of the water balance components (especially runoff) at ungauged locations. Probability and Statistical Theory is used to estimate runoff characteristics related to various scales in time and space. A river basin was studied as a system consisting of interacting entities without splitting it into these entities—looking for patterns rather than isolated steps of causality, i.e. letting the data speak for themselves. Links between the water balance statistics and the way they develop in time and/or in space, i.e. along the river network, are analysed and these dependences are introduced explicitly in interpolation procedures. Thus the approach preserves the statistical properties in terms of covariances and semi-variograms at a basin scale as well as satisfying water balance constraints. A review of the basic concepts is given first. Second, applications of the concepts are presented with examples of their application to different extensive datasets (France and Costa Rica) and for different hydrological variables. Mean annual and monthly discharges and flow duration curves are investigated in this article.

key words empirical orthogonal function; flow duration curve; runoff; scale; stochastic interpolation

INTRODUCTION

A fundamental problem in applied hydrology is to estimate runoff parameters at an ungauged location. Various methods have been developed for this purpose and the results are often summarized by maps of runoff characteristics over the studied region. Maps produced manually (Gannett, 1912) were published first, since computation facilities were limited at the time they were made. The procedures involved in producing these maps needed lots of attention and were highly time consuming and usually subjective. Empirical relationships among average streamflow, land use, geomorphology and climate have received wide attention for several decades (Solomon *et al.*, 1968; Liebscher, 1972; Hawley & McCuen, 1982; Gustard *et al.*, 1989; Vogel *et al.*, 1999). They have usually been established by multivariate regional regression. Drainage area and precipitation are by far the most explanatory variables. Other basin characteristics may be incorporated, but their inclusion in relationships is not warranted. One critical point is that when values are required at ungauged locations, each gauged basin has the same weight in the calculation. Our experience is that geographical proximity criteria give more reliable estimates and this is the background for the approach developed herein.

Runoff is, by definition, an integrator of a random spatially distributed process. The spatial variability of runoff characteristics is mostly governed by the river network. From upstream to downstream, the basins collect water from tributaries with sometimes different river flow regimes. One consequence is that the hydrological information cannot be propagated in all directions but in a preferred direction along the river network. The size of the basin as well as the topology of the river network influence most parameters like variance-covariance, skewness of the original data series, and all statistics of low flow and floods (Gottschalk, 1993).

The objective of this paper is to present an approach that we will call the "hydrostochastic" approach. This approach attempts to incorporate the previously mentioned properties of runoff characteristics in an interpolation framework:

- a river basin is studied as a system consisting of interacting blocks without decomposing it into blocks—looking for patterns rather than identifying steps of causality, i.e. letting data speak for themselves;
- a river basin is studied in its context looking at it from outside in search of contextual rules of its functioning (water balance, consistency with statistical laws);
- the theoretical background is given by Probability and Statistical Theory to analyse and model results of integration in space over a drainage basin and in time over an interval following the structure of the river network.

These concepts have been applied in a number of studies (Gottschalk & Krasovskaia, 1998; Sauquet *et al.* 2000a,b; Skøien *et al.*, 2005). The following sections briefly describe the developments. Applications of these concepts to map monthly discharges and to estimate flow duration curves are presented using different extensive datasets in France and in Costa Rica.

MAPPING MEAN ANNUAL RUNOFF

Starting with the most fundamental characteristic of runoff—the long-term mean $qa(A)$ for a basin A—it must be so that the values over M sub-basins ΔA_i, $i = 1, \ldots, M$ should sum up to the value at the outlet for total basin A. This is the water balance equation for the lateral flow in a basin or in a statistical sense the average value for the whole basin should be consistent with averages over its parts. Mapping the mean value is a rather straightforward task and basically it is a problem of stochastic interpolation with local support (or block kriging) with an added water balance constraint. Gottschalk (1993) introduced a method for stochastic interpolation of runoff along the river network with a constraint preserving the water balance, i.e. at each downstream point in the river the runoff is the sum of the upstream inflow. Sauquet *et al.* (2000) developed this methodology further and combined it with a system for structuring hydrographical networks in a hierarchical way called HydroDem (Leblois & Sauquet, 2000). It allows an effective reconstruction of the variation of mean annual runoff (first order moment) along the river network in a basin from discharge observations and a DEM. The resolution of the underlying DEM defines the size of computational units (grid cells, sub-basins). The applications below use maps of the long-term annual mean derived in this manner.

MAPPING MEAN MONTHLY RUNOFF PATTERN

Variables under study

Traditionally, the variable considered is runoff observed at the outlet at gauging stations. When measurements at a gauging station are directly used in the interpolation scheme, redundant information is introduced due to partial overlapping drainage areas resulting in bias in the spatial analysis. Here, runoff generated by each portion of the basin between two or more gauging stations or uppermost headwaters are considered instead. These runoff values are calculated by subtracting the discharge(s) measured upstream from the value observed downstream. If N gauging stations are available, N basins or sub-basins A_i ,i = 1, ..., N can be defined and N related runoff values computed. To eliminate scale effects within the dataset due to the size of the basin, mean annual runoff $qa(A_i)$ and the 12 monthly discharges $qm(A_i, t)$, t = 1, …, 12 are expressed in mm year^{-1} and mm month^{-1}, respectively.

The pattern of monthly values is estimated following a procedure that accounts for the dependence between consecutive monthly discharges imposed by groundwater storage, flow routing in addition to scale effect, and links to topological river patterns. Temporal disaggregation of the mean annual discharge is considered to ensure consistency with the map of mean annual runoff. The interpolation will deal with dimensionless discharges, i.e. the 12 coefficients:

$$Z(A,t) = \frac{qm(A,t)}{qa(A)} \tag{1}$$

where $qa(A)$ is the mean annual runoff. $\overline{qm}(A)$ is the average of the 12 long-term monthly flows and an approximation is given by:

$$\overline{qm}(A) = qa(A)/12 \tag{2}$$

and:

$$\overline{Z}(A) \approx 1/12 \tag{3}$$

To keep the month-to-month dependence within the flow pattern, the time-series of Z is interpreted as linear combinations of L temporal functions β_i invariant in space:

$$Z(A,t) = \overline{Z}(A) + \sum_{i=1}^{L} \alpha_i(A)\beta_i(t), t = 1, ..., 12 \tag{4}$$

where $L \leq 12$ and the weight coefficients $\alpha_i(A)$ vary in space but are constant in time. The Empirical Orthogonal Function (EOF) analysis (Holmström, 1963) is a convenient method to extract the L significant amplitude functions from a dataset of observed times-series. α_i, i = 1, …, L are the eigenvectors of the covariance matrix between all the sub-basins. The number of variables to be interpolated is reduced from 12 to L.

An adaptation of kriging

With kriging the estimated value $z(\boldsymbol{u})$ at location $\boldsymbol{u}$ is a linear combination of observed values $z(\boldsymbol{u}_i)$, $i = 1, \ldots, N$ located in the neighbourhood of $\boldsymbol{u}$:

$$z(\boldsymbol{u}) = \sum_{i=1}^{N} \lambda_i z(\boldsymbol{u}_i) \tag{5}$$

where the weights $\lambda_i, i = 1, \ldots, N$ are found by minimizing the expected error, under an unbiasedness constraint (i.e. the expected bias is equal zero). Under the assumption that the process is homogeneous, this leads to the resolution of the following linear system (Matheron, 1965):

$$\begin{cases} \sum_{j=1}^{N} \lambda_j(\boldsymbol{u})\gamma(|\boldsymbol{u}_i - \boldsymbol{u}_j|) - \mu(\boldsymbol{u}) = \gamma(|\boldsymbol{u}_i - \boldsymbol{u}|), i = 1,\ldots, N \\ \sum_{j=1}^{N} \lambda_j(\boldsymbol{u}) = 1 \end{cases} \tag{6}$$

where: $|\mathbf{u}_i - \mathbf{u}_j|$ denotes the Euclidian distance between $\mathbf{u}_i$ and $\mathbf{u}_j$, μ is a Lagrangian multiplier and γ is the theoretical model function to the distance fitted to the empirical semi-variogram. This approach is widely used in the interpolation of meteorological fields (e. g. Creutin & Obled, 1982) but needs to be adapted for runoff features z_q that are related to areas. In particular, a relevant distance between pairs of basins has to be defined. Huang & Yang (1998) used the distance between the centres of gravity. The main drawback in this approach is that several basins may have the same centre of gravity. That is why the average of all possible distances between two sub-basins A and B, as suggested by Ghosh (1951), is used here instead. This distance is determined as:

$$d(A,B) = \frac{1}{AB} \iint_{A,B} |\boldsymbol{u}_A - \boldsymbol{u}_B| d\boldsymbol{u}_A d\boldsymbol{u}_B \tag{7}$$

Possible theoretical models of the semi-variogram are tested and compared graphically to the empirical semi-variogram. The selected function for γ is the one giving the best fit. The runoff characteristic $z_q(A)$ related to the element A is calculated using the weighted linear combination of N observed values $z_q(A_i)$, $i = 1, \ldots, N$:

$$z_q(A) = \sum_{i=1}^{N} \lambda_i z_q(A_i) \tag{8}$$

When spatial homogeneity is rejected, an empirical formula linking the area-related runoff values z_q to K basin characteristics X_i, $i = 1, \ldots, K$ is fitted:

$$z_q{}^*(A) = f(X_i(A)) \tag{9}$$

$\varepsilon_q(A) = z_q(A) - z_q{}^*(A)$ is the residual interpolated by kriging under the assumption of a second-order stationarity random field. Including basin descriptors in the empirical formulas is a way of accounting for the fact that streamflow data result from processes operating over the whole basin. The combination of the map of the residuals and the map of z_q* gives an estimate of z_q for each area ΔA_i, $i = 1, \ldots, M$. A special case is when runoff is estimated on elements of a partition of the study area ΔA_i, $i = 1, \ldots, M$. Estimated runoff values can be provided along the river network at any point **u** considered as the outlet of the upstream area. The annual and monthly discharges are the sums of the runoffs generated in all fundamental units ΔA_i flowing into that location **u**.

$$qa(\boldsymbol{u}) = \frac{1}{A} \sum_{\Delta A_i \subset A} qa(\Delta A_i)\Delta A_i \tag{10}$$

$$qm(\boldsymbol{u},t) = \frac{1}{A} \sum_{\Delta A_i \subset A} qm(\Delta A_i)\Delta A_i \text{ , } t = 1, \ldots, 12 \tag{11}$$

where A is the drainage area at location **u** and discharge are expressed in mm. Equation (11) can be developed using equation (3) and equation (4):

$$qm(\boldsymbol{u},t) = \frac{1}{A} \sum_{\Delta A_i \subset A} qa(\Delta A_i)Z(\Delta A_i, t)\Delta A_i = \frac{1}{A} \sum_{\Delta A_i \subset A} qa(\Delta A_i)\left[\frac{1}{12} + \sum_{i=1}^{L} \alpha_i(A)\beta_i(t)\right]\Delta A_i \tag{12}$$

REGIONALIZATION OF FLOW DURATION CURVES

Flow Duration Curves (FDC) are widely used in applied hydrology. A common practice for regionalization of FDCs is to derive a dimensionless FDC (DFDC) from observed FDCs valid within a homogeneous region, assuming that this DFDC can be transferred to a site with no data. This DFDC is obtained by dividing all observations by the mean annual flow (Yu & Yang, 1996, Singh *et al.*, 2001). Here the same standardization procedure is used, but the regionalization is supported by scaling properties, i.e. a relationship between the parameters of the DFDC and the first moments of the time series. First we consider the normalized daily runoff, in the same way as before (equation (1)):

$$Z(A,t) = q(A,t)/\,qa(A) \tag{13}$$

and then model the covariance between the values $Z(A_1)$ and $Z(A_2)$ over two (basin) averages with areas A_1 and A_2, respectively, by a theoretical function:

$$\mathrm{Cov}[Z(A_1); Z(A_2)] = \frac{1}{A_1 A_2} \int_{A_1}\int_{A_1} Cov(\boldsymbol{u}', \boldsymbol{u}'')d\boldsymbol{u}'d\boldsymbol{u}'' = \frac{1}{A_1 A_2} \int_{A_1}\int_{A_1} \sigma(\boldsymbol{u}')\sigma(\boldsymbol{u}'')\rho(\boldsymbol{u}', \boldsymbol{u}'')d\boldsymbol{u}'d\boldsymbol{u}'' \tag{14}$$

where $Cov(\mathbf{u}', \mathbf{u}'')$ and $\rho(\boldsymbol{u}', \boldsymbol{u}'')$ is the point covariance and correlation functions, $\sigma(\mathbf{u})$ the point standard deviation and with **u′** representing a point within the basin area A_1 and **u″** within the basin area A_2. The variance is formalized by:

$$\mathrm{V}ar[Z(A)] = \frac{1}{A^2} \int_A\int_A Cov(\boldsymbol{u}', \boldsymbol{u}'')d\boldsymbol{u}'d\boldsymbol{u}'' = \frac{1}{A^2} \int_A\int_A \sigma(\boldsymbol{u}')\sigma(\boldsymbol{u}'')\rho(\boldsymbol{u}', \boldsymbol{u}'')d\boldsymbol{u}'d\boldsymbol{u}'' \tag{15}$$

Aggregated values were assigned to each couple of gauged basins applying equation (15). The choice of the best point function is based on a comparison between theoretical values and empirical values. As we are dealing with normalized values $Z(A)$ the standard deviation equals the coefficient of variation of the original variable $q(A)$. These calculations allow mapping of the coefficient of variation along the river network. The next step is to establish the dependency between the DFDCs and use them to calculate percentiles of the normalized flow duration curve. The interpolated mean annual runoff (obtained by the procedure detailed in the previous section) is introduced to finally calculate the percentiles of the FDC (for more details see Krasovskaia *et al.*, 2005).

APPLICATION

The study areas

The study areas cover Costa Rica and France, except for the rivers flowing into the Mediterranean Sea. In France, the snowmelt-fed regimes are found in the mountainous part (high altitude rivers in the Pyreneans) in contrast to the northern and western part of the study area under Atlantic climate influences, where the pluvial regime is governed by rainfall and evaporation dominates (alluvial plains of the Seine basin and rivers from Brittany). In Costa Rica, tropical river flow regimes dominate with high flows in autumn and a temporal variability closely connected to the El Nino Southern Oscillation Index. The French and Costa Rican data sets will be used to illustrate mapping mean monthly runoff and flow duration curves, respectively.

Database

A total of 726 French stations and 70 Costa Rican stations with minor human impact were considered for this application. Mean annual runoff for the gauged basins ranges from 100 to more than 1500 mm in the French area. Mean annual runoff can be higher than 5 m year^{-1} in the centre of Costa Rica, varying on average around 2000–4000 mm year^{-1}.

The interpolation scheme requires a well-defined hierarchical structure of the river network to identify links with runoff variability and to be utilized thereafter in the interpolation procedure. The river network has been extracted from a raster digital elevation model DEM with 1 × 1 km cells. Basin attributes are calculated by combining GIS layers with the drainage pattern. The mean basin elevation was a single basin characteristic utilized as at large scale relief, is the most important physiographic factor. The French study area was divided into five major hydrological sub-areas based on topography. In Costa Rica the hydrological conditions are different at either side of the water divider between the Pacific and the Caribbean basins. Separate calculations have therefore been made for the Caribbean and Pacific regions.

Mapping mean monthly runoff, an example from France

Results are detailed for one region which includes the Dordogne basin, the Garonne basin and coastal rivers in the southwest of France. Runoff estimates were computed for more than 5100 elements of a partition of the study area ΔA_i, $i = 1, \ldots, M$, considered as fundamental units (i.e. M non-overlapping small target elements that form the whole study area).

EOF analysis was achieved using a sub dataset of 205 gauging stations. Eleven independent amplitude functions were identified and arranged in descending order according to their contribution to the explained variance. The first four of them $\beta_i, i = 1, \ldots, 4$ (Fig. 1) explained more than 98% of the total variance within the dataset, and consequently four weight coefficients were considered for the mapping. The amplitude function β_1 stands for the largest portion of the explained variance (84%) and describes the most common monthly pattern within the dataset.

The regionalization procedure is illustrated on the example of the mean annual runoff qa and the first weight coefficient α_1. Empirical relationships were established between runoff characteristics and H, the mean elevation:

$$qa^* = -0.0423H + 261.73 + 0.0735\ (R^2 = 0.40) \tag{16}$$

$$\alpha_1^* = -0.0263(H/1000)^2 + 0.0144(H/1000) + 0.0735\ (R^2 = 0.55) \tag{17}$$

The same procedure is applied to the three other weight coefficients α_i, $i = 2, \ldots, 4$. Residuals are calculated at the gauging stations and an exponential model is fitted to the experimental variogram (Fig. 2). Figure 3 displays the final map of the first weight coefficient α_1, which is a combination of the map of residuals ε_1 and the empirical formula application α_1^*. α_1 is close to zero for rivers with an even flow regime whereas a high positive value for α_1 shows a contrasted regime, with pronounced seasonal variation.

All the results from the interpolation were combined to estimate the long-term mean monthly pattern (Fig. 4). Low flows in January are noted in the Pyrenean sector

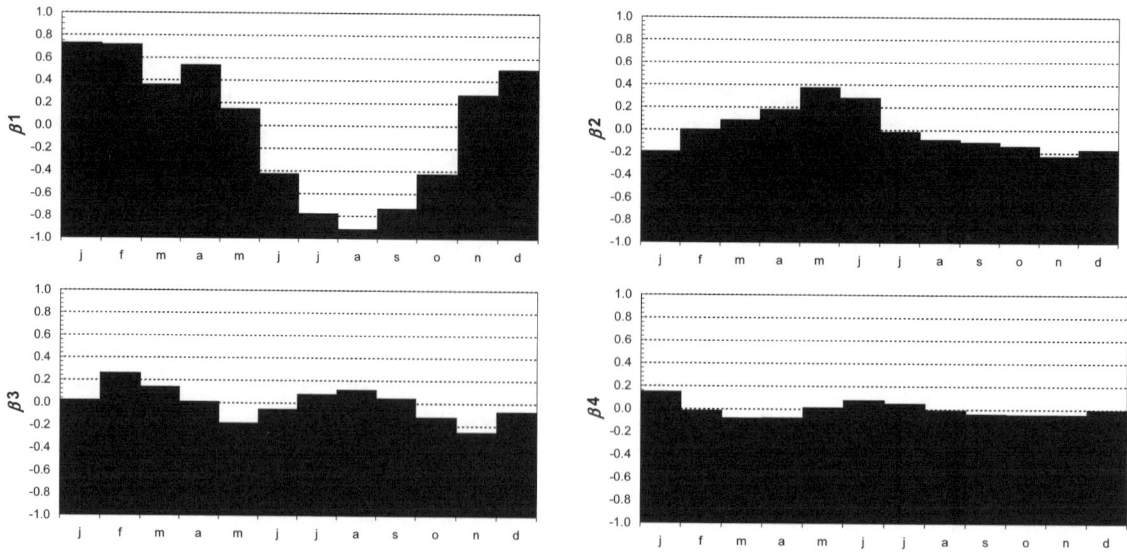

Fig. 1 Amplitude functions for southwestern France.

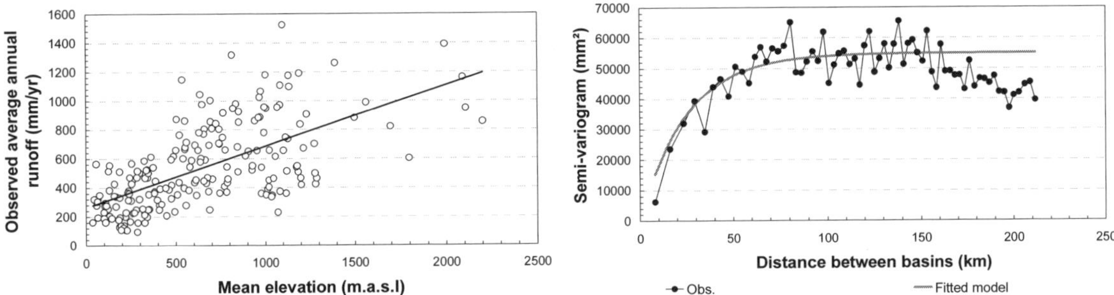

Fig. 2 Empirical relationship between mean annual runoff and mean elevation and semi-variograms for the residual of this relationship for the southwestern part of France.

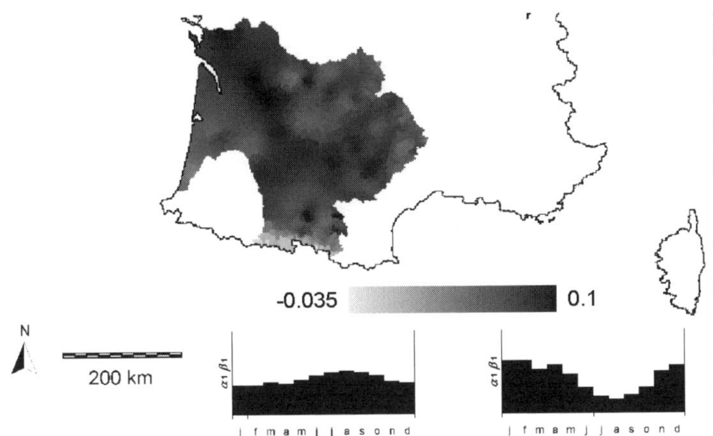

Fig. 3 Map of the first spatial component for the southwestern part of France.

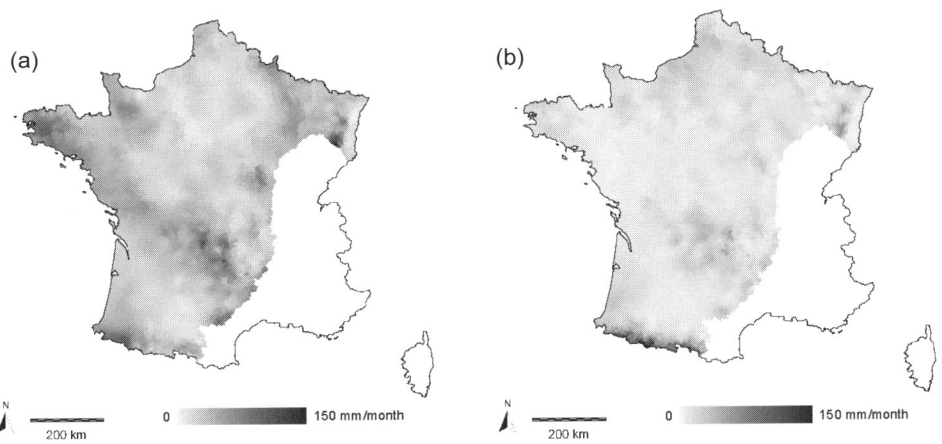

Fig. 4 Map of mean runoff for (a) January and (b) July for France.

while water is abundant in the northwestern part of France affected by oceanic rains. High values are found in the mountainous sector with moderate altitude. In July, the air temperature is above zero in the Pyrenean sector. Melting snow at high altitude in the Pyreneans in July generates monthly flow peaks, elsewhere evaporation processes are predominant and cause low flow in the rivers.

Regionalization of flow duration curves, an example from Costa Rica

The DFDC were mapped along the river net using the established strong dependency between the percentiles of the normalized runoff and the coefficient of variation, shown in Fig. 5, together with the interpolation procedures described in the previous chapter for interpolating the mean annual runoff and the coefficient of variation along the river net.

The dependence shown in Fig. 5 allows establishing a linear regression between the values of the variation coefficient and percentiles of the normalized runoff. As example, the relationship between the normalized percentile $p = 1\%$ and the coefficient of variation is linear:

$$Z_{p=1\%} = 4.99Cv + 0.12 \quad (R^2{=}0.93) \tag{18}$$

It is interesting to note that at the duration level of 80% there is almost no dependence on the coefficient of variation, while it is high for, for example, 50%, i.e. the median.

The map of the coefficient of variation over Costa Rica is shown in Fig. 6. This map is consistent with the statistical law for variance reduction of the mean annual runoff (not shown in this paper). Figure 7 shows the Dimensionless Flow Duration Curves (DFDC) for the Caribbean region. It can be noticed that almost all DFDCs

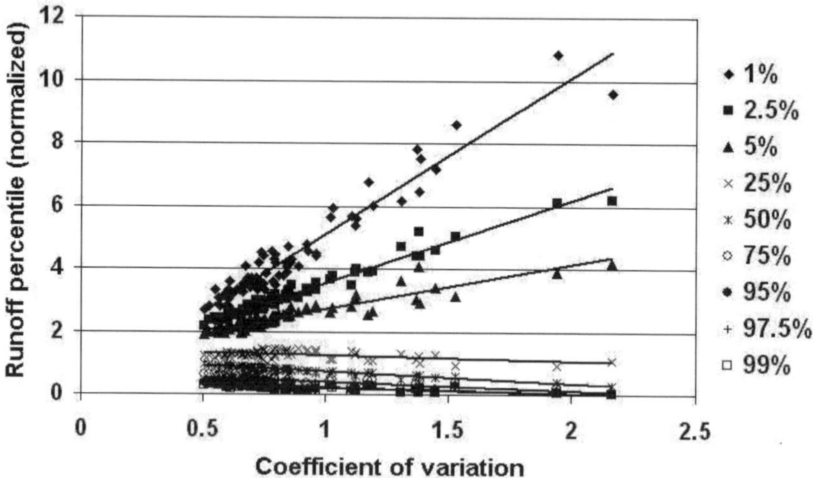

Fig. 5 Relationship between the coefficient of variation and normalized percentiles in Costa Rica.

have a value equal to one when the duration is close to 30%. This property is also observed under other climate conditions.

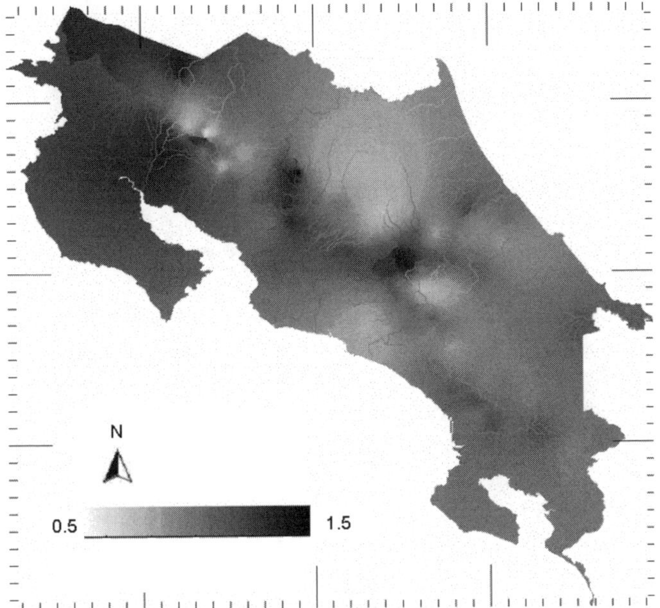

Fig. 6 Map of the coefficient of variation of daily runoff for Costa Rica. The map shows basin values along rivers.

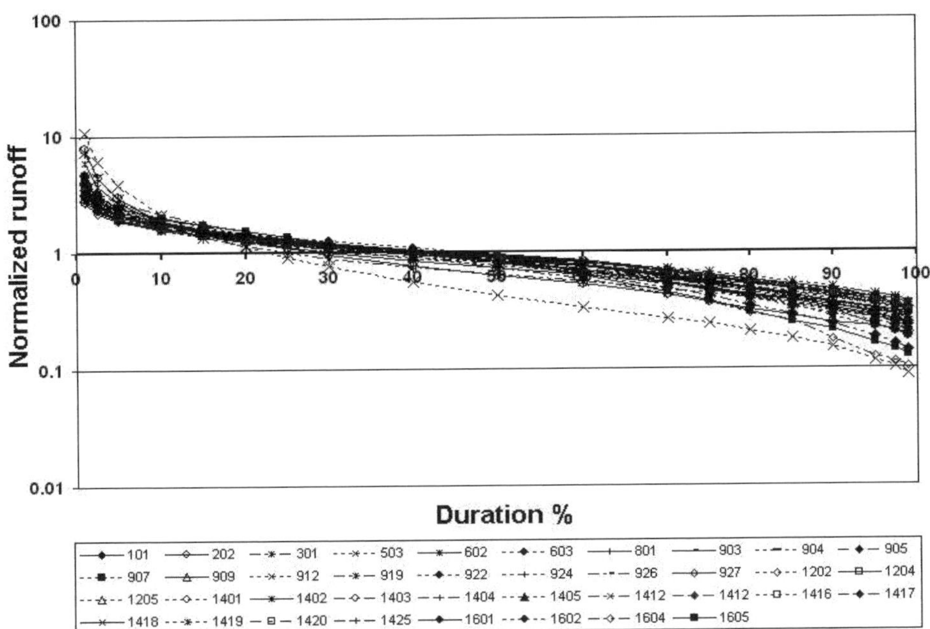

Fig. 7 Observed Dimensionless Flow Duration Curves (Caribbean area, Costa Rica).

Thus, the procedure for estimating the daily runoff corresponding to the percentiles p, is as follows: extract the value of the coefficient of variation and the mean annual runoff from the map, apply the relationships between the coefficient of variation and the normalized percentiles and multiply the obtained value by the extracted mean annual runoff value.

CONCLUSIONS

An approach to estimate runoff characteristics at ungauged locations based on a hydrostochastic concept is presented. Runoff is the specific focus for the approach with due consideration that this variable represents an integrated value over a basin, i.e. in theoretical terms a generalized random process in time and space. Furthermore the approach benefits from the fact that different statistical parameters are internally linked by statistical and physical laws. The basic parameters are the long-term mean and the variance. The developments have been demonstrated on two examples from France and Costa Rica. These developments have one common point: the map of mean annual runoff from which monthly pattern and DFDC can be straightforwardly derived. In practice, the same normalized variable is studied at different time scale and its spatial properties are introduced to estimate monthly runoff and percentiles.

The hydrostochastic approach is a "top-down" alternative to "bottom-up" rainfall–runoff modelling and even complements the latter for validation purposes, as rainfall is not involved in the interpolation framework. Comparison of maps derived by each method, respectively, is a convenient way to identify errors in measurements as well as in interpolation or modelling. Noticeable differences between the interpolated values and those deduced from the models may be a sign of deficiency of one method or of inconsistency in the data. Modelling should benefit from these developments, as one major difficulty is to improve the estimate of the parameters of lumped conceptual models.

Acknowledgements The authors wish to express their thanks to the French data base HYDRO and Instituto Costarricense de Electricidad for supplying the discharge data used in this study.

REFERENCES

Creutin, J. D. & Obled C. (1982) Objective analysis and mapping techniques for rainfall fields an objective comparison. *Water Resour. Res.* **18**, 413–431.

Gannett, H. (1912) Map of the United States showing mean annual runoff. In surface water supply of the United States, 1911. *US Geol. Survey Water Supply Papers* 301–312, Government Printing Office, Washington, DC, USA, pt. II.

Ghosh, B. (1951) Random distances within a rectangle and between two rectangles. *Bull. Calcutta Math. Soc.* **43,** 17–24.

Gottschalk, L. & Krasovskaia, I (1998) Development of grid-related estimates of hydrological variables. *Report of the WCP-Water Project B.3, Geneva, WCP/WCA, February 1998.*

Gottschalk, L. (1993) Interpolation of runoff applying objective methods. *Stochast. Hydrol. Hydraul.* **7**, 269–281.

Gustard, A, Roald, L. A., Demuth, S., Lumadjeng, H. S. & Gross R. (1989) *Flow Regimes from Experimental and Network Data (FREND)*, Vol. 1. UNESCO, Paris, France.

Hawley, E. M. & MacCuen, R.H. (1982) Water yield estimation in western United States. *J. Irrig. Drain. Div.* **108,** 25–34.

Holmström, I. (1963) On a method for parametric representation of the state of the atmosphere. *Tellus* **15**, 127–149.

Huang, W. C. & Yang, F. T. (1998) Streamflow estimation using Kriging. *Water Resour. Res.* **34**, 1599–1608.

Krasovskaia, I., Gottschalk, L, Leblois, E. & Pacheco, A. (2006) Regionalization of flow duration curves. In: *Climate Variability and Change—Hydrological Impacts* (ed. by S. Demuth, A. Gustard, E. Planos, F. Scatena & E. Servat (Proc. Fifth FRIEND World Conference held at Havana, Cuba, November 2006), 105–110. IAHS Publ. 308. IAHS Press, Wallingford, UK.

Leblois, E. & Sauquet, E. (2000) Grid elevation models in hydrology—Part 1: Principles and literature review; Part 2: HydroDEM. User's manual. Cemagref, Technical notes, Lyon, France.

Liebscher, H. (1972) A method for runoff-mapping from precipitation and air temperature data. In: *Proc. Symp. World Water Balance* (Reading, 1970), vol. 1, 115–121. IAHS Publ. 92. IAHS Press, Wallingford, UK.

Matheron, G. (1965) Les variables régionalisées et leur estimation. Une application de la théorie des fonctions aléatoires aux sciences de la nature. Ed. Masson, Paris, France (in French).

Singh, R. D., Mishra, S. K. & Chowdhary, H. (2001) Regional flow-duration models for large number of ungauged Himalayan catchments for planning microhydro projects. *J. Hydrol. Engng* **6**(4), 310–316.

Sauquet, E., Gottschalk, L. & Leblois, E. (2000a) Mapping average annual runoff: a hierarchical approach applying a stochastic interpolation scheme. *Hydrol. Sci. J.* **45**, 799–816.

Sauquet E., Krasovskaïa, I. & Leblois, E. (2000b) Mapping mean monthly runoff pattern using EOF analysis. *Hydrol. Earth System Sci.* **4**, 79–93.

Skøien, J. O., Merz, R. & Blöschl G. (2005) Top-kriging geostatistics on stream networks. *Hydrol. Earth System Sci.*, Discussions, **2**, 2253–2286.

Solomon, S. I., Denouvilliez, J. P., Chart, E. J., Woolley, J. A. & Cadou C. (1968) The use of a square grid system for computer estimation of precipitation, temperature, and runoff. *Water Resour. Res.* **4**, 919–929.

Vogel, R. M., Wilson, I. & Daly, C. (1999) Regional regression models of annual streamflow for the United States. *J. Irrig. Drain. Engng*, **125**, 148–157.

Yu, P. S. & Yang, T. C. (1996) Synthetic regional flow duration curve for southern Taiwan. *Hydrol. Processes* **10**(3), 373–391.

Accounting for spatial variability: a way to improve lumped modelling approaches? An assessment on 3300 chimera catchments

M. BOURQUI[1], C. LOUMAGNE[1], N. CHAHINIAN[2] & M. PLANTIER[1]

1 *CEMAGREF, Parc de Tourvoie BP 44, F-92163 Antony cedex, France*
marie.bourqui@cemagref.fr

2 *Institut de Recherche pour le Développement(IRD), UMR HydroSciences–Université Montpellier II, Case Courrier MSE, Place Eugène Bataillon, 34095 Montpellier cedex 5, France*

Abstract Recent progress in collecting spatialized data with remote sensing techniques should allow the accounting for: (i) the spatial variability of rainfall and (ii) the basins' physical characteristics in rainfall–runoff models. To benefit from this spatial information, lumped approaches can quite easily be replaced by semi-distributed approaches. However, two questions need to be investigated. Does integrating additional information into a semi-distributed approach successfully improve the performance of flow simulations at the basin outlet? Which type of heterogeneity should first be taken into account to yield the most significant improvements? This paper presents a method to account for basin heterogeneity in lumped and semi-distributed models through the use of indices. Given the requirement for a large database to produce statistically significant results, "chimera" basins (virtual aggregation of two real basins) were used. We characterized 212 French basins using approximately 50 indices of pedology, geology, morphology and land use. Lumped and semi-distributed versions of a rainfall–runoff were compared on 3300 chimera basins. Results indicate that integrating "useful" spatial data in a lumped model can improve its performance without altering its parsimonious structure. Some indices correlated with rainfall confirm that the semi-distributed approach is more advantageous than the lumped approach for basins with high spatial variability of precipitation. The possible relations between physical characteristics and model parameters are investigated to help regionalization attempts and hence improve modelling abilities in ungauged basins.

Key words basin heterogeneities; disaggregation; rainfall–runoff modelling

INTRODUCTION

For water resources management and flood forecasting lumped rainfall–runoff models are well adapted to the requirements of operational applications. Simple models can efficiently represent the rainfall–runoff transformation while using a limited number of parameters (Perrin *et al.*, 2001).

However, estimating even a small number of parameters remains a major problem in the case of ungauged basins. Estimating parameters based on the basin physical characteristics is even more difficult when these characteristics are variable in space and time (Beven *et al.*, 1988; Diermanse, 1998; McDonnell, 2004). In addition, the spatial and temporal variability of the rainfall distribution can affect the runoff distribution (Wilson, 1979; Arnaud *et al.*, 2002). Taking this variability into account should be easier today, thanks to the high availability of spatialized data.

Many studies have attempted to take these variabilities into account through different lumped and distributed approaches. (Ambroise, 1995; Refsgaard & Knudsen, 1996; Krysanova *et al.*, 1999). However, most results are based on a limited number of basins, which leaves several questions open on the advantages of taking variability into account in rainfall–runoff modelling:

(a) Starting with a lumped approach, does taking spatial heterogeneities into account by dividing the basin into sub-basins effectively improve performance of flow simulations on a wide range of basins and climates?
(b) To improve runoff simulation, do we need to take the spatial variability into account by treating it in a distributed way?

The objective of this paper is first to compare the efficiency of different lumped and semi-distributed modelling strategies. Then the relationships between the temporal and spatial variations observed within a basin are analysed to investigate how model efficiency can be improved by semi-distribution of the inputs and the basin characteristics.

To meet these objectives, a database of French basins was built. Physical attributes (morphology, geology, vegetation and pedology) were collected for all basins.

For semi-distributed approaches, we used a methodology that uses virtual basins called chimeras (Andreassian *et al.*, 2004), which are the combination of two real, similarly sized basins that are located in different geographical areas. The exaggerated heterogeneity of these basins should allow us to understand which basin characteristics are interesting to use for the semi-distribution. To evaluate the amount of heterogeneity of each virtual basin, we used some global indices computable on each basin.

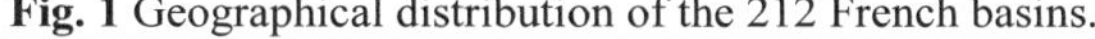

Fig. 1 Geographical distribution of the 212 French basins.

MATERIAL AND METHODS

Database

The streamflow data was taken from a sample of 212 basins in France representing a wide range of area, geomorphology and climatic conditions. Mean daily rainfall, streamflow and mean potential evapotranspiration data were available at a daily time-step. The distribution of the basins is illustrated in Fig. 1 and the main hydrological characteristics are shown in Table 1. This database has a good geographical diversity to guarantee the generality of our results. For every basin we used a GIS database and a digital terrain model (DTM) to calculate a number of geological, pedological and geomorphological characteristics that could be potentially useful to explain the hydrological features of the basin. The databases and the distribution of each type of information are summarized in Table 2. For each basin, approximately 30 indices, representing a majority of the invariant physical characteristics that could influence runoff formation, were calculated.

Table 1 Annual hydro-climatic characteristics of the real sample.

	Min.	Median	Max.
Mean annual rainfall (mm)	620	940	2300
Mean annual ETP (mm)	640	720	1250
Mean annual runoff (mm)	20	420	1960
Basin area (km^2)	7	136	43800
Time series length (years)	5	13	36

Table 2 Type, origin and scale of data bases used to establish the physiographical characteristics of the basins and classes of attributes chosen for each type.

Genre	Origin of the data bases and scale	Defined types et represented classes	
Pedology	SGBDE The Soil Geographical Database of Europe at Scale 1:1000 000, 1996	SOIL TYPE (5 CLASSES) [Cambisol], [Podzoluvisol], [Rendzina], [Lithosol], [Fluvisol]	TEXTURE (4 classes) [Coarse], [Medium], [Medium fine], [Fine]
Geology	BRGM Numerical geological map at 1:100 000 000 (6' edition,1996) + diverses geologicals maps	LITHOLOGY (6 CLASSES) [Alluvial deposits, Ice deposits and Sends], [Massives limestones], [Chalks, Molasses], [Marls], [Basaltic crystalline magmatic rocks], [Shists and metamorphous,détrital rocks]	BED ROCK PERMEABILITY (4 classes) [Impermeables], [Permeables with cracks], [Permeables with interstice], [Few permeables]
Land Use	Corine Land Cover (Source IFEN, 2000) at scale 1:500 000	7 CLASSES [Urban areas], [Arable and irrigated lands and permanent crops], [Heterogeneous agricultural zones], [Prairies], [Forests], [Végétation arbusive], [Natural areas without vegetation]	
Morphology	Logiciel River Tools cuple with a DTM at mesh of 75 meters	5 CLASSES [saturability], [slope and form], [arborescence of hydrological network], [hydrological response]	

Model and methodology

The lumped model used for this study is the GR4J model (Perrin *et al.*, 2003). It is a daily four-parameter model. It was tested on more than 400 basins in several countries. The methodology adopted here to create chimera basin was developed by Andreassian *et al.* (2004). The chimera construction and the model parameterization strategies are illustrated in Fig. 2. Based on 212 real basins, this technique provided nearly 3300 highly heterogeneous virtual basins with known intermediate flows.

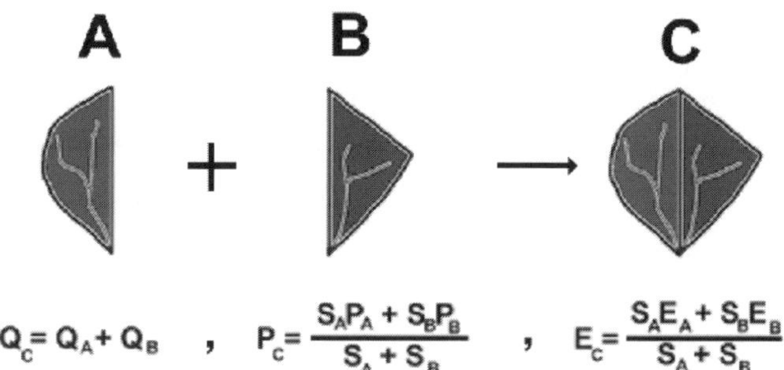

Fig. 2 Construction of the chimera basin C from sub-basins A and B with a surface area of S_A and S_B, respectively. Computation of data input for the sub-basin C from P_A and P_B precipitations Q_A ,Q_B streamflows, and E_a E_b potential evapotranspiration.

Our goal here was to evaluate the efficiency of three modelling approaches differentiated by their disaggregation level: a classical initial lumped approach (IL), an intermediate semi-distributed approach (rainfall SD) where only the mean rainfall of each sub-basin is distributed and finally a true semi-distributed approach (true or total SD).

For the lumped approach, a single parameter vector is optimized with a single rainfall P_c and a single evapotranspiration E_c variable as input data. For the intermediate approach, we used two sub-models for sub-basins A and B, each one having as input data P_a, E_a and P_b, E_b, respectively. However, the same parameter vector was used in calibration for the two sub-models. Consequently, this approach is also defined as semi-lumped. The simulated flow Q_c was obtained by the sum of two simulated intermediary flows Q_b and Q_a. For the last approach, called the true semi-distributed approach, we had two models running in parallel, each one fed by rainfall and the potential evapotranspiration of sub-basins A and B. Two parameter vectors, for each sub-basin, were optimized. This gave a semi-distribution of the input data and the parameters.

Optimization and performance evaluation The optimization algorithm is the step by step method (Edijatno *et al.*, 1999). We used the split-sample test framework. (Klemeš, 1986) to assess models. The objective function chosen is the *C2M* criterion (Mathevet, 2005) defined by:

$$C2M(\%) = \frac{Nash}{(200 - Nash)} * 100 \quad (1)$$

in which Nash (%) is the Nash & Sutcliffe (1970) criterion. The *C2M* is bounded in the [–100; 100] domain.

The performance of the three approaches was compared in validation mode on periods different from those used in calibration mode. The $C2M_Q$ criterion used in validation mode, is a criterion based on the quadratic error between flows as the evaluation criterion, whereas in calibration mode we used this criterion based on the errors between the $\sqrt{Q}$. We will compare the average values of $C2M_Q$ obtained for the three approaches in validation for the whole sample of chimera basins.

Quantifying the heterogeneities of the chimeras To evaluate the level of heterogeneity of the different physical characteristics observed within each chimera, we developed a simple heterogeneity index. We found it necessary to choose a heterogeneity index that can be calculated on each basin and for physical each characteristic. We used the following index, applied to each physical characteristic described and each chimera. We define the distance d_x between basins by:

$$d_X = \frac{|I_A - I_B|}{I_{\max} - I_{\min}} \tag{2}$$

where I_A and I_B are the values of the physical descriptor *X* for sub-basins *A* and *B*, and $I_{\max}$ and $I_{\min}$ the maximum and minimum values, respectively, of descriptor *X* on the initial sample of 212 basins. This heterogeneity index varies between 0 for the homogeneous chimera basins and 1 for the most heterogeneous basins of the sample.

Relation between heterogeneities and improvement We originally hypothesized that the semi-distributed approach would benefit more to the basins with the highest variability indices. To test the validity of this hypothesis, we sought to relate the performance improvement from lumped to semi-distributed approaches to the level of heterogeneity for different physical characteristics. Exaggerating the natural heterogeneity of the basins using the chimera method allowed us to highlight the physical characteristics that, in the case of high variability, made semi-distribution advantageous. Relations between the model performance considering only the semi-distribution of the parameters (parameter SD) called here $\Delta C2M$ (equation (3)) and the heterogeneity index of the different physical characteristics evaluated have thus been established:

$$\Delta C2M = [C2M_Q(SDtotally) - C2M_Q(SDrain)] \tag{3}$$

RESULTS AND DISCUSSION

The difference in performance provided by the criterion $C2M_Q$ between the lumped approach and the totally semi-distributed (SD) approach on the 3300 chimera basins is presented in Fig. 3. The points above the (1:1) line represent the basins whose performance was improved by semi-distribution. It should be noted that approximately 70% of the basins had significantly improved (greater than 1% of $C2M_Q$) and approximately 14% were degraded (greater than –1%). However, there was enormous variability in the performance improvements. For the group of basins whose

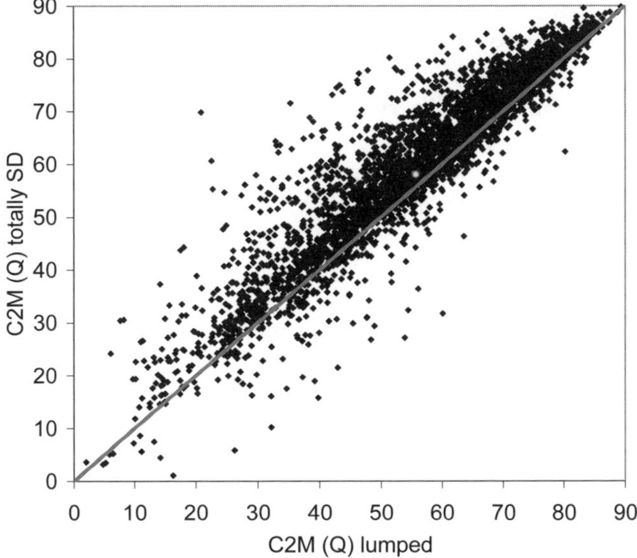

Fig. 3 Model mean efficiency ($C2M_Q$) in validation mode on the whole sample: IL *vs* totally SD approach.

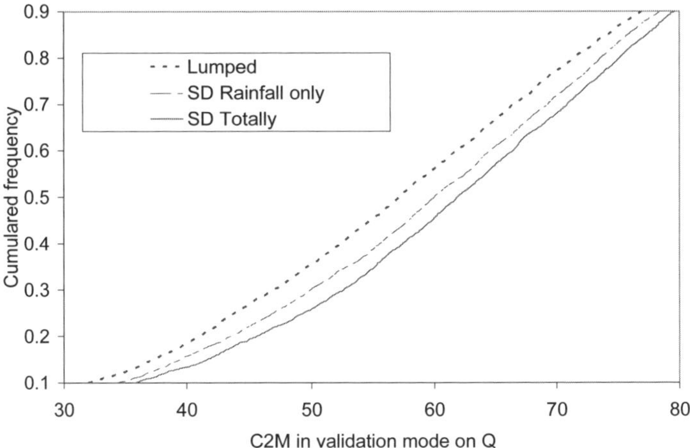

Fig. 4 Cumulative frequency curves of the mean efficiency ($C2M_Q$) in validation mode for the 3300 chimera with IL and the two SD approaches.

performance in lumped mode was already high (over 75%), no major improvement was observed. In fact a high criterion may indicate that the lumped approach is already well suited to these basins. Thus, it will be more difficult to improve results. For the group of basins whose criterion was lower, clear improvements due to semi-distribution were observed, as well as a greater number of degradations.

In terms of cumulative frequency of the mean efficiency, both semi-distributed versions significantly overtouch the initial lumped version (Fig. 4). Table 3 indicates the average performances for the whole sample obtained by the three approaches. The true SD version was the best, with a mean difference of 4% in validation. The rainfall SD approach provided a mean gain over the lumped approach of approximately 2.4%. The improvement attributable only to the distribution of the parameters, independent from the improvement caused by the semi-distribution of rainfall, was (4–2.4%) thus 1.6%. Semi-distribution of rainfall alone was responsible in average of 60% of the improvement. Moreover, the greatest improvements can be explained by the non-correlation of rainfall.

Table 3 Average *C2M(Q)* for the whole sample of chimera for the three different approaches.

C2M(Q)	Lumped approach	SD Rainfall only	SD totally
Average	55.8	58.1	59.8
Median	57.1	60.0	61.9
Percentile 10%	31.9	34.3	35.8
Percentile 90%	79.9	78.6	79.17

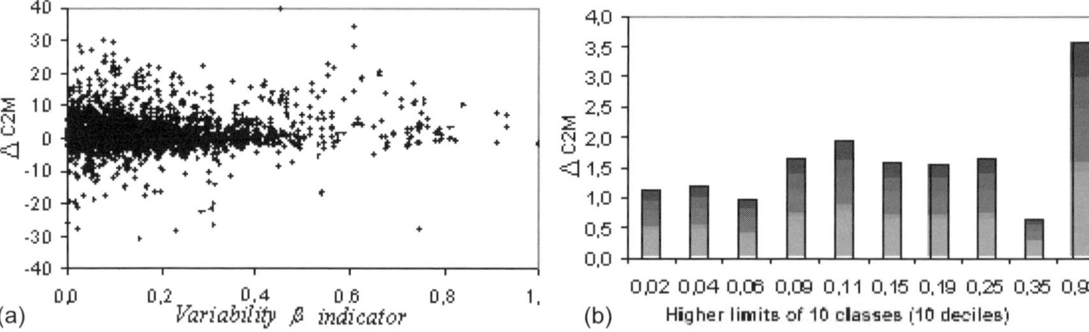

Fig. 5 (a) Mean efficiency gain obtained with parameters SD approach in function of heterogeneity level of β indicator on the whole sample; (b) same result but presented in function of increasing classes of the distribution of β heterogeneity indicator.

Then we compared the different types of heterogeneity and the gains obtained by the SD parameter approach ($\Delta C2M$). In some cases, "positive" trends were observed, i.e. the higher the heterogeneity, the higher the gain obtained by the SD parameter approach.

For the indexes on the geomorphology and the hydrographic network, the types of heterogeneity presenting a "positive" relation were β form factor (Moussa & Bocquillon, 1993), the drainage network density (DD) (Horton, 1945) and the hypsometric integral (HI) (Chow, 1964).

Figure 5(a) shows the relationship obtained for the heterogeneity indicator of the β form index for all the chimeras of the sample. It can be observed that even though virtual basins were used, the majority of these indicators had values lower than 0.5, and, although improvements for the criterion are present, substantial degradations also exist. On the other hand, the basins whose heterogeneity was greater than 0.5 were systematically improved. Figure 5(b) summarizes these results through the means of the gains obtained on ten crescent classes of this index. These classes represent the ten crescent deciles of the index's distribution. This leads to the conclusion that on average for this index, a "positive" relationship is obtained between the improvement using the SD parameter approach and the basin heterogeneity index. The same types of observation were made for the indexes representing the heterogeneity of the drainage network density (Fig. 6) and the hypsometric integral (Fig. 7). The other indicators involving the basin morphology showed no significant relationships.

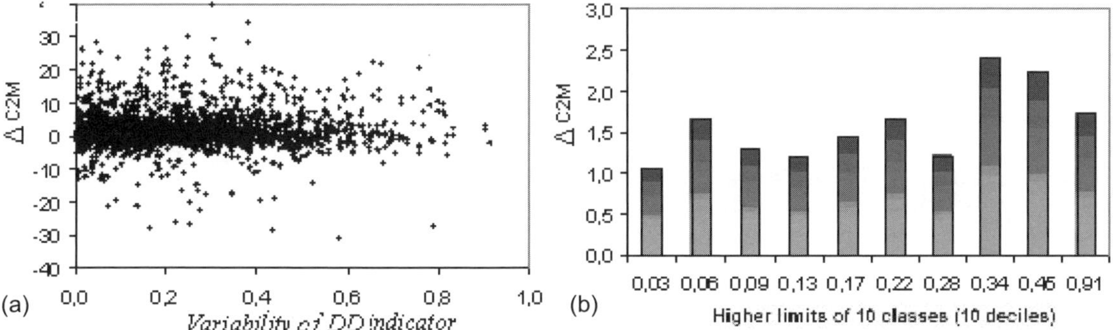

Fig. 6 (a) Mean efficiency improvement obtained with the SD parameters approach as a function of the heterogeneity level of the DD indicator on the whole sample; (b) Same result but presented as a function of increasing classes of the distribution of DD heterogeneity indicator.

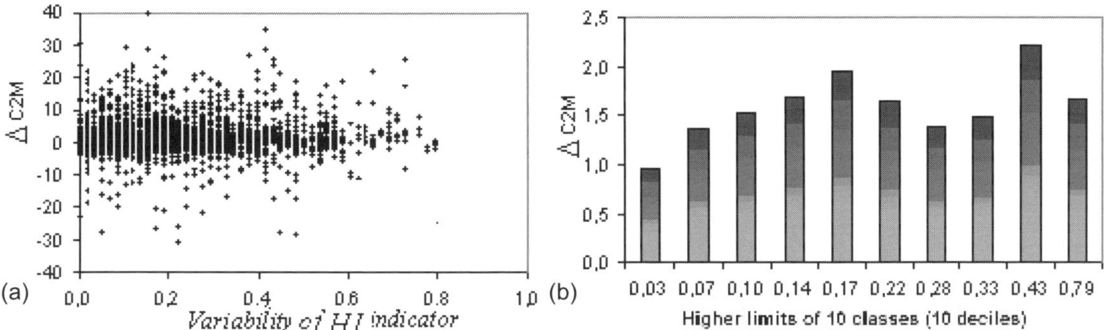

Fig. 7 (a) Mean efficiency improvement obtained with the SD approach as a function of the heterogeneity level of the HI indicator on the whole sample; (b) same result but presented as a function of increasing classes of this indicator.

For the geological indexes, the limestone indexes show a positive trend between the mean gains obtained and their heterogeneity rate within the chimeras (Fig. 8(a)), as for the permeable rock indicator (Fig. 8(b)). For land use, only the combined indicator of arable land and heterogeneous agricultural land described the same type of relation (Fig. 8(c)). For pedology, a positive trend is observed only on the amount of heterogeneity of Cambisol (Fig. 8(d)) and Podzoluvisol (Fig. 6(e)) type soils.

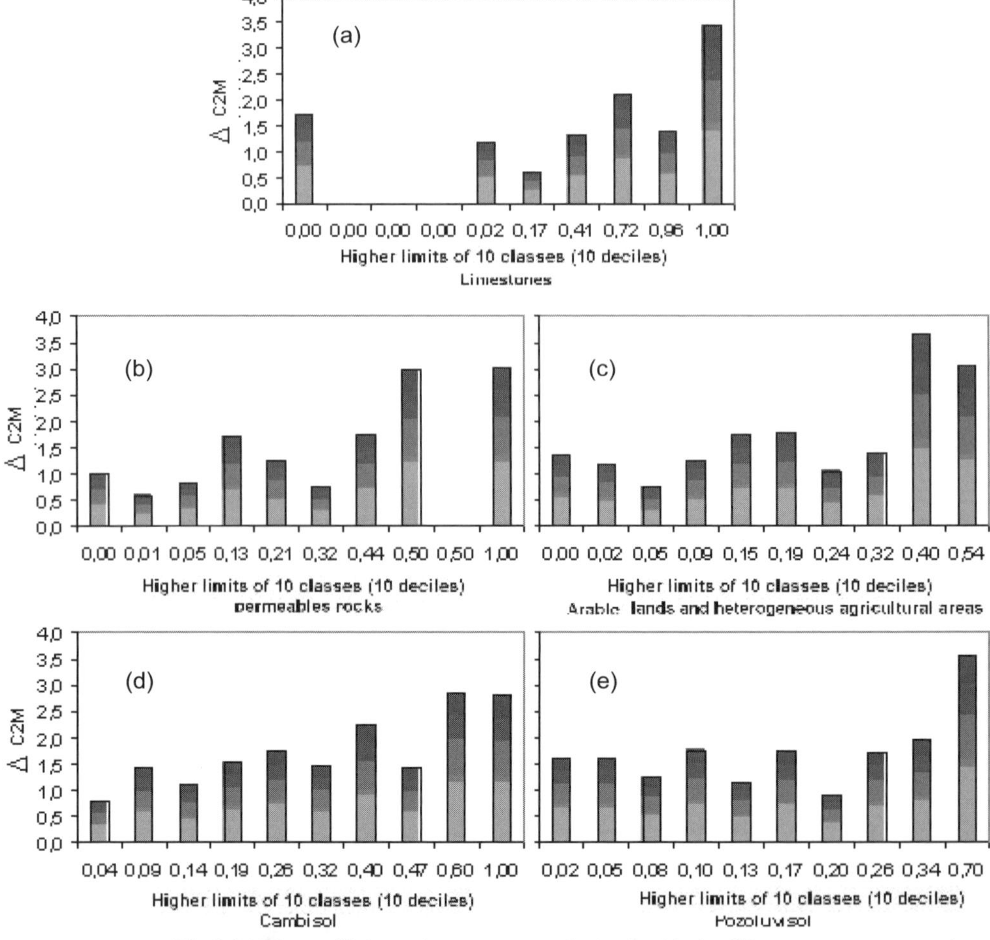

Fig 8 (a) Mean efficiency improvement obtained with the SD parameters approach as a function of increasing classes of the distribution obtained for the limestones indicator; (b) for permeable rocks; (c) for arable land added to heterogeneous agricultural area; (d) for Cambisol; (e) for Podzoluvisol.

The positive trends between the heterogeneity index of the basins and the gains obtained using the semi-distributed parameter approach only appeared for only a few types of the descriptors tested.

Nevertheless, the significance of this heterogeneity index is complex. It is difficult to physically interpret its value. For example, a value of 0.5 for the forest indicator means that the basin is covered 50% by forest and 50% by another land use type. But, there is no information on the composition of this other part.

These results are in line with those obtained by Merz & Blöschl (2004) who tried to regionalize the parameters of the HBV model (Bergström & Forsman, 1973) from basin attributes such as the percentage of various geological types, soil types, land uses and topographic indices on 308 Austrian catchments. This research shows the poor relationships between these attributes and the parameters, although the model parameters seem to represent the physics of the basins.

CONCLUSION

The objective of this study was to bring out the differences in performance between a lumped approach and semi-distributed approaches as well as to relate the basin variability to model performance improvements for artificially heterogeneous basins. This investigation has confirmed the relative superiority of the semi-distributed approaches. However, this result was not systematically observed on basins. The greater part of the improvement stems from taking the rainfall distribution into account. This is observed especially in the cases where the rainfall timeseries of each sub-basin are weakly correlated in time.

While analysing the impact of parameter semi-distribution for given indicators, relationships were found between the basin heterogeneities and the performance gains of the semi-distributed approach. This variation in performance is conditioned by the initial performance of the lumped approach: the basins with the weakest initial performance obtained the greatest gains. In fact it is easier to improve the modelling of the basins when a lumped approach is not suitable.

Similar tests at the hourly time-step should be interesting to study to continue this work. Indeed at this time step, the rainfall field is more decorrelated in time for a given distance. In addition, the relationships that emerged from the physical heterogeneities may be used in the definition of similar basins for the modelling of ungauged basins (Rojas Serna, 2005).

REFERENCES

Ambroise, B., Perrin, J. L. & Reutenauer, D. (1995) Multicriterion validation of a semidistributed conceptual model of the water cycle in the Fecht Catchment (Vosges Massi, France). *Water Resour. Res.* **31**(6), 1467–1481.

Andreassian, V. *et al.* , Oddos A. Michel C., Anctil F., Perrin C. and Loumagne, C.(2004) Impact of spatial aggregation of inputs and parameters on the efficiency of rainfall–runoff models: a theoritical study using chimera watersheds. *Water Resour. Res.* **40**(5),W05209, doi: 10.1029/2003WR002854.

Arnaud, P., Bouvier, C., Cisneros, L. & Dominguez, R. (2002) Influence of rainfall spatial variability on flood prediction. *J. Hydrol.* **260**(1–4), 216–230.

Bergström, S. & Forsman, A. (1973) Development of a conceptual deterministic rainfall–runoff model. *Nordic Hydrol.* **4**, 147–170.

Beven, K. J., Wood, E. F. & Sivapalan, M. (1988) On hydrological heterogeneity—catchment morphology and catchment response. *J. Hydrol.* **100**(1–3), 353–375.

Chow, V. T. (1964) *Handbook of Applied Hydrology*. McGraw-Hill Book Co, New York, USA.

Diermanse, F. L. M. (1998) Representation of natural hétérogeneity in rainfall–runoff models. Physics and chemistry of the Earth, Part B. *Hydrol. Oceans Atmos.* **24**(7), 787–792.

Edijatno Nascimento, N. O., Yang, X., Makhlouf, Z. & Michel, C. (1999) GR3J: a daily watershed model with three free parameters. *Hydrol. Sci. J.* **44**(2), 263–277.

Horton, R. E. (1945) Erosional development of streams and their drainage basins: hydrophysical approach to quantitative morphology. *Geol. Soc. Am. Bull.* **56**, 275–370.

Klemeš, V. (1986) Diletantism in hydrology: transition or destiny? *Water Resour. Res.* **22**(9), 177S–188S.

Krysanova, V., Bronstert, A. & Muller-Wohlfeil, D. (1999) Modelling river discharge for large drainage basins: from lumped to distributed approach. *Hydrol. Sci. J.* **44**(2), 313–331.

Mathevet, T. (2005) Quels modèles pluie-débit globaux pour le pas de temps horaire? Développements empiriques et intercomparaison de modèles sur un échantillon de 313 bassins versants. PhD Thesis, ENGREF Paris, France.

McDonnell, J. & Ross, J. W. (2004) On the need of classement classification. *J. Hydrol.* **299**, 2–4.

Merz, R. & Bloschl, G. (2004) Regionalisation of catchment model parameters. *J. Hydrol.* **287**(1–4), 95–123.

Moussa, R. & Bocquillon, C. (1993) Morphologie fractale du réseau hydrographique. *Hydrol. Sci. J.* **38**(3), 187–201.

Nash, J. E. & Sutcliffe, V. (1970) River flow forecasting through conceptual models. I. A discussion of principles. *J. Hydrol.* **10**, 282–290.

Perrin, C., Michel, C. & Andreassian, V. (2001) Does a large number of parameters enhance model performance? Comparative assessment of common catchment model structures on 429 catchments. *J. Hydrol.* **242**(3–4), 275–301.

Perrin, C., Michel, C. & Andreassian, V. (2003) Improvement of a parsimonious model for streamflow simulation. *J. Hydrol.* **279**(1–4), 275–289.

Refsgaard, J. C. & Knudsen, J. (1996) Operational validation and intercomparison of different types of hydrological models. *Water Resour. Res.* **32**(7), 2189–2202.

Rojas Serna, C. (2005). Quelle connaissance hydrométrique minimale pour définir les paramètres d'un modèle pluie-débit? PhD Thesis, ENGREF, Paris, France.

Wilson, C. B., Valdes, J. D. & Rodriguez-Iturbe, I. (1979) On the influence of the spatial distribution rainfall on storm runoff. *Water Resour. Res.* **15**(2), 321–328.

5 Perspectives

Compilation of the MOPEX 2004 results

N. CHAHINIAN[1,2], V. ANDRÉASSIAN[1], Q. DUAN[3], V. FORTIN[4,5], H. GUPTA[6], T. HOGUE[7], T. MATHEVET[1,8], A. MONTANARI[9], G. MORETTI[10], R. MOUSSA[11], C. PERRIN[1], J. SCHAAKE[12], T. WAGENER[6,13] & Z. XIE[14]

1 *Hydrology group, Cemagref Antony, Parc Tourvoie BP 44, 92163 Antony cedex, France*
nanee.chahinian@agrocampus-rennes.fr

2 *Now at: Institut de Recherche pour le Développement(IRD), UMR HydroSciences–Université Montpellier II, Case Courrier MSE, Place Eugène Bataillon, 34095 Montpellier cedex 5, France*

3 *Lawrence Livermore National Laboratory, 7000 East Avenue, Livermore, California 94550, USA*

4 *Institut de Recherche d'Hydro Qubec 1800, boulevard Lionel-Boulet, Varennes, Québec J3X 1S1, Canada*

5 *Now at: Numerical Prediction Research, Meteorological Research Division, Environment Canada, 2121, North Service Road, Trans-Canada Highway, Dorval, Quebec H9P 1J3, Canada*

6 *University of Arizona, NSF STC SAHRA, Marshall Building Room 532A, 845 North Park Avenue, Tuscon, Arizona 85721, USA*

7 *University of California Los Angeles, Department of Civil and Environmental Engineering, 5732C Boelter Hall, California 90095-1593, USA*

8 *Now at: EDF-DTG, BP 41, 38040 Grenoble cedex, France*

9 *University of Bologna, Viale del Risorgimento 2, 40136 Bologna, Italy*

10 *Ingenieurbüro Winkler und Partner GmbH, Schloßstr. 59A, 70176 Stuttgart, Germany*

11 *Institut National de la Recherche Agronomique, UMR LISAH, 2 place Pierre Viala, 34060 Montpellier cedex, France*

12 *NOAA/NWS/OHD, Hydrology Laboratory, Silver Spring, Maryland 20910, USA*

13 *Penn State University, Department of Civil and Environmental Engineering, 212 Sackett Building, University Park, Pennsylvania 16802, USA*

14 *Institute of Atmospheric Physics, Chinese Academy of Sciences, Beijing 100029, China*

Abstract As part of the MOPEX 2004 workshop, the participants were asked to submit simulations using the common database provided for the workshop (see Chahinian *et al.*, this issue). The simulations were then analysed and the evaluation criteria computed to compare the models' performance for the gauged and ungauged modes using six criteria describing the model's performance in both high and low flow conditions. The comparisons were undertaken for all three participation levels (i.e. 3, 12 and 40 catchments). The results indicate that on the 3-catchment level model ranking may vary according to the tested criterion and catchment. Hence a larger number of catchments are necessary to evaluate the models' performance. Among the 10 models tested on the 12 and 40 catchment samples in gauged mode, GR5H, Mordor and SAC-SMA rank as the top three models. When analysing model results in the ungauged mode, SAC-SMA ranks as the best of the four tested models. The analysis of the submitted files highlights the need for continuing efforts to develop model parameterization strategies for ungagued catchments in order to improve prediction in ungauged basins (PUB).

Key words model comparison; MOPEX; parameter estimation; PUB; rainfall–runoff modelling; ungauged catchments

INTRODUCTION

In view of the focus on hydrological predictions on ungauged catchments as highlighted by the IAHS PUB decade, the MOPEX (Model Parameter Estimation Experiment) extended its scope to the parameterization of hydrological models on ungauged catchments. This paper summarizes the MOPEX 2004 workshop results and provides insights into the simulation capacities of the tested models on both gauged and ungauged catchments.

Model parameter estimation is a bigger problem for ungauged catchments where traditional calibration techniques (Hendrickson *et al.*, 1988; Duan & Gupta 1992; Gupta *et al.*, 2003) cannot be used because of the lack of data. Hence some authors turn to regionalization methods in order to link model parameters to the catchments' descriptors (Micovic & Quick, 1999; Seibert, 1999; Perrin, 2000; Kokkonen *et al.*, 2003; Van der Linden & Woo, 2003; Croke *et al.*, 2004; Lee *et al.*, 2005; Merz & Blöschl, 2005) or use *a priori* model estimates based on the catchments' physical properties (Koren *et al.*, 2003; Leavesley *et al.*, 2003). In order to succeed these methods need a strong relationship between the catchment characteristics and the model's parameters as well as a large data set to extract statistically significant results. These two conditions are rarely satisfied. Hence trying to infer post-calibration relationships between calibrated parameters and catchment properties is, in most instances, a perilous exercise because of the inter-dependencies between the various parameters (Johnston & Pilgrim, 1976; Beven, 1993). In addition, even the most "physically based" models need some calibration (Abbott *et al.*, 1986) to account for the discrepancies between the parameter measurement and model application scales (Haverkamp *et al.*, 1998). Finally, it is often difficult to find long time series of rainfall–runoff data to ensure a sound calibration and large databases containing both hydrological and morphological attribute data to derive these relations. In this setting, model parameter estimation experiments such as MOPEX, involving different models and large databases, provide a creative interaction framework where modellers can share their experiences and benefit from the latest advances in model parameter estimation techniques. In this perspective, the MOPEX 2004 workshop was a clear success marked by high participation and the use of new hourly data.

LIMITS OF THE PRESENT COMPARISON

Even if we believe the Paris MOPEX workshop results to be extremely valuable, we deem it necessary to mention that, for the most part, the models have only been tested on either 3 or 12 catchments, and that this is necessarily too small a data set to provide results of general value. For example, in his comparison of 20 rainfall–runoff models over 429 catchments, Perrin (2000) found several combinations of 3 or 12 catchments showing that 19 of the 20 models were performing best. Nonetheless, it is obvious that the results of this comparison will help participating modellers to improve their models.

METHODS

For the first time since the start of the MOPEX programme, three distributed models took part in the comparison (Afdeff, Hydrotel, Modspa). In total 13 models were tested (Table 1). Four models were used at the daily time step (ModSpa, SAC-SMA,

Table 1 Participant information and general model characteristics.

Participant	Model name	Reference	Runoff/rainfall model	Channel routing function
UCLA University of Arizona	SAC-SMA	Burnash *et al.*, 1973	SAC-SMA	Kinematic wave
Hydro-Quebec	Hydrotel	Fortin *et al.* (1995); (2001)	3 Layer Richards 1D	Kinematic or diffusive wave
INRA Montpellier Cemagref Antony	ModSpa	Moussa (1991) ; (1993)	Diskin-Nazimov (1995)	Diffusive wave
Cemagref Antony	GR4J	Edijatno *et al.* (1999) Perrin *et al.*, (2001)	Non-linear reservoir	Unit hydrograph
	GR5H	Mathevet (2005)	Non-linear reservoir	Unit hydrograph
	HBV0	Perrin *et al.* (2001) Bergström & Forsman (1973)	Linear reservoir/modified HBV	Unit hydrograph
	IHAC	Jakeman *et al.* (1990) Perrin *et al.* (2001)	Linear reservoir/modified IHACRES	Pure delay
	Mordor	Mathevet (2005)	Non-linear reservoirs	Unit hydrograph
	Topmo	Beven & Kirkby (1978) Perrin *et al.* (2001)	Modified TopModel	Pure delay
University of Bologna	Affdef	Brath *et al.* (2002) Moretti & Montanari (2005)	SCS Reservoir model	Maskingham-Cunge
Laurence Livermore Laboratory	NOAH Land Surface Model (NLS)	Chen *et al,* (1996)	Multi-layer soil and energy balance	Linearized St Venant
National Weather Service	SAC-SMA	Burnash *et al.* (1973)	SAC-SMA	Kinematic wave
	SWB	Schaake *et al.* (1996)	Reservoir model	
Institute of Atmospheric Physics, Chinese Academy of Sciences	VIC	Liang *et al.* (1994) Liang & Xi (2001)	Multi-layer soil and energy balance	One parameter simple routing

NOAH, SWB) while the remaining ran hourly simulations. The model structures are presented briefly in this issue (see catalogue of models). It should be noted that although the production stores and transfer functions vary in complexity and nature, all the production functions use reservoir analogies.

Some models needed information that was not provided in the database (Hydrotel) while others did not use any of the morphometric data provided (GR4J; GR5H; HBV0, etc.). The participants were free to seek for any additional data through their own means.

Ideally, the comparison should have been made on the whole data set of 40 catchments. However, in order to encourage participation, the contributors were given the possibility to run their models on either 3, 12 or 40 catchment samples (no research funds are provided by MOPEX). As can be seen in Table 1, the two distributed models and the SAC-SMA model at the hourly time step were run on the 3-catchment (3C) sample. GR4J, GR5H, HBV0, IHAC, Mordor and Topmo were tested on the 40-

catchment (40C) sample but only in the gauged mode. Whereas NOAH, SAC-SMA and SWB were tested on the 12C sample at the daily time step. VIC was the only model tested on the 12C sample at the hourly time step. However, there are disparities in the original submissions, which render the comparisons more difficult. One of the participating teams replaced one of the catchments from the 12C set with another from the 40C. Thus, in order to present consistent results, we carried out the model inter-comparison analysis on the "common" 11 catchments(11C). This slight problem could easily be overcome in the future by providing clearer data submission formats and guidelines.

Hydro-meteorological data consisted of hourly (and daily) measurements of discharge and hourly estimates of rainfall, evapotranspiration, downward solar and infrared radiation, specific air humidity, air temperature and wind speed. For more details on the database, see Chahinian *et al* (this issue).

The participants were free to chose the calibration and validation criteria of their choice. However, they were informed that the model comparison would be carried out on a number of criteria based on the Nash and Sutcliffe efficiency, the bias and the root mean square error. In order to evaluate the importance of low flows the NSE and RMSE criteria were also computed on the square root of the discharge. The RMSE on the log of the discharge was also computed.

$$NS_Q = 1 - \frac{\sum_{t=1}^{N}\left(Q_{obs,t} - Q_{sim,t}\right)^2}{\sum_{t=1}^{N}\left(Q_{obs,t} - \overline{Q}_{obs,t}\right)^2} \quad (1)$$

$$NS_{\sqrt{Q}} = 1 - \frac{\sum_{t-1}^{N}\left(\sqrt{Q_{obs,t}} - \sqrt{Q_{obs,t}}\right)^2}{\sum_{t-1}^{N}\left(\sqrt{Q_{obs,t}} - \overline{\sqrt{Q_{obs,t}}}\right)^2} \quad (2)$$

$$RMSE_Q = \sqrt{\frac{1}{N}\left(\sum_{t=1}^{N}\left(Q_{sim,t} - Q_{obs,t}\right)^2\right)} \quad (3)$$

$$RMSE_{\sqrt{Q}} = \sqrt{\frac{1}{N}\left(\sum_{t=1}^{N}\left(\sqrt{Q_{sim,t}} - \sqrt{Q_{obs,t}}\right)^2\right)} \quad (4)$$

$$RMSE_{\ln Q} = \sqrt{\frac{1}{N}\left[\sum_{t=1}^{N}\ln\left(Q_{sim,t} + \overline{Q}_{obs,t}/10\right) - \ln\left(Q_{obs,t} + \overline{Q}_{obs,t}/10\right)\right]^2} \quad (5)$$

$$Bias = \left[\frac{\sum_{t=1}^{N}\left(Q_{sim,t} - Q_{obs,t}\right)}{\sum_{t=1}^{N} Q_{obs,t}}\right] * 100 \quad (6)$$

where N, number of time steps used; t, time step index; Q, streamflow (*sim* and *obs* refer to simulated and observed flows); $\overline{Q}$, mean annual discharge.

High flows and floods will influence the criteria calculated on the discharge values, whereas those calculated on the ln(Q) will be more influenced by low flows. The criteria calculated on $\sqrt{Q}$ will represent all the discharge values, giving equal weight to both high and low flows. To include the greatest number of models, in the analysis the criteria were computed both at the hourly and daily time steps. However, given the redundancy in the results, only the daily time step will be presented and discussed in this paper.

This analysis is based on the digital files submitted by the participants from June 2004 until March 2005 only. As some modellers chose to improve their results by carrying additional work after the deadline, some discrepancies might occur between their results (companion papers, this issue) and the ones presented herein.

No analysis will be carried out on the evolution of parameter values between the ungauged and gauged modes as companion papers in this issue cover this aspect.

RESULTS AND DISCUSSION

The results are organized by mode and participation level. First, model results will be analysed in the gauged mode for increasing participation levels and then in the ungauged mode.

Table 2 Participation level.

Author	Model	Participation level	Gauged Mode	Ungauged Mode	Parameterization method
Fortin	Hydrotel	3 Catchments	X	X	Pedotransfer+ vegetation classification
Hogue *et al.*,	SAC-SMA	3 Catchments	X	X	Sister catchment
Montanari & Moretti	Affdeff	3 Catchments	X	X	Single flood event duration < 600 hrs
Moussa (*)	ModSpa*	3 Catchments	X	X	Pedotransfer+ vegetation classification
Xie	VIC	12 Catchments	X	X	Vegetation classification
Schaake & Duan	SAC-SMA*	12 Catchments	X	X	Ensemble simulations
	NOAH-LSM*	12 Catchments	X	X	
	SWB*	12 Catchments	X	X	
Mathevet *et al.*	GR4J	40 Catchments	X		
	GR5H	40 Catchments	X		
	HBV0	40 Catchments	X		
	IHAC	40 Catchments	X		
	Mordor	40 Catchments	X		
	Topmo	40 Catchments	X		

(*) Daily time step.

Table 3 Main hydrological characteristics of the selected catchments based on the "Banque Hydro" database.

Participation level	Code	Name	Area (km^2)	Instantaneous peak discharge (m^3s^{-1})	Mean annual discharge (m^3s^{-1})
3C	J3024010	Le Guillec à Trézilidé	43.0	12.40	0.67
	V6035010	Le Toulourenc à Malaucène	150.0	81.00	1.32
	Y5615030	Le Loup à Villeneuve-Loubet	279.0	228	4.47
12C	A1522020	La Lauch à Guebwiller	68.1	41	1.65
	H2001020	L'Yonne à Corancy	98.0	45.90	2.89
	H3613020	Le Lunain à Épisy	252.0	12.10	0.74
	H5723011	L'Orgeval à Boissy-le-Châtel	104.0	33.10	0.63
	J2034010	Le Guindy à Plouguiel	125.0	27.60	1.21
	J4124420	La Rivière de Pont-l'Abbé à Plonéour-Lanvern	32.1	4.52	0.53
	K0744010	L'Anzon à Débats-Rivière-d'Orpra	181.0	72.30	2.54
	K0753210	Le Lignon du Forez à Boën	371.0	285	5.70
	Y3514020	Le Vistre à Bernis	291.0	43.30	2.11
40C	A5723010	L'Ingressin à Toul	54.7	9.930	0.44
	H2513110	Le Tholon à Champvallon	131.0	17.90	0.86
	H3613010	Le Lunain à Paley	163.0	18.30	0.56
	H3923010	Le ru d'Ancoeur à Blandy	181.0	23.90	0.59
	H4252010	L'Orge à Morsang-sur-Orge	922.0	41.20	3.96
	H7853010	Le Sausseron à Nesles-la-Vallée	101.0	3.320	0.56
	H7913030	La Mauldre à Aulnay-sur-Mauldre	369.0	28.50	2.15
	J4712010	L'Éllé au Faouët	142.0	59.20	2.75
	K0100020	La Loire à Goudet	432.0	1600	5.70
	K0253020	La Borne occidentale à Espaly-Saint-Marcel	375.0	261	3.66
	K0550010	La Loire à Bas-en-Basset	3234.0	3500	38.60
	K0614010	Le Furan à Andrézieux-Bouthéon	178.0	142	2.54
	K0813020	L'Aix à Saint-Germain-Laval	193.0	195	3.02
	K0974010	Le Gand à Neaux	85.0	58.40	0.90
	K1173210	L'Arconce à Montceaux-l'Étoile	599.0	147	5.77
	K2724210	L'Artière à Clermont-Ferrand	49.0	8.660	0.27
	K2783010	La Morge à Maringues	713.0	103	4.29
	K5623010	L'Auron au Pondy	199.0	29.80	0.98
	K5653010	L'Auron à Bourges	585.0	83.80	3.77
	P3245010	Le Mayne à Saint-Cyr-la-Roche	49.0	22.90	0.70
	U4305410	La Denante à Davayé	11.1	8.300	0.13
	U4525210	Le Morgon à Villefranche-sur-Saône	68.0	17.90	0.49
	V3315010	La Valencize à Chavanay	36.0	17.30	0.36
	V3517010	Le Ternay à Savas	25.5	16.00	0.34
	V6052010	L'Ouvèze à Vaison-la-Romaine	585.0	1000	6.07
	X2414030	L'Artuby à la Bastide	91.0	104	1.04
	Y5615010	Le Loup à Tourrettes-sur-Loup	206.0	147	3.67
	Y5625020	La Cagne à Cagnes-sur-Mer	95.0	160	0.82

Validation mode

All 14 models were tested on the 3 catchments in validation mode, 10 were further tested on the 11 catchments and 6 on the complete set of 40 catchments.

Validation mode 3C level

The results for the validation period of the gauged mode are presented in Table 4(a)–(e). All the models produce high NS values for the first and last catchments. The lowest ranking model is SAC-SMA calibrated by Hogue *et al.*, with a NS criterion of 67%, and the highest is SWB with 87%. With the exception of GR4J, Topmo and Mordor, the results are poorer for the second catchment. In some instances the difference in performance ranges 40–50% (Afdeff, NOAH, SAC-SMA and VIC). It is interesting to note that results of Schaake & Duan on the SAC-SMA model are overall more stable than those of Hogue *et al.* As the model structure and data are identical, the difference in performance is likely to be caused by the differences in the calibration procedure.

The NS values obtained on the third catchment are of the same order of magnitude as the first one except for Afdeff and HBV0. However, the two catchments are in

Table 4 Daily Nash and Sutcliff, *Bias* and RMSE on ln(Q) $\sqrt{Q}$ and Q criteria per model and per catchment for the gauged mode (validation period; 01 August 1999–31 July 2002; daily time step).

Model	Nash and Sutcliffe efficiency (%)		
(a)	J3024010	V6035010	Y5615030
Affdeff	73	10	43
GR4J	75	77	79
GR5H	82	76	81
HBV0	76.5	66	53
Hydrotel	83	82	NA
IHAC	74	70	76
ModSpa	82	77	84
Mordor	84	88	88
NOAH–LSM	72	28	78
SAC–SMA[1]	67	17	81
SAC–SMA	85	82	88
SWB	87	75	82.5
Topmo	70	76	84
VIC	79	38	84
(b)	*Bias* (%)		
Affdeff	–23.71	56.31	56.46
GR4J	–13.37	–6.63	–5.14
GR5H	–9.18	–7.64	–3.70
HBV0	–6.69	5.07	–1.43
Hydrotel	–0.11	1.06	NA
IHAC	–9.78	–2.71	–3.61
ModSpa	–8.41	–6.62	7.61
Mordor	–14.44	–1.37	–0.98
NOAH–LSM	–14.95	50.43	0.74
SAC–SMA[1]	–7.75	–0.25	0.01
SAC–SMA	–11.02	12.52	1.35
SWB	–13.13	4.95	–6.76
Topmo	13.30	6.12	–2.13
VIC	22.68	87.45	16.44

Table 4 (*cont.*)

(c)	RMSE *ln(Q)* (mm day^{-1})		
Affdeff	0.16	0.87	0.41
GR4J	0.25	0.59	0.38
GR5H	0.26	0.45	0.50
HBV0	0.28	0.60	0.42
Hydrotel	0.14	0.20	NA
IHAC	0.30	0.56	0.41
ModSpa	0.09	0.44	0.18
Mordor	0.24	0.43	0.37
NOAH–LSM	0.41	0.55	0.27
SAC–SMA[1]	0.91	1.46	1.43
SAC–SMA	0.05	0.27	0.14
SWB	0.06	0.64	0.23
Topmo	0.30	0.49	0.40
VIC	0.10	0.92	0.28
(d)	RMSE √*(Q)* (mm day^{-1})		
Affdeff	0.27	0.50	0.58
GR4J	*0.17*	0.24	0.25
GR5H	0.15	0.23	0.24
HBV0	0.17	0.24	0.28
Hydrotel	0.22	0.22	NA
IHAC	0.19	0.25	0.24
ModSpa	0.21	0.32	0.22
Mordor	0.15	0.20	0.21
NOAH–LSM	0.35	0.42	0.35
SAC–SMA[1]	0.21	0.35	0.30
SAC–SMA	0.11	0.25	0.25
SWB	0.17	0.36	0.32
Topmo	0.24	0.20	0.20
VIC	0.24	0.49	0.34
(e)	RMSE(*Q*) (mm day^{-1})		
Affdeff	0.84	1.52	2.35
GR4J	*0.69*	0.63	1.12
GR5H	0.55	0.60	1.03
HBV0	0.66	0.75	1.78
Hydrotel	0.66	0.72	NA
IHAC	0.71	0.71	1.26
ModSpa	0.68	0.80	0.75
Mordor	0.60	0.50	0.99
NOAH–LSM	0.86	1.36	1.55
SAC–SMA[1]	0.91	0.68	1.43
SAC–SMA	0.62	0.81	1.13
SWB	0.59	0.61	1.38
Topmo	0.75	0.61	1.01
VIC	0.45	1.27	1.30

NA, not available.

different climatic zones and are not prone to the same overland flow processes, i.e. Dunn type mechanism for the first one and Hortonian overland flow for the second and third catchments. Therefore, model performance cannot be related to the process type, nor can it be linked to the model structure.

A comparison between NS_Q and $NS_{\sqrt{Q}}$ indicates that model ranking is not affected by high flows (Table 4(e)). Table 4(b) further indicates that the bias values are higher than the NS values;. $|Bias| \leq 10\%$ for six models for the first catchment and 10 for the second and third catchments. Afdeff and VIC produce the highest bias values. In the case of the latter, the associated NS criteria are high. This indicates that the model can accurately reproduce the response time of the catchments but tends to overestimate the peak discharges, i.e. the rising and falling limbs of the simulated and observed hydrographs are well aligned, but the peaks are over estimated (Fig. 1). In comparison Table 4(c) indicates that low flows and recession curves are well reproduced.

In conclusion, based on the 3C sample, SAC-SMA (Schaake & Duan) has the best performance, followed by Hydrotel and Mordor. Afdeff and SAC-SMA (Hogue *et al.*) have the poorest performance. At this stage, with the exception of SAC-SMA, we cannot establish whether the difference is due to the model structure or the calibration strategy and effort. Indeed, some models were calibrated by trial and error (Afdeff, Modspa) while others used automatic search algorithms (GR5H, SAC-SMA, NOAH); some authors specified the time spent on the calibration process while others did not.

No clear difference can be seen between the distributed and lumped models. However, in comparison with Hydrotel and Modspa, Afdeff seems to have a poorer performance. This can be explained by the difference between the model's grid size and the coarse data resolution (for further details see Fortin *et al.*, this issue).

Validation mode 11C level

The validation results of the 11C models are summarized in Figures 2 and 3. They represent the cumulative Nash & Sutcliffe efficiencies on respectively discharge and its square root value. Figure 3 indicates that Noah and VIC yield the poorest results: 50% of the catchments are simulated with a Nash & Sutcliffe efficiency <30% and < 60% respectively. In comparison GR5H, SAC-SMA and Mordor have the best performances as 50% of the catchments have Nash and Sutcliffe efficiencies >75%. The five models calibrated by Mathevet *et al.*, exhibit very similar performances over the validation period, highlighting the robustness of their simple calibration procedure (Mathevet, 2005). The rankings, with regard to low flows, show that NOAH and VIC perform comparatively less well than the other models, while GR5H and Mordor yield the best results, followed closely by SAC-SMA.

With regard to the bias criterion (Table 5(b)), NOAH has the overall highest bias values for seven of the catchments with VIC having the other four. No clear ranking can be seen for the lowest bias values (i.e. the best model with regard to bias) as with the exception of GR5H and NOAH, all the models rank first on at least one catchment. In addition, GR4H, GR5H and Mordor tend to underestimate discharge values for all but one catchment (K0753210). The latter's flows are overestimated by all the models. This is not surprising as it has the highest mean annual and instantaneous peak discharge values of the sample (Table 3).

The model rankings regarding $RMSE_Q$, $RMSE_{\sqrt{Q}}$, $RMSE_{\ln Q}$ are similar to those of the NS_Q and $NS_{\sqrt{Q}}$ (Table 5). Mordor, GR5H and SAC-SMA rank in the top three with

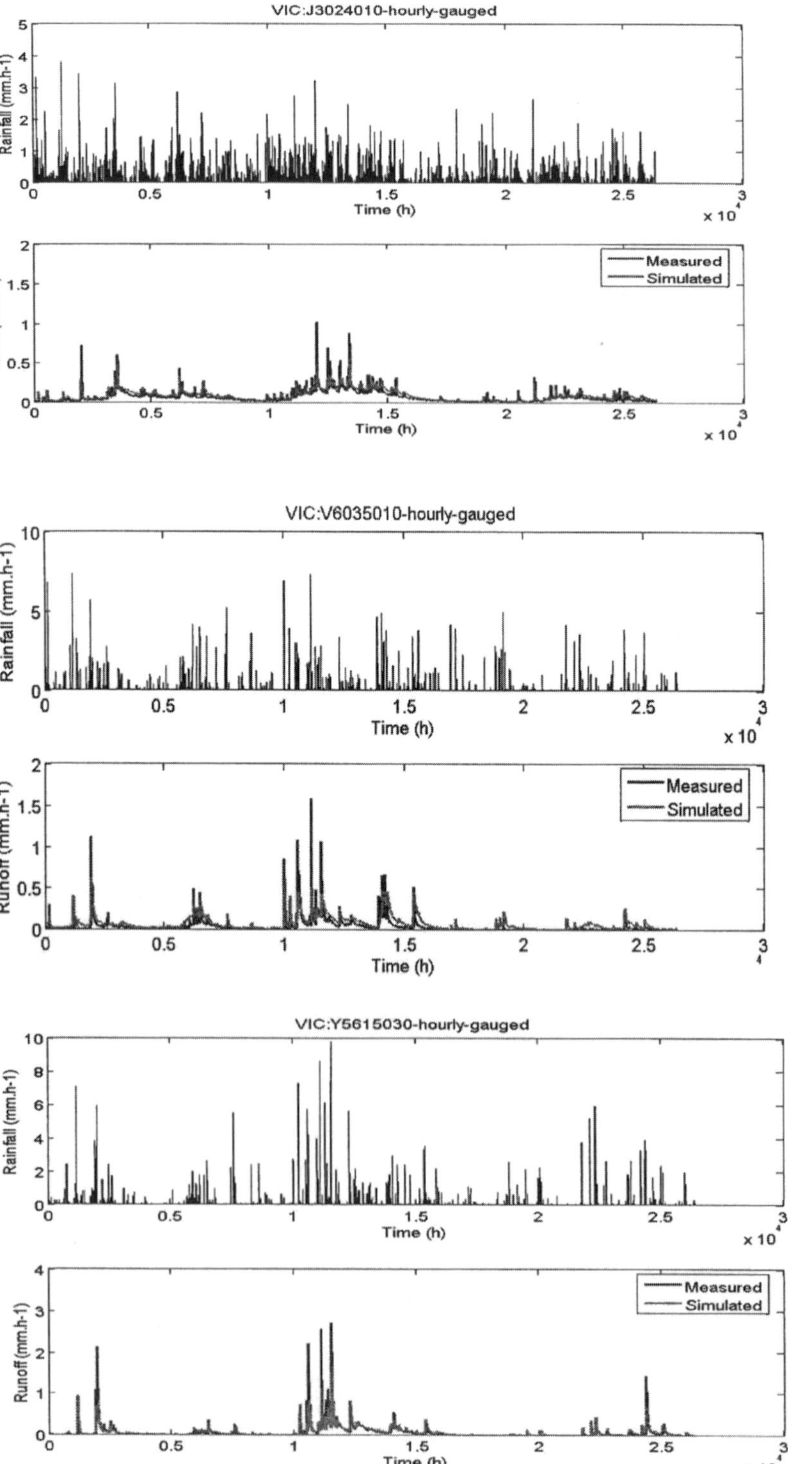

Fig. 1 Simulated and observed hydrographs for VIC model.

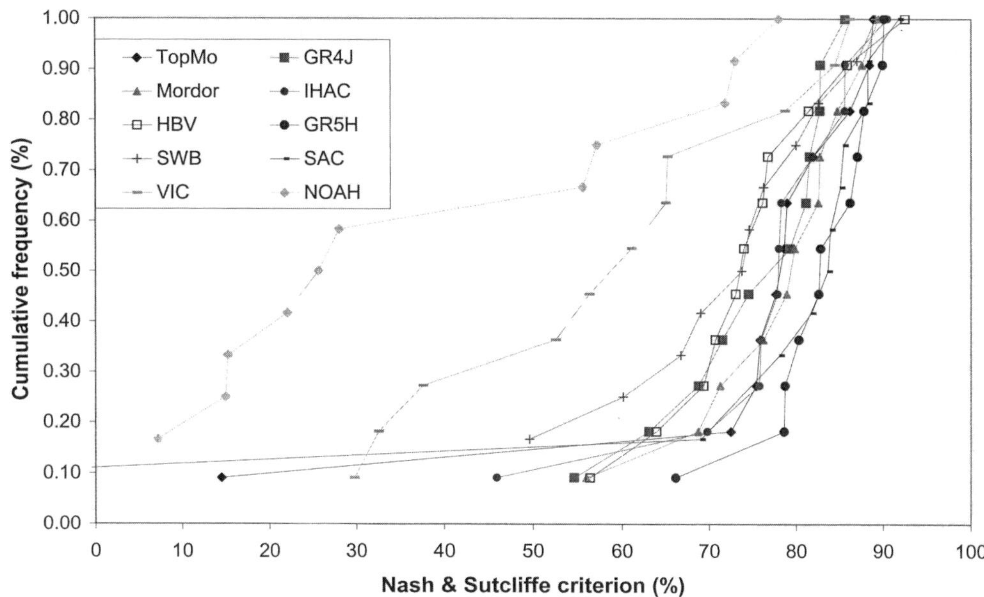

Fig. 2 Nash and Sutcliffe criteria distribution for the GR4J, GR5H, HBV0, IHAC, Mordor NOAH, SAC-SMA, SWB, Topmo and VIC (validation).

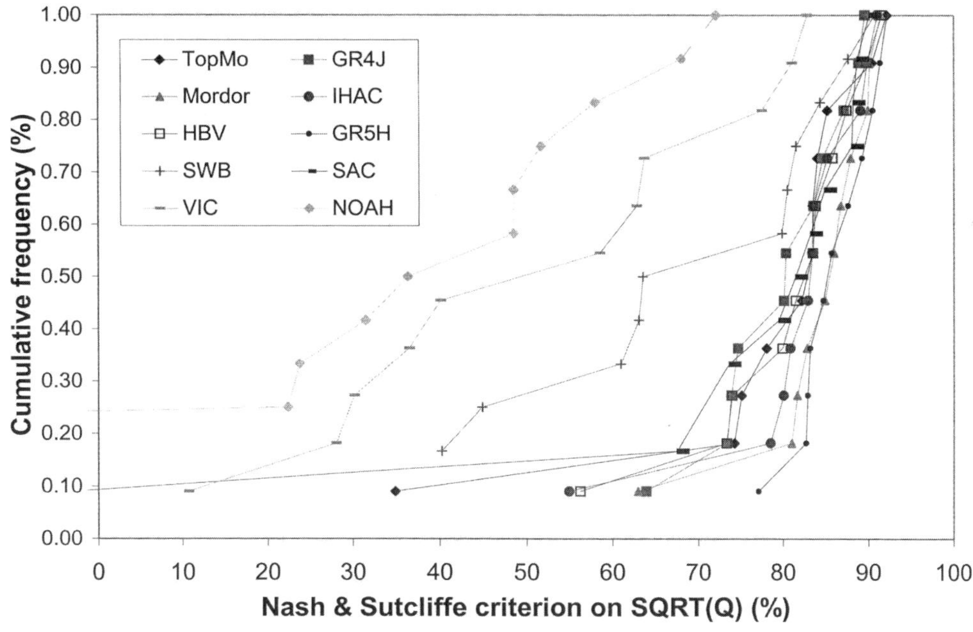

Fig. 3 Nash and Sutcliffe on SQRT (Q) criteria distribution for the GR4J, GR5H, HBV0, IHAC, Mordor NOAH, SAC-SMA, SWB, Topmo and VIC (validation).

Mordor having the best performance. NOAH ranks last once again. Table 6 summarizes the overall model performance for all 11 catchments and six criteria.

Table 5 *Bias*, RMSE on Q, $\sqrt{Q}$ and ln(Q) values per model and catchment for the gauged mode (validation period) for the 11 catchment sample.

Catchment	GR4H	GR5H	HBV	IHAC	Mordor	NOAH	SAC–SMA	SWB	Topmo	VIC
(a) *Bias* (%)										
a1522020	–5.61	–2.08	29.25	0.63	–1.98	67.91	35.42	5.65	29.44	22.68
h2001020	–4.86	–4.29	–11.07	–3.79	–3.14	14.07	48.21	–11.45	–10.00	–6.52
h3613020	–19.64	–8.40	11.74	–25.40	–20.84	50.50	–8.40	21.03	26.50	–4.91
j2034010	–14.07	–9.22	3.21	–9.97	–11.61	–8.54	4.07	–5.35	4.07	30.58
j3024010	–13.38	–9.19	–6.70	–9.79	–12.44	16.73	–11.02	–13.13	–13.31	6.46
j4124420	–13.39	–6.94	5.18	–5.70	–10.23	26.30	3.25	–19.26	5.16	16.44
k0744010	–5.67	–8.68	4.31	–3.84	–3.07	63.44	7.54	2.50	3.63	34.54
k0753210	0.03	2.18	0.33	3.10	2.08	51.88	5.88	9.70	2.68	35.51
v6035010	–6.66	–7.67	5.07	–2.74	–1.38	46.28	12.52	4.95	6.12	65.39
y3514020	–5.46	–5.02	–4.01	–5.48	–4.63	4.61	–4.19	–2.80	–2.36	70.01
y5615030	–5.15	–3.70	–1.43	3.62	–0.99	1.25	1.35	–6.76	–2.13	87.45
(b) RMSE (Q) (mm day^{-1})										
a1522020	1.23	1.10	1.60	1.38	1.10	2.81	1.37	1.31	1.46	2.01
h2001020	1.39	1.21	1.45	1.28	1.09	1.21	0.94	1.30	1.37	0.87
h3613020	0.18	0.12	0.18	0.20	0.17	1.66	0.32	1.28	0.25	0.23
j2034010	0.43	0.33	0.39	0.40	0.35	0.72	0.97	0.36	0.36	0.92
j3024010	0.69	0.55	0.66	0.71	0.60	0.86	0.62	0.59	0.75	0.75
j4124420	0.73	0.54	0.47	0.53	0.51	1.27	0.54	0.93	0.79	1.26
k0744010	0.79	0.76	0.66	0.71	0.67	1.06	0.68	0.71	0.61	0.76
k0753210	0.71	0.59	0.62	0.60	0.58	1.00	0.46	0.63	0.58	0.67
v6035010	0.63	0.60	0.75	0.71	0.50	1.36	0.68	0.81	0.61	1.27
y3514020	0.37	0.31	0.36	0.28	0.30	0.75	0.28	0.51	0.28	0.48
y5615030	1.12	1.04	1.78	1.26	1.10	1.55	1.13	1.38	0.99	1.30
(c) RMSE on $\sqrt{Q}$ (mm day^{-1})										
a1522020	0.11	0.26	0.38	0.30	0.26	0.61	0.42	0.32	0.38	0.57
h2001020	0.25	0.25	0.28	0.25	0.23	0.34	0.25	0.29	0.31	0.28
h3613020	0.16	0.07	0.12	0.12	0.11	0.32	0.18	0.43	0.14	0.14
j2034010	0.23	0.12	0.14	0.13	0.12	0.49	0.27	0.13	0.13	0.42
j3024010	0.24	0.15	0.17	0.20	0.15	0.29	0.15	0.17	0.24	0.24
j4124420	0.29	0.16	0.16	0.18	0.16	0.35	0.21	0.27	0.24	0.51
k0744010	0.30	0.23	0.20	0.22	0.20	0.34	0.24	0.29	0.19	0.30
k0753210	0.22	0.18	0.18	0.18	0.18	0.41	0.17	0.26	0.18	0.26
v6035010	0.17	0.23	0.24	0.25	0.20	0.37	0.25	0.36	0.21	0.49
y3514020	0.25	0.13	0.16	0.14	0.12	0.42	0.15	0.26	0.13	0.20
y5615030	0.14	0.24	0.28	0.24	0.26	0.35	0.25	0.32	0.21	0.34
(d) RMSE on ln(Q) (mm day^{-1})										
a1522020	0.41	0.38	0.44	0.36	0.37	0.47	0.33	0.15	0.43	0.50
h2001020	0.43	0.34	0.34	0.36	0.31	0.18	0.09	0.11	0.39	0.17
h3613020	0.43	0.30	0.50	0.47	0.44	0.61	0.19	0.11	0.48	0.13
j2034010	0.31	0.34	0.39	0.36	0.39	0.55	0.19	0.71	0.33	0.59
j3024010	0.59	0.26	0.28	0.30	0.24	0.37	0.08	0.07	0.30	0.10
j4124420	0.37	0.25	0.29	0.29	0.25	0.41	0.05	0.10	0.33	0.49
k0744010	0.41	0.46	0.37	0.44	0.40	0.55	0.08	0.33	0.39	0.33
k0753210	0.31	0.31	0.31	0.29	0.30	0.35	0.18	0.22	0.31	0.19
v6035010	0.25	0.45	0.60	0.56	0.43	0.55	0.27	0.64	0.49	0.92
y3514020	0.38	0.39	0.48	0.46	0.40	0.44	0.16	0.52	0.38	0.21
y5615030	0.36	0.51	0.42	0.41	0.37	0.27	0.14	0.23	0.40	0.28

Table 6 Synthesis of overall model classification for the gauged mode (validation period) for the 11 catchment sample.

Criteria	Best Model	Worst Model
Nash(Q)	– (SAC–SMA on 4 catchments)	NOAH
Nash($\sqrt{Q}$)	– (GR5H on 5 catchments)	NOAH
Bias	–	NOAH
RMSE	Mordor	NOAH
RMSE($\sqrt{Q}$)	Mordor	NOAH
RMSE(ln(Q))	SAC_SAM	–

(–) No model scores best on at least 6 catchments.

Table 7 *Bias*, RMSE values for the gauged mode (validation period) for the 40 catchment sample (continues next page).

Catchment	GR4J	GR5H	HBV0	IHAC	Mordor	Topmo
(a) *Bias*						
A1522020	–5.61	–2.08	29.25	0.63	–1.98	29.44
A5723010	0.76	1.82	18.40	4.59	6.01	19.02
H2001020	–4.86	–4.29	–11.07	–3.79	–3.14	–10.00
H2513110	–6.96	–4.21	–4.71	–16.81	–14.03	–1.41
H3613010	–27.01	–21.25	14.33	–32.87	–27.97	24.53
H3613020	–19.64	–8.40	11.74	–25.40	–20.84	26.50
H3923010	–33.14	–29.54	19.25	–34.24	–34.61	22.66
H4252010	–0.01	5.51	8.39	–5.20	–8.25	28.23
H5723011	–37.06	–32.84	–12.26	–35.45	–28.42	–8.80
H7853010	16.10	27.58	8.96	–4.15	3.72	43.37
H7913030	1.74	14.34	8.23	–0.10	–5.79	19.99
J2034010	–14.07	–9.22	3.21	–9.97	–11.61	4.07
J3024010	–13.38	–9.19	–6.69	–9.78	–12.44	–13.30
J4124420	–13.46	–7.00	5.16	–5.66	–10.23	5.04
J4712010	–13.61	–12.85	–11.46	–9.65	–13.39	–10.15
K0100020	–5.64	–7.90	–10.81	–7.20	–5.61	62.32
K0253020	–4.94	–4.06	10.07	–7.45	–1.27	8.93
K0550010	–11.78	–12.33	–2.92	–12.51	–9.15	–2.33
K0614010	–8.23	–7.51	–7.21	–12.90	–6.34	–2.32
K0744010	–5.67	–8.68	4.31	–3.84	–3.07	3.63
K0753210	0.03	2.18	0.33	3.11	2.08	2.68
K0813020	–3.25	–8.05	6.66	–2.11	–1.50	5.02
K0974010	–18.87	–21.29	–13.32	–23.37	–19.17	–8.65
K1173210	–17.69	–18.79	0.95	–19.99	–16.48	–4.72
K2724210	–8.68	–2.23	22.45	–9.54	–5.82	21.95
K2783010	0.92	4.06	5.17	–4.53	–0.29	17.18
K5623010	–13.65	–14.55	7.07	–17.84	–15.28	6.01
K5653010	–12.17	–15.29	–1.74	–18.34	–15.38	–0.71
P3245010	–16.02	–15.78	2.50	–12.55	–12.71	–5.88
U4305410	–6.90	–7.44	–7.03	–11.52	–7.25	–9.17
U4525210	–9.43	–7.51	–3.48	–14.28	–10.43	–1.57
V3315010	–7.59	–7.18	5.98	–6.54	–5.35	3.62
V3517010	3.69	2.47	–2.11	8.07	3.67	1.17

Table 7 (*cont.*)

Catchment	GR4J	GR5H	HBV0	IHAC	Mordor	Topmo
V6035010	−6.64	−7.65	5.08	−2.71	−1.38	6.13
V6052010	17.01	11.75	12.18	9.00	16.48	13.28
X2414030	10.29	11.80	18.09	19.47	9.97	13.77
Y3514020	−5.45	−5.02	−4.00	−5.48	−4.63	−2.36
Y5615010	0.98	9.78	5.26	14.65	7.11	9.52
Y5615030	−5.15	−3.71	−1.44	3.61	−0.99	−2.13
Y5625020	−0.11	−0.17	9.33	7.43	4.94	10.96
(b) RMSE(Q)						
A1522020	1.23	1.10	1.60	1.38	1.10	1.46
A5723010	0.41	0.34	0.41	0.36	0.28	0.45
H2001020	1.39	1.21	1.45	1.28	1.09	1.37
H2513110	0.22	0.18	0.25	0.27	0.23	0.21
H3613010	0.32	0.25	0.30	0.32	0.29	0.32
H3613020	0.18	0.12	0.18	0.20	0.17	0.25
H3923010	0.60	0.49	0.48	0.58	0.57	0.42
H4252010	0.27	0.22	0.27	0.19	0.19	0.30
H5723011	1.32	1.18	1.08	1.28	1.25	0.88
H7853010	0.18	0.22	0.19	0.14	0.15	0.45
H7913030	0.26	0.23	0.28	0.20	0.23	0.34
J2034010	0.43	0.33	0.39	0.40	0.35	0.36
J3024010	0.69	0.55	0.66	0.71	0.60	0.75
J4124420	0.73	0.54	0.47	0.53	0.51	0.79
J4712010	1.29	0.90	1.15	1.16	1.19	0.83
K0100020	1.14	1.12	1.13	1.22	1.03	1.35
K0253020	0.44	0.38	0.40	0.43	0.37	0.38
K0550010	0.64	0.59	0.61	0.66	0.56	0.55
K0614010	0.75	0.69	0.69	0.69	0.64	0.62
K0744010	0.79	0.76	0.66	0.71	0.67	0.61
K0753210	0.71	0.59	0.62	0.60	0.58	0.58
K0813020	0.88	0.85	0.75	0.76	0.77	0.73
K0974010	0.84	0.81	0.76	0.88	0.80	0.66
K1173210	0.70	0.63	0.53	0.68	0.57	0.49
K2724210	0.33	0.31	0.35	0.33	0.32	0.71
K2783010	0.17	0.14	0.17	0.17	0.13	0.20
K5623010	0.56	0.48	0.41	0.59	0.46	0.33
K5653010	0.52	0.46	0.43	0.56	0.45	0.35
P3245010	0.82	0.71	0.73	0.76	0.64	0.68
U4305410	0.80	0.71	0.74	0.73	0.63	0.73
U4525210	0.42	0.37	0.43	0.44	0.38	0.44
V3315010	0.50	0.43	0.39	0.46	0.41	0.36
V3517010	0.63	0.56	0.55	0.63	0.57	0.55
V6035010	0.63	0.60	0.75	0.71	0.50	0.61
V6052010	0.89	0.81	0.92	0.99	0.84	0.84
X2414030	1.20	1.07	1.23	1.26	0.91	1.23
Y3514020	0.37	0.31	0.36	0.28	0.30	0.28
Y5615010	1.16	1.48	1.21	1.21	0.87	1.35
Y5615030	1.12	1.04	1.78	1.26	1.01	0.99
Y5625020	0.90	0.83	1.12	1.09	0.97	0.92
(c) RMSE $\sqrt{Q}$						
A1522020	0.30	0.26	0.38	0.30	0.26	0.38

Table 7 (*cont.*)

Catchment	GR4J	GR5H	HBV0	IHAC	Mordor	Topmo
A5723010	0.17	0.14	0.17	0.14	0.11	0.19
H2001020	0.29	0.25	0.28	0.25	0.23	0.31
H2513110	0.11	0.09	0.14	0.14	0.12	0.13
H3613010	0.18	0.14	0.21	0.19	0.17	0.20
H3613020	0.11	0.07	0.12	0.12	0.11	0.14
H3923010	0.25	0.20	0.23	0.24	0.22	0.18
H4252010	0.15	0.12	0.17	0.11	0.10	0.17
H5723011	0.35	0.31	0.29	0.34	0.34	0.23
H7853010	0.11	0.14	0.12	0.08	0.09	0.22
H7913030	0.12	0.11	0.14	0.09	0.10	0.17
J2034010	0.14	0.12	0.14	0.13	0.12	0.13
J3024010	0.17	0.15	0.17	0.20	0.15	0.24
J4124420	0.22	0.16	0.16	0.18	0.16	0.24
J4712010	0.25	0.19	0.21	0.23	0.21	0.18
K0100020	0.26	0.23	0.25	0.26	0.23	0.39
K0253020	0.19	0.16	0.18	0.18	0.16	0.17
K0550010	0.21	0.18	0.19	0.20	0.17	0.17
K0614010	0.24	0.23	0.23	0.23	0.21	0.22
K0744010	0.25	0.23	0.20	0.22	0.20	0.19
K0753210	0.23	0.18	0.18	0.18	0.18	0.18
K0813020	0.27	0.25	0.23	0.23	0.23	0.23
K0974010	0.31	0.29	0.25	0.32	0.28	0.24
K1173210	0.23	0.20	0.19	0.23	0.19	0.20
K2724210	0.18	0.18	0.21	0.18	0.18	0.27
K2783010	0.11	0.09	0.12	0.11	0.09	0.14
K5623010	0.22	0.18	0.17	0.22	0.17	0.15
K5653010	0.17	0.14	0.16	0.19	0.14	0.14
P3245010	0.26	0.23	0.23	0.23	0.20	0.22
U4305410	0.28	0.26	0.25	0.25	0.23	0.26
U4525210	0.16	0.14	0.18	0.17	0.15	0.18
V3315010	0.20	0.17	0.17	0.19	0.16	0.16
V3517010	0.25	0.22	0.20	0.24	0.22	0.20
V6035010	0.24	0.23	0.24	0.25	0.20	0.21
V6052010	0.28	0.24	0.27	0.30	0.25	0.25
X2414030	0.30	0.22	0.32	0.31	0.24	0.28
Y3514020	0.16	0.13	0.16	0.14	0.12	0.13
Y5615010	0.25	0.23	0.26	0.25	0.19	0.22
Y5615030	0.25	0.24	0.28	0.24	0.21	0.21
Y5625020	0.26	0.23	0.27	0.26	0.22	0.23
d) RMSE ln(Q)						
A1522020	0.41	0.38	0.44	0.36	0.37	0.43
A5723010	0.39	0.37	0.38	0.37	0.37	0.40
H2001020	0.37	0.34	0.34	0.36	0.31	0.39
H2513110	0.31	0.27	0.30	0.28	0.28	0.28
H3613010	0.55	0.52	0.74	0.58	0.54	0.62
H3613020	0.41	0.30	0.50	0.47	0.44	0.48
H3923010	0.53	0.48	0.54	0.59	0.53	0.48
H4252010	0.50	0.46	0.55	0.42	0.37	0.50
H5723011	0.52	0.44	0.49	0.52	0.62	0.48
H7853010	0.29	0.29	0.34	0.33	0.29	0.30
H7913030	0.23	0.20	0.23	0.23	0.23	0.24

Table 7 (*cont.*)

Catchment	GR4J	GR5H	HBV0	IHAC	Mordor	Topmo
J2034010	0.36	0.34	0.39	0.36	0.39	0.33
J3024010	0.25	0.26	0.28	0.30	0.24	0.30
J4124420	0.31	0.25	0.29	0.29	0.25	0.33
J4712010	0.33	0.29	0.35	0.33	0.26	0.31
K0100020	0.48	0.42	0.34	0.45	0.41	0.46
K0253020	0.43	0.40	0.45	0.47	0.41	0.43
K0550010	0.38	0.37	0.39	0.40	0.39	0.40
K0614010	0.21	0.19	0.20	0.20	0.19	0.21
K0744010	0.43	0.46	0.37	0.44	0.40	0.39
K0753210	0.31	0.31	0.31	0.29	0.30	0.31
K0813020	0.44	0.46	0.39	0.44	0.42	0.39
K0974010	0.61	0.61	0.54	0.68	0.59	0.54
K1173210	0.44	0.39	0.49	0.48	0.37	0.45
K2724210	0.54	0.55	0.63	0.51	0.49	0.58
K2783010	0.47	0.40	0.52	0.47	0.44	0.54
K5623010	0.67	0.60	0.50	0.70	0.57	0.55
K5653010	0.49	0.40	0.50	0.50	0.43	0.46
P3245010	0.42	0.37	0.44	0.38	0.39	0.40
U4305410	0.46	0.46	0.44	0.43	0.43	0.47
U4525210	0.38	0.38	0.50	0.44	0.37	0.49
V3315010	0.44	0.42	0.45	0.46	0.44	0.44
V3517010	0.44	0.41	0.37	0.45	0.42	0.38
V6035010	0.59	0.45	0.60	0.56	0.43	0.49
V6052010	0.58	0.46	0.54	0.54	0.49	0.51
X2414030	0.48	0.48	0.50	0.52	0.48	0.51
Y3514020	0.43	0.39	0.48	0.46	0.40	0.38
Y5615010	0.48	0.44	0.49	0.46	0.42	0.46
Y5615030	0.38	0.51	0.42	0.41	0.37	0.40
Y5625020	0.68	0.55	0.64	0.60	0.52	0.65

Validation mode 40C level

Although no clear ranking can be established on average regarding the NS_Q and $NS_{\sqrt{Q}}$ criteria, Mordor, GR5H and SAC-SMA perform better than the other models. NOAH seems to have the poorest performance. These findings highlight the fact that model performance in validation cannot be linked to the number of calibrated parameters as GR5H has 5 free parameters, Mordor 10 and SAC-SMA 13.

The results for the validation mode on the 40C sample are summarized in Figs 4, 5 and 6 and Table 8(a)–(d).

Analysis of Figs 4 and 5 indicates that similar to the 11C sample, Mordor and GR5H have the best overall performance with regard to the Nash and Sutcliffe criterion on both (Q) and $\sqrt{Q}$. In comparison Topmo has the worst performance. However, all five models yield Nash and Sutcliffe efficiencies >70% for 50% of the catchments, therefore their performances should be considered as good.

Figure 6 shows a general similarity in the models' behaviour regarding bias. However, for five catchments (H3613010, H3613020, H3923010, J4124420,

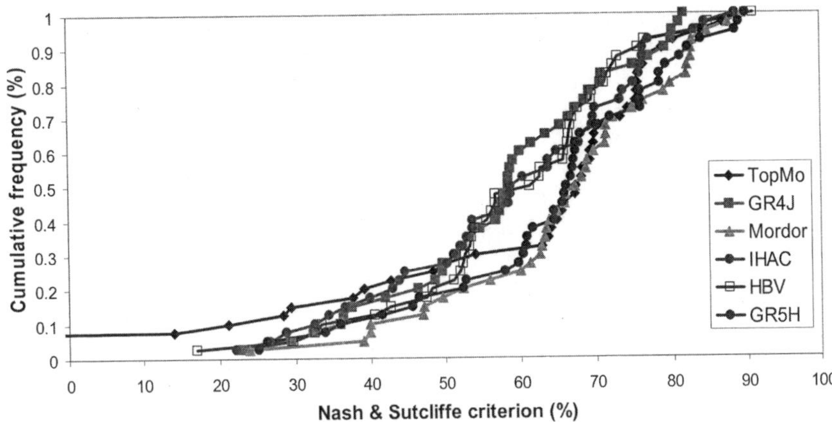

Fig. 4 Nash and Sutcliffe on (Q) criteria distribution for the GR4J, GR5H, HBV0, IHAC, Mordor and Topmo models on the 40 catchments in gauged mode (validation).

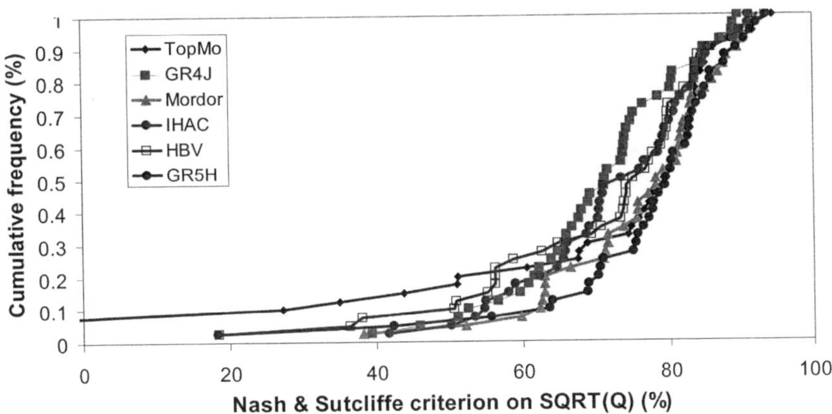

Fig. 5 Nash and Sutcliffe on SQRT(Q) criteria distribution for the GR4J, GR5H, HBV0, IHAC, Mordor and Topmo models on the 40 catchments in gauged mode (validation).

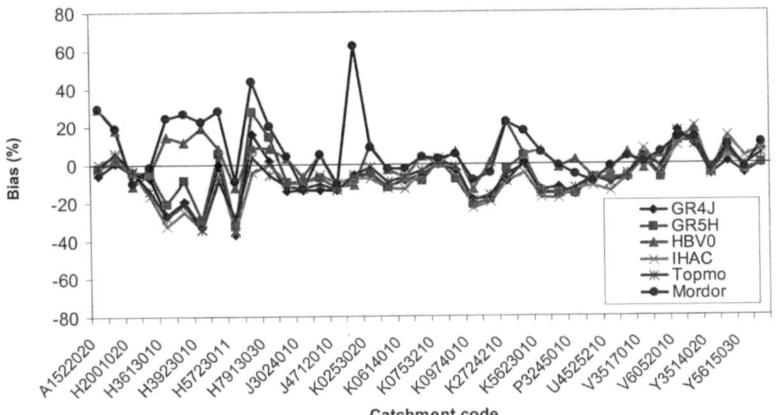

Fig. 6 Bias values for the GR4J, GR5H, HBV0, IHAC, Mordor and Topmo models on the 40 catchments in gauged mode (validation).

Table 8 Nash and Sutcliffe and Bias criteria per model and per catchment for the ungauged mode (01 August 1996–31 July 2002; daily time step).

Model	Nash and Sutcliffe efficiency (%)		
	J3024010	V6035010	Y5615030
Affdeff	76	–188	–88
GR4J	*NA*	*NA*	*NA*
GR5H	*NA*	*NA*	*NA*
HBV0	*NA*	*NA*	*NA*
Hydrotel	–1	53	50
IHAC	*NA*	*NA*	*NA*
ModSpa	65	64	69
Mordor	*NA*	*NA*	*NA*
NOAH–LSM	37	16	77
SAC–SMA[1]	64	43	39
SAC–SMA	71	78	78
SWB	66	71	80
Topmo	NA	NA	NA
VIC	72	45	82
	Bias (%)		
Affdeff	–16.44	104.87	52.88
GR4J	NA	NA	NA
GR5H	NA	NA	NA
HBV0	NA	NA	NA
Hydrotel	2.11	30.50	–17.05
IHAC	NA	NA	NA
ModSpa	1.80	10.20	12.38
Mordor	NA	NA	NA
NOAH–LSM	–48.20	93.53	7.55
SAC–SMA1	–24.85	–18.92	–46.90
SAC–SMA	–11.79	10.91	–22.96
SWB	–4.88	26.84	–15.58
Topmo	NA	NA	NA
VIC	26.24	94.73	21.91
	RMSE ln(Q) (mm day^{-1})		
Affdeff	0.16	1.14	0.58
GR4J	NA	NA	NA
GR5H	NA	NA	NA
HBV0	NA	NA	NA
Hydrotel	0.44	0.30	0.59
IHAC	NA	NA	NA
ModSpa	0.11	0.58	0.34
Mordor	NA	NA	NA
NOAH–LSM	0.51	0.79	0.33
SAC–SMA1	0.35	0.55	0.52
SAC–SMA	0.21	0.30	0.14
SWB	0.47	0.42	0.29
Topmo	NA	NA	NA
VIC	0.12	1.20	0.32
	RMSE $\sqrt{Q}$ (mm day^{-1})		
Affdeff	0.24	0.63	0.61
GR4J	NA	NA	NA
GR5H	NA	NA	NA

Table 8 (*cont.*)

	J3024010	V6035010	Y5615030
HBV0	NA	NA	NA
Hydrotel	0.39	0.29	0.48
IHAC	NA	NA	NA
ModSpa	0.24	0.35	0.39
Mordor	NA	NA	NA
NOAH–LSM	0.38	0.48	0.36
SAC–SMA1	0.32	0.38	0.52
SAC–SMA	0.27	0.25	0.29
SWB	0.35	0.32	0.34
Topmo	NA	NA	NA
VIC	0.24	0.51	0.36
	RMSE(Q) (mm day^{-1})		
Affdeff	0.66	2.59	4.16
GR4J	NA	NA	NA
GR5H	NA	NA	NA
HBV0	NA	NA	NA
Hydrotel	1.34	1.15	2.25
IHAC	NA	NA	NA
ModSpa	0.80	0.92	1.68
Mordor	NA	NA	NA
NOAH–LSM	1.07	1.40	1.46
SAC–SMA1	0.81	1.15	2.37
SAC–SMA	0.73	0.72	1.43
SWB	0.79	0.82	1.34
Topmo	NA	NA	NA
VIC	0.72	1.14	1.28

NA, not available.

K2724210) Mordor and HBV stand out from the rest of the tested models. No physical or hydrometric characteristic can be found to explain this behaviour as the database contains other catchments having similar morphometric and hydrological properties for which the models give similar trends.

With regard to RMSE on Q, Mordor and Topmo rank first and second, respectively. GR4H is the only model that does not rank first at all; all remaining four models rank first at least once. For RMSE on SQRT(Q) Mordor and GR5H rank first and second and Topmo ranks third. However, the RMSE values on both (Q) and SQRT(Q) calculated for all five models are very similar i.e. the greatest difference between the highest and the lowest value for a given catchment is 0.16 mm day^{-1}. Hence one cannot conclude on a difference in the models' performances based on these criteria. This is also the case for the RMSE values on ln(Q).

Ungauged mode

Eight models were used to submit simulations for the ungauged mode on the 3C database and four models were tested on the 11C sample. Table 2 synthesizes the parameterization strategies used by the various participants; the outcomes are presented in detail in the companion papers (this issue).

Ungauged mode 3C Level

Table 8 indicates that all models performed less well in the ungauged mode than in the gauged mode, all catchments and criteria considered. This finding is not surprising in itself and stresses the extreme difficulty of the exercise.

With regard to the Nash and Sutlciffe criterion on (Q) SAC-SMA, SWB and Modspa have the most stable overall behaviour as the criteria vary within a maximum bound of 10% on all three catchments. VIC seems to perform rather well in the ungauged mode; it ranks first in terms of NS on the third catchment and second on the first one.

The most unstable behaviour, in terms of Nash and RMSE variation, is reported for Afdeff. It is interesting to note that the NS value obtained using Afdeff on the first catchment (76%) is very high both with regard to the other models and to the model's own performance on the other two catchments. It is within close range of the NS criterion calculated for the ungauged mode. The model was parameterized using a single flood event of duration <600 hours, hence the good result achieved on this catchment could be linked to the quality and representativeness of the selected flood event. In contrast, the method seems poorly adapted to the two other catchments (V6035010 and Y5615030), which are subject to more Mediterranean semiarid conditions highlighted by a high temporal variability of both precipitation and runoff.

None of the selected models clearly out-performs the rest with regard to both low and high flows, respectively, RMSE on $\ln(Q)$ and $\sqrt{Q}$ as the model ranking changes according to the test catchments. However, SAC-SMA (Schaake & Duan) ranks first twice for both criteria and shows an overall more stable behaviour than the other models. The best performance in terms of bias is reported for Modspa, respectively, first twice and second once, while the worst performances are reported for Afdeff, NOAH and VIC.

Ungauged mode 11C Level

Figures 7 and 8 present the Nash and Sutcliffe criterion distributions, both using Q and SQRT(Q), for the 11C samples. The results are once again poorer than for the gauged mode, and clearly worse for NOAH than for the other three models as NOAH scores a Nash criterion <–4% on 50% of the catchments. In comparison VIC, SWB and SAC-SMA have Nash criteria that are respectively >55%, >66% and >70%. SAC-SMA's performance is the best.

The models' ranking is not modified with regard to the NS on SQRT(Q). However, NOAH's performance improves drastically as 50% of the catchments have now a Nash criterion >27%. This would mean that NOAH is more prone to errors on high flows. The difference in performance is less clear for the other three models.

Table 9 indicates that the bias values are high for all four models. Furthermore, the models' behaviour is quite similar when considering their variation per catchment,

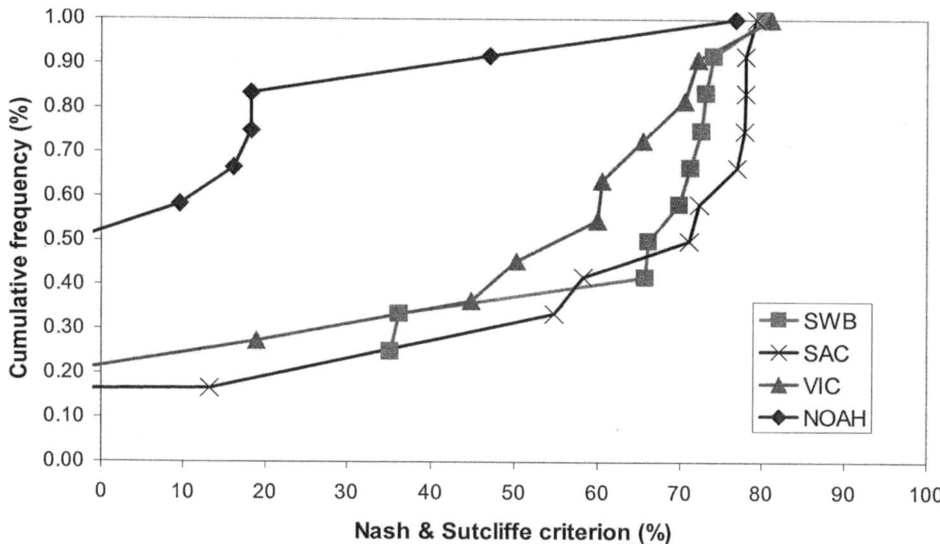

Fig. 7 Nash and Sutcliffe criteria distribution for the NOAH, SAC-SMA, SWB, and VIC models in ungauged mode.

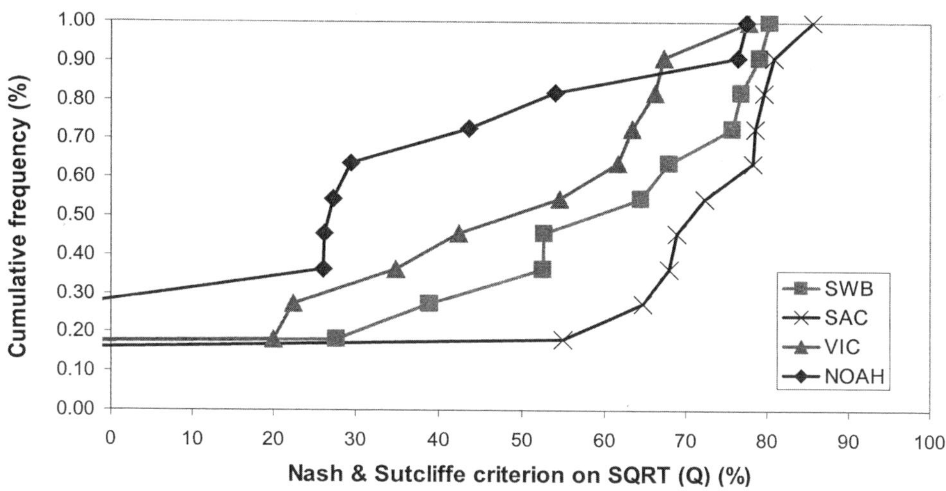

Fig. 8 Nash and Sutcliffe on SQRT(Q) criteria distribution for the NOAH, SAC-SMA, SWB, and VIC models in ungauged mode.

with the exception of catchments V6035010 and J4124420 for VIC. The bias values for the ungauged mode are higher than those of the gauged mode, when considering the same length of record (Fig. 9). This is also the case of the RMSE criteria both on Q, SQRT(Q) and ln(Q) (Fig. 10).

Table 10 synthesizes the models' overall performance, taking into account all the catchments and criteria. It indicates that the best results are obtained by SAC-SMA and the poorest by NOAH. It is not clear however, whether the difference is due to the models' performance or the modellers' longer experience with the model.

Table 9 *Bias*, RMSE on Q, $\sqrt{Q}$ and $\ln(Q)$ values per model and catchment for the ungauged mode for the 11 catchment sample.

Catchment	NOAH	SAC–SMA	SWB	VIC
(a) *Bias* (%)				
a1522020	68.39	37.19	43.71	63.08
h2001020	17.52	–21.46	–15.91	–10.59
h3613020	129.96	108.00	143.42	181.81
j2034010	38.51	1.02	18.04	77.09
j3024010	34.69	23.63	34.85	26.24
j4124420	–4.82	–11.79	–4.88	65.35
k0744010	46.11	25.28	34.58	24.72
k0753210	90.02	–0.91	13.80	19.38
v6035010	68.51	–10.82	1.65	94.73
y3514020	93.53	10.91	26.84	30.56
y5615030	19.86	–26.78	–18.87	21.91
(b) RMSE (Q) (mm day^{-1})				
a1522020	4.11	1.80	1.62	1.96
h2001020	1.78	1.29	1.27	1.18
h3613020	0.84	0.68	0.81	0.78
j2034010	1.37	0.91	1.10	1.01
j3024010	0.85	0.63	0.75	0.72
j4124420	1.07	0.73	0.79	1.40
k0744010	1.79	1.42	1.23	0.78
k0753210	1.29	0.58	0.70	0.73
v6035010	1.25	0.62	0.67	1.14
y3514020	1.40	0.72	0.82	0.60
y5615030	0.95	0.44	0.48	1.28
(c) RMSE $\sqrt{Q}$ (mm day^{-1})				
a1522020	0.73	0.40	0.43	0.55
h2001020	0.43	0.33	0.31	0.29
h3613020	0.41	0.32	0.42	0.44
j2034010	0.48	0.27	0.37	0.37
j3024010	0.31	0.19	0.28	0.24
j4124420	0.38	0.27	0.35	0.45
k0744010	0.47	0.36	0.37	0.31
k0753210	0.53	0.21	0.23	0.26
v6035010	0.47	0.21	0.21	0.51
y3514020	0.48	0.25	0.32	0.23
y5615030	0.31	0.21	0.32	0.36
(d) RMSE $\ln(Q)$ (mm day^{-1})				
a1522020	0.53	0.19	0.27	0.45
h2001020	0.23	0.16	0.17	0.13
h3613020	0.77	0.46	0.91	1.17
j2034010	0.78	0.31	0.49	0.41
j3024010	0.38	0.08	0.28	0.12
j4124420	0.51	0.21	0.47	0.35
k0744010	0.35	0.17	0.25	0.38
k0753210	0.94	0.16	0.15	0.18
v6035010	0.58	0.12	0.11	1.20
y3514020	0.79	0.30	0.42	0.28
y5615030	0.59	0.33	0.93	0.32

Table 10 Synthesis of overall model classification for the ungauged mode for the 11 catchment sample.

Criteria	Best Model	Worst Model
Nash(Q)	SAC-SMA	NOAH
Nash($\sqrt{Q}$)	SAC-SMA	NOAH
Bias	SAC-SMA	NOAH
RMSE	SAC-SMA	NOAH
RMSE($\sqrt{Q}$)	SAC-SMA	NOAH
RMSE(ln(Q))	SAC_SAM	NOAH

(a) **SAC**

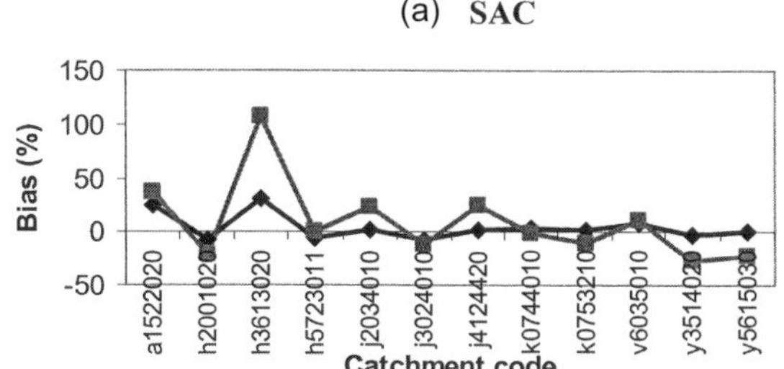

(b) **SWB**

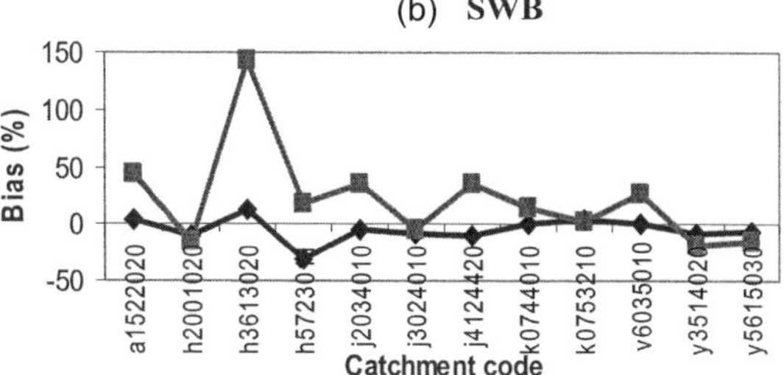

(c) **NOAH**

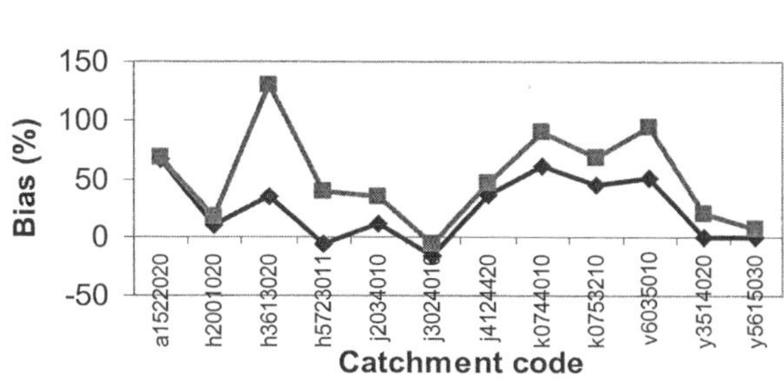

Fig. 9 Comparison between the bias values of the ungauged and gauged modes for NOAH, SAC-SMA and SWB.

The failure of the models in the ungauged mode highlights the need to investigate new parameterization strategies and methods in order to find alternatives for model calibration. Currently three mainstream strategies are followed by the MOPEX community to estimate model parameters on ungauged catchments. The first believes in the physical meaning of the model parameters and hence tries to infer these from soil and vegetation data (Modspa, VIC, Hydrotel, …). In this instance the results depend directly on the quality and spatial resolution of the data, two conditions that are not easily met in operational hydrology. The second group tries to investigate new methods to transfer parameter values from other model applications (see paper on the sister catchment approach by Hogue *et al.*, and papers on ensemble simulations by Schaake *et al.*, and Andréassian *et al.*, this issue) of course these methods then need to

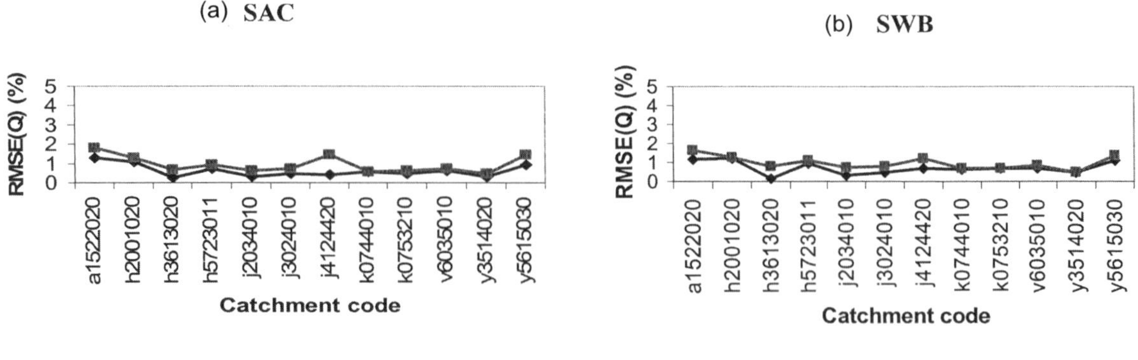

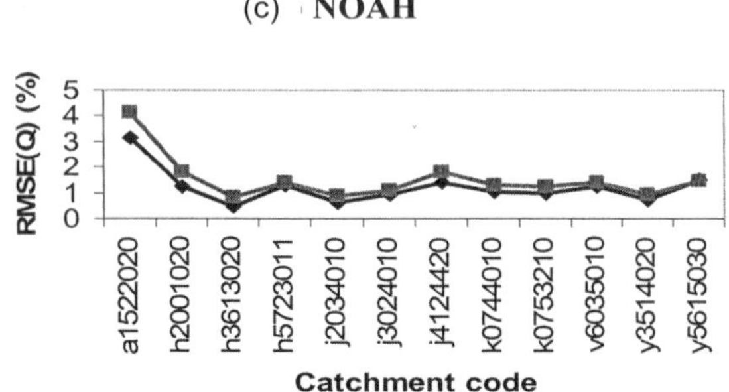

Fig. 10 Comparison between the RMSE values of the ungauged and gauged modes for NOAH, SAC-SMA and SWB.

determine a similarity index to quantify the resemblance between catchments. The third alternative consists of calibrating the model on fewer data points such as a single flood event and using the parameters for the rest of the time series (see papers by Fortin *et al.*, and Rojas-Serna *et al.*, this issue). This approach assumes a temporal stability of the models' parameters and still needs some minimal hydrological data to ensure model calibration.

In light of the simulation files that were submitted it is not possible to choose a single method as being the most efficient. Further tests should be undertaken to investigate these issues. The participation level and enthusiasm generated by the MOPEX 2004 workshop is a clear indication of the necessity and collective will to pursue these efforts both as individual scientists and as a research group. Further collaboration with the distributed modellers' community, through programmes such as DMIP, could provide an enriching forum in this sense.

CONCLUSION AND PERSPECTIVES

As part of the MOPEX 2004 workshop, the participants were asked to submit simulations using the database provided by Météo-France and Cemagref (see Chahinian *et al.*, this issue). For the first time since the launch of the MOPEX programme, the data set contained hourly hydrometeorological data. Fourteen models were tested on 3 catchments, 10 on 11 catchments and 6 on 40 catchments in gauged mode. The participation was slightly lower for the ungauged mode: 8 models were tested on 3 catchments and 4 on 11 catchments.

The objective of this paper was to analyse and compare the results produced by various models in order to see the advantages and shortcomings of their parameter estimation strategies. The results indicate that all the models succeeded in yielding acceptable results when calibrated using discharge data. The differences in calibration critera did not alter these results, although some models had high bias values despite good Nash & Sutcliffe efficiencies, this may be related to the individual effort modellers put in to the calibration process, although the latter is difficult to quantify. However, all the models' performance dropped in the ungauged mode, all criteria and catchments considered. This highlights a need to improve alternative parameter estimation techniques that do not rely heavily on calibration. Interesting approaches based on similarity indices between catchments, ensemble simulations and limited discharge data mining were presented. In order to achieve good results these techniques need large data samples to establish statistically significant relationships. Hence more experiments should be carried out on a sample of a large number of catchments, assuming that data providers are willing to share their data. Of course modellers will also have to agree to put their models through "blind tests". With its open access databases and experience in knowledge transfer, MOPEX can be an important forum between modellers, experimentalists, and water agencies. Its growing interest in predictions in ungauged catchments renders it an essential part of the IAHS's PUB decade.

Acknowledgements The MOPEX workshop organization funds were provided by the Cemagref's International Relations' Department. The authors would like to thank Météo-France, Cemagref and the National Weather Service for providing the comparison data and ENGREF for hosting the 2004 workshop.

REFERENCES

Abbott, M., Bathurst, J., Cunge, J., O'Connel, P. & Rasmussen, J. (1986) An introduction to the European Hydrological System-Système Hydrologique Européen, "SHE", 1: History and philosophy of a physically-based, distributed modelling system. *J. Hydrol.* **87**, 45–59.

Bergström, S. & Forsman, A. (1973) Development of a conceptual deterministic rainfall–runoff model. *Nordic Hydrol.* **4**, 147–170.

Beven, K. (1993) Prophecy, reality and uncertainty in distributed hydrological modelling. *Adv. Water Resour.* **16**: 41–51.

Beven, K. & Kirkby, M. (1978) A physically based, variable contributing area model of basin hydrology. *Hydrol. Sci.* **24**(1), 43–69.

Brath, A., Montanari, A & Moretti, G. (2002). On the use of simulation techniques for the estimation of peak river flows. In: *Proc. Int. Conf. on Flood Estimation, Berna, 6-8 March 2002, Rep II-17*), 587–599. CHR/KHR, International Commission for the Hydrology of the Rhine basin, Lelystad, The Netherlands.

Burnash, R., Ferral, R. & McGuire, R. (1973) *A Generalized Streamflow Simulation System: Conceptual Modeling for Digital Computers*. Joint Federal-State River Forecast Center, Sacramento, Calicornia, USA.

Chen, F., Mitchell, K., Schaake, J., Xue, Y., Pan, H., Koren, V., Duan, Q., Ek, M. & Betts, A. (1996) Modeling of land--surface evaporation by four schemes and comparison with FIFE observations. *J. Geophys. Res.* **101**, 7251–7268.

Croke, B., Merritt, W. & Jakeman, A. (2004) A dynamic model for predicting hydrologic response to land cover changes in gauged and ungauged catchments. *J. Hydrol.* **291**(1–2), 115–131.

Duan, Q. & Gupta, V. (1992) Effective and efficient global optimization for conceptual rainfall–runoff models. *Water Resour. Res.* **28**(4), 1015–1031.

Edijatno (1991) Mise au point d'un modèle élémentaire pluie-débit au pas de temps journalier. PhD Thesis, University Louis Pasteur/ENGEES, Strasbourg, France.

Fortin, J. P., Moussa, R., Bocquillon, R. & Villeneuve, J. P. (1995) Hydrotel un modèle hydrologique distribué pouvant bénéficier des données fournies par la télédétection et les systèmes d'informations géographiques. *Rev. Sci. Eau* **8**(1): 97–124.

Fortin, J. P., Turcotte, R., Massicotte, S., Moussa, R., Fitzback, J. & Villeneuve, J. P. (2001) Distributed watershed model compatible with remote sensing and GIS data. I: Description of model. *J. Hydrol. Engng* **6**(2), 91–99.

Gupta, H., Soorooshian, S., Hogue, T. & Boyle, D. (2003) Advances in automatic calibration of watershed models. In *Calibration of Watershed Models* (ed. by Q. Duan, H. Gupta, S. Sorooshian, A. Rousseau & R. Turcotte), 9–28. American Geophysical Union, Washington, DC, USA.

Haverkamp, R., Parlange, J. Y., Cuenca, R. & Ross, P. (1998) Scaling of the Richards equation and its application to watershed modelling. In *Scale Dependence and Scale Invariance in Hydrology* (ed. by G. Sposito), 190–243, Cambridge University Press, New York, USA.

Hendrickson, J., Sorooshian, S. & Brazil, L. (1988) Comparison of Newton-type and direct search algorithms for calibration of conceptual rainfall–runoff models. *Water Resour. Res.* **24**(5), 691–700.

Jakeman, A. J., Littlewood, I. G. & Whitehead, P. G. (1990) Computation of the instantaneous unit hydrograph and identifiable component flows with application to two small upland catchments. *J. Hydrol.* **117**, 275–300.

Johnston, P. & Pilgrim, D. (1976) Parameter optimization for watershed models. *Water Resour. Res.* **12**(3), 477–186.

Kokkonen, T., Jakeman, A., Young, P., Koivusalo, H. (2003) Predicting daily flows in ungauged catchments: model regionalization from catchment descriptors at the Coweeta Hydrologic Laboratory, North Carolina. *Hydrol. Processes* **17**(11), 2219–2238.

Koren, V., Smith, M. & Duan, Q. (2003) Use of a priori parameter estimates in the derivation of spatially consistent parameter sets of rainfall–runoff models In: *Calibration of Watershed Models* (ed. by Q. Duan, H. Gupta, S. Sorooshian, A. Rousseau & R. Turcotte), 239–254. American Geophysical Union, Washington, DC, USA.

Liang, X. & Xie, Z. (2001) A new surface runoff parameterisation with subgrid-scale soil heterogeneity for land surface models. *Adv. Water Resour.* **24**, 1173–1193.

Liang, X., Lettenmaier, D., Wood, E. & Burges, S. (1994) A simple hydrologically based model of land surface water and energy fluxes for general circulation models. *J. Geophys. Res.* **99**(D7), **14**, 415–414, 428.

Leavesley, G., Hay, L., Viger, R. & Markstorm, S. (2003) Use of *a priori* parameter-estimation methods to constrain calibration of distributed-parameter models. In: *Calibration of Watershed Models* (ed. by Q. Duan, H. Gupta, S. Sorooshian, A. Rousseau & R. Turcotte), 255–266. American Geophysical Union, Washington, DC, USA

Lee, H., McIntyre, N., Wheater, H. & Young, A. (2005) Selection of conceptual models for regionalization of the rainfall–runoff relationship. *J. Hydrol.* **315**(1–4), 71–92.

Mathevet T. (2005) Quels modèles pluie-débit au pas de temps horaire? Développements empiriques et intercomparaison de modèles sur un large échantillon de bassins versants. PhD Thesis, ENGREF-Cemagref, Paris, France.

Mertz, R. & Bloschel, G. (2005) Flood frequency regionalization—spatial proximity vs. catchment attributes. *J. Hydrol.* **302**(1–4), 283–306.

Micovic, J. & Quick, M. (1999) A rainfall and snowmelt runoff modelling approach to flow estimation at ungauged sites in British Columbia. *J. Hydrol.* **226**(1–2), 101–120.

Moretti, G. & Montanari, A. (2005) AFFDEF: a spatially distributed grid based rainfall–runoff model for continuous time simulations of river discharge. *Environ. Modelling & Software* (submitted).

Moussa, R. (1991) Variabilité spatio-temporelle et modélisation hydrologique. Application au bassin du Gardon d'Anduze. PhD Thesis, University of Montpellier II, France.

Moussa, R. (1993) Modélisation hydrologique spatialisée et système d'information géographique. *La Houille Blanche* **5**, 293–301.

Perrin, C. (2000) Vers une amélioration d'un modèle global pluie-débit au travers d'une approche comparative. PhD thesis, INPG (Grenoble) / Cemagref (Antony), France.

Perrin, C., Michel, C. & Andréassian, V. (2001) Does a large number of parameters enhance model performance? Comparative assessment of common catchment model structures on 429 catchments. *J. Hydrol.* **242**(3–4), 275–301.

Schaake, J., Koren, V., Duan, Q., Mitchell, K. & Chen, F. (1996) Simple water balance model for estimating runoff at different spatial and temporal scales. *J. Geophys. Res.* **101**(D3), 7461–7475.

Seibert, J. (1999) Regionalization of parameters for a conceptual rainfall–runoff model. *Agr. For. Met.* **98–99**, 279–293.

Van der Linden, S. & Woo, M. (2003) Transferability of hydrological model parameters between basins in data-sparse areas, subarctic Canada. *J. Hydrol.* **270**(3–4), 182–194.

The Model Parameter Estimation Experiment (MOPEX): its structure, connection to other international initiatives and future directions

THORSTEN WAGENER[1], TERRI HOGUE[2], JOHN SCHAAKE[3], QINGYUN DUAN[4], HOSHIN GUPTA[5], VAZKEN ANDREASSIAN[6], ALAN HALL[7] & GEORGE LEAVESLEY[8]

1 *Department of Civil & Environmental Engineering, Pennsylvania State University, University Park, Pennsylvania, USA*
thorsten@engr.psu.edu

2 *Department of Civil & Environmental Engineering, University of California, Los Angeles, California, USA*

3 *Office of Hydrologic Development, NOAA/National Weather Service, 1325 East–West Avenue, Silver Spring, Maryland 20910, USA*

4 *Lawrence Livermore National Laboratory, Energy and Environment Directorate, 7000 East Avenue, Livermore, California 94550, USA*

5 *Department of Hydrology and Water Resources, University of Arizona, Tucson, Arizona, USA*

6 *Groupe Hydrologie, Cemagref, Antony, France*

7 *Water Resources Application Project/GEWEX, Cooma, Australia*

8 *US Geological Survey, Denver, Colorado, USA*

Abstract The Model Parameter Estimation Experiment (MOPEX) is an international project aimed at developing enhanced techniques for the *a priori* estimation of parameters in hydrological models and in land surface parameterization schemes connected to atmospheric models. The MOPEX science strategy involves: database creation, *a priori* parameter estimation methodology development, parameter refinement or calibration, and the demonstration of parameter transferability. A comprehensive MOPEX database has been developed that contains historical hydrometeorological data and land surface characteristics data for many hydrological basins in the United States (US) and in other countries. This database is being continuously expanded to include basins from various hydroclimatic regimes throughout the world. MOPEX research has largely been driven by a series of international workshops that have brought interested hydrologists and land surface modellers together to exchange knowledge and experience in developing and applying parameter estimation techniques. With its focus on parameter estimation, MOPEX plays an important role in the international context of other initiatives such as GEWEX, HEPEX, PUB and PILPS. This paper outlines the MOPEX initiative, discusses its role in the scientific community, and briefly states future directions.

Key words *a priori*; calibration; hydrological models; parameters; regionalization; transferability; uncertainty

INTRODUCTION

Hydrological and land surface models are tools of increasing importance for water resources management worldwide: they support decision making and strategic planning in a context of possible climate and land use change in basins. They are important to assess the impact of these changes on water resources. Requirements to successfully perform these modelling tasks are that we can reliably model a wide range of hydroclimatic regimes and understand how changes in the basin are reflected as changes in the model parameters or in its structure. Current model structure requires that at least some key parameters are adjusted according to the fit of the model predictions to observations of the response variable of interest (usually streamflow), a process called calibration. However, these measurements are not always available and where available, they may be very sparse. Thus, other approaches are needed to derive *a priori* (before calibration) parameter estimates. Current procedures for *a priori* parameter estimation are often based on relationships between model parameters and basin characteristics—that is, soils, vegetation, topography, climate, geology, etc. These developed relationships have not been fully validated through rigorous testing using retrospective hydrometeorological data and corresponding land surface characteristics. This is partly because the necessary database needed for such testing has not been available. Moreover, there still exists a gap in our understanding of the links between model parameters and land surface characteristics. Generally, available information about soils and vegetation typically only indirectly relates to model parameters that conceptualize aspects such as the hydraulic properties of soils and rooting depths of vegetation. It is also not clear how heterogeneity associated with spatial land surface properties affects those characteristics at the scale of a basin or a grid cell. Consequently, there is a considerable degree of uncertainty associated with the parameters derived using current procedures, which is propagated into the model predictions and into the subsequent decision making process.

The Model Parameter Estimation Experiment (MOPEX) is an ongoing international project to help develop techniques for the *a priori* estimation of parameters used in land surface parameterization schemes of atmospheric and hydrological models. MOPEX is affiliated with both the Predictions in Ungauged Basins (PUB) and the Global Energy and Water Cycle Experiment (GEWEX), and is supported by the GEWEX Americas Prediction Project (GAPP) as well as by individual participants. MOPEX evolved to address the parameter estimation problem and to promote and guide the development of improved *a priori* parameter techniques applicable to both gauged and ungauged basins. The MOPEX project has been an international collaborative effort since 1996, with the involvement of international scientists and data sets assembled from different countries.

The scope of this paper includes an outline of the MOPEX initiative, a discussion of its role in the scientific community and closes with a statement of future directions.

MOPEX SCIENCE STRATEGY

The MOPEX science strategy involves three major steps (Fig. 1):

- To develop the necessary data sets from a range of hydroclimatic regimes.
- To use these data to develop *a priori* and calibrated parameters and then develop and test new *a priori* techniques.
- To demonstrate transferability to other basins.

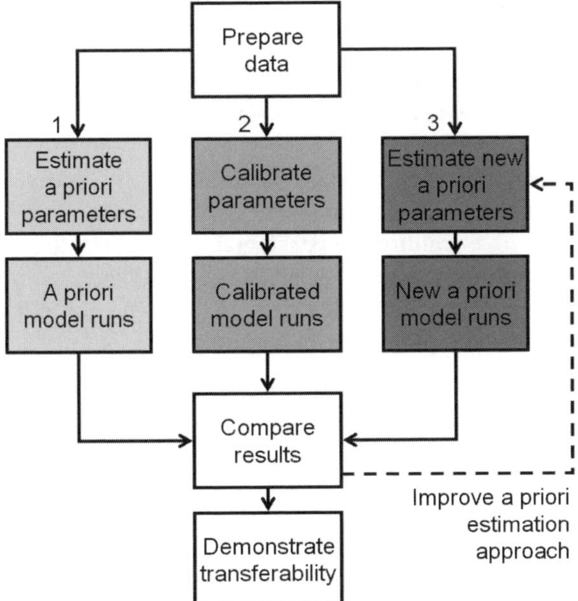

Fig. 1 MOPEX implementation strategy.

Figure 1 outlines the three-path strategy for the second step in the MOPEX science strategy. The benchmark for any development is *a priori* parameter estimation made with the currently available techniques (path 1). Any newly developed technique has to improve on this benchmark. The second path includes the calibration of the model parameters against observations of the variable of interest. These parameter estimates therefore provide the optimal (at least from an optimization point of view, not necessarily from a hydrological point of view) parameter estimates for the basin under investigation. So, while path 1 provides the lower bound of performance that any *a priori* parameter estimation method has to achieve, path 2 provides the upper bound since it is (or at least should be) not feasible that any *a priori* estimates perform better than optimized parameters. Within the path 2, the calibrated parameters are analysed to improve our understanding of relationships between model parameters and basin characteristics (i.e. climate, soils, vegetation and topographic features). This improved understanding can then be used to derive new methods and to make new model runs in path 3. The success of step two is measured by how much improvement in model performance is achieved in path 3 compared with results from the benchmark runs in path 1.

A range of objectives or focus areas has been defined by the MOPEX participants over the years in order to foster the research and activities required to achieve improved *a priori* estimates. These can be listed as follows:

- Objective 1: To evaluate the transferability of *a priori* model parameters between watersheds;
- Objective 2: To develop and test improved *a priori* model parameter estimation techniques for ungauged basins;

- Objective 3: To develop improved strategies to classify basins and to measure hydrological similarity;
- Objective 4: To develop diagnostic tools to foster improved understanding of natural hydrological processes at basin scales and of the related behaviour of hydrological models;
- Objective 5: To promote the standard assessment of uncertainty, including assessing the uncertainty in *a priori* parameter estimates;
- Objective 6: To improve calibration strategies, including the use of *a priori* parameter estimates within these strategies;
- Objective 7: To develop objective measures to evaluate the parameter estimate techniques and to understand parameter uncertainty.

The research will be facilitated through the development and use of an international database of retrospective hydrometeorological data and basin characteristics data for a wide range of climate and geophysical conditions. This database should include data for a wide range of Intermediate Scale Area (ISA) river basins (500–10 000 km^2) throughout the world. MOPEX has already assembled retrospective hydrometeorological data and basin characteristics data for many US basins. Data for select basins from other countries such as Australia, France, UK, Germany, and China have also been collected. Details about this database effort can be found in Schaake *et al.* (this issue). The MOPEX initiative will continue to promote and facilitate the exchange of ideas and experiences on approaches to model parameter estimation for different climatic regimes.

MOPEX WORKSHOPS

MOPEX participants have organized a series of workshops (Table 1) as one of the main instruments in advancing MOPEX science. Workshops, which started in 1999, have been held in each of the past three years to investigate various issues related to parameter estimation and regional model applications. Prior to the workshops, model participants are asked to apply their model and parameter estimation procedures to (preferably) a nominated set of MOPEX basins for prior analysis. The main goal of these workshops has been to investigate some of the MOPEX science strategies using various hydrological and land surface models on long-term hydrometeorological data. Details of past and currently planned workshops are given in Table 1. Past workshops centred particularly around the following questions:

- How do we define the relationships between model parameters and basin characteristics?
- How can model calibration be used to refine the *a priori* parameters?
- How do we evaluate the uncertainty due to model structure, calibration data and model parameters?

While the preceding workshops have been very successful, it was felt that a more concentrated effort is required to increase the science output derived from the cooperative efforts of the workshop participants. Starting with the 2004 workshop in France, workshops will be based on a new format with the following underlying principles:

Table 1 MOPEX workshops and symposia.

Date	Location	Sponsor	Focus
Past			
1999	Birmingham, UK	IUGG[1]/IAHS[2]	Initiating workshop
2001	Maastricht, The Netherlands	IAHS	SVAT symposium with examples of parameter estimation techniques, IAHS Publ. 270
2002	Tucson, USA	NSF STC SAHRA[3]	US Data
2003	Sapporo, Japan	IUGG/IAHS	US Data and review of techniques
2004	Paris, France	CEMAGREF	French Data
2005	Foz do Iguassu, Brazil	IAHS	Use of large data sets
Future			
2007	Perugia, Italy	IUGG/IAHS	Link to atmospheric science

[1] International Union of Geodesy and Geophysics. [2] International Association of Hydrological Sciences. [3] National Science Foundation Science and Technology Center for Sustainability of Semi-Arid Hydrology and Riparian Areas.

- Each workshop will have a specific focus, either in terms of hydro-climatic region (for example, humid or semiarid) or in terms of a specific application (for example, flood forecasting).
- Each workshop will allow different levels of participation; that is, different numbers of basins that have to be simulated, to enable every participant to contribute to the collaborative science investigation. A minimum number of basins will be specified that each participant has to simulate in order to take part in the workshop. A larger number of basins is available for those who want to contribute more.
- The data sets of 12 basins from each workshop will be archived to create a database of benchmark basins that will be taken forward in time and used for comparison studies.
- This benchmark database is in addition to the general MOPEX database, which is aimed at creating a high-quality, historical hydrometeorological and river basin characteristic data sets for a wide range of ISA river basins throughout the world. High-quality data sets have been obtained for Australia from the University of Melbourne and for the United Kingdom from the Institute of Hydrology, in addition to a large number of US data sets.
- A small, but well-defined set of science objectives will be listed for each workshop to allow for a coordinated and meaningful analysis of workshop results.

INTERNATIONAL INITIATIVES AND THEIR RELATION TO MOPEX

The MOPEX initiative has its place in the international arena of initiatives, with its clear focus on improving parameters for hydrological and land surface models. Here we discuss briefly how MOPEX relates to some of the main international initiatives that are currently ongoing.

The Project for Intercomparison of Land-surface Parameterization Schemes (PILPS) has shown that different land–surface schemes can provide widely different simulation results, although driven by the same meteorological forcing data and required to use the same values for commonly named parameters (such as soil hydraulic properties and vegetation phenology parameters). The large scattering of model results can be partially explained by the uncertainty in the values of the parameters used in each scheme. Development of enhanced *a priori* parameter estimation methodologies is therefore necessary to improve the performance of hydrological models and land surface schemes.

MOPEX has the endorsement of several international organizations and projects including: the World Meteorological Organization (WMO) Commission on Hydrology, International Association of Hydrological Sciences (IAHS) Predictions in Ungauged Basins (PUB) Initiative (Sivapalan, 2003; Wagener *et al.*, 2004) and the Global Energy and Water Cycle Experiment (GEWEX). The Office of Global Programs in the National Oceanic and Atmospheric Administration (NOAA) and funding agencies in different countries have all provided financial support for scientists to participate in MOPEX activities.

The Hydrologic Ensemble Prediction Experiment (HEPEX) focuses on bringing meteorological and hydrological scientists from research, operational and user communities together to work on advancing probabilistic hydrological forecasting (Franz *et al.*, 2005). The HEPEX initiative was launched in the spring of 2004. With its focus on connecting the hydrological and meteorological communities, and with a strong emphasis on prediction, this initiative looks to MOPEX to develop improved parameterizations of the terrestrial components of the overall modelling system and to represent uncertainty in these parameterizations.

Within the PUB initiative of IAHS, MOPEX has an important position as the working group that focuses on the problems of parameter estimation in gauged and ungauged basins. Therefore it has a central role and the MOPEX database is likely to be of great value within PUB where comparison studies are a major component. Two other working groups with which we anticipate particularly strong interaction are the Top-Down Modelling Working Group (http://www.stars.net.au/tdwg/; Littlewood *et al.*, 2003)—focusing on the development of parsimonious data-based modelling approaches—and the Working Group on Uncertainty Estimation for Hydrological Modelling (Wagener *et al.*, 2006)—focusing on the development of an uncertainty framework for hydrological predictions.

LESSONS LEARNED SO FAR

Ultimately, models are (by definition) simplified representations of the real world and are therefore imperfect. This means that their parameters are also simplified representations of the real world characteristics they represent, and we will therefore never find a 1:1 correlation between model parameters and basin characteristics. It is likely that there will remain some dependency on fitting the model to our intended real-world application considering the model purpose, data availability, hydro-climatic region, etc. This also means that there will always be some degree of uncertainty that has to be considered in this process and that has to be communicated to any decision maker. How to estimate and represent this uncertainty in the context of imperfect models is still unclear.

Results to-date of the MOPEX initiative have recently been summarized by Duan *et al.* (2006). Here we restate the results in the form of testable hypotheses that can be corroborated or refuted through our future workshop activities:

- Result: Study results confirm earlier statements that the existing *a priori* parameter estimation procedures are problematic and are in need of improvement. This means that there is still large uncertainty regarding how parameter values would change under altered conditions, e.g. climate and land use.
- Hypothesis 1: *Current a priori estimates of hydrologic model parameters are very uncertain.*
- Hypothesis 2: *Our lack of understanding regarding how model parameters relate to physical basin characteristics precludes reliable predictions of how watershed response would change with altered conditions (e.g. land-use change).*
- Result: Calibration results clearly demonstrate the potential for improvement in *a priori* parameter estimation. The difference between the benchmark and optimized modelling results shows that there is still a clear need for calibration to be included in modelling studies and that it should be possible to improve currently available *a priori* approaches. In addition, different models seem to represent hydrological processes differently and all of them are imperfect. This means that parameters will in most cases be model dependent, even if they have the same name and units. Understanding this model dependence should help in understanding what level of correlation between model parameters and basin characteristics can be expected.
- Hypothesis 3: *We are currently not able to achieve reliable hydrological predictions without relying on the calibration of at least some key parameters.*
- Hypothesis 4: *Model parameters are model structure dependent and any regionalization or transfer approach will necessarily be model specific.*

Corroborating or refuting these hypotheses requires repeatable tests on a sufficiently large number of basins to allow for statistically results. However, possible sample sizes will—at least for the near future—be limited for some more physically-based model approaches that cannot be applied easily to new basins.

MOPEX thrives on the contributions of a heterogeneous group of scientists. Its experiments/tests have to be sufficiently flexible to allow for the participation of everybody who wants to contribute. Our intention is to provide a platform to perform tests on the above mentioned and other hypotheses utilising as wide a range of model structures, methods and expertise as possible.

FUTURE OUTLOOK

The work of MOPEX will continue contributing to the improved understanding of *a priori* and *a posteriori* (calibrated) model parameters for hydrological and land surface models. This will be implemented through workshops with different foci, and an increased interaction with other initiatives such as HEPEX, and other working groups, such as the PUB Top-down modelling and Uncertainty working groups.

A particular strength of MOPEX is the continuing growth and use of an international database of basin data for the comparison of techniques and models. This enables us to create an increasing knowledge base of methods applied to the same data. Connecting this database to others created within the context of the PUB initiative will provide international benchmark data sets.

We invite discussion and input from interested researchers on the following aspects:

- *MOPEX initiative*: We invite anybody interested in the above subjects to participate in the MOPEX initiative. Additional information about the MOPEX initiative can be found at http://www.seas.ucla.edu/~thogue/MOPEX/ and in the article by Hogue *et al.* (2004).
- *MOPEX workshops*: We invite suggestions for geographical or topical foci of future workshops. Of particular interest would be a group willing to host a MOPEX workshop. The most effective workshops so far have lasted around 2½–3 days.
- *MOPEX database*: We are continuously expanding the MOPEX database and are always looking for new data sets of different hydroclimatic and geographical regions (or additions to data sets of regions already covered). Please contact any of the authors if you have a data set that fulfils the MOPEX quality and content requirements, and could be added to our database.

The next MOPEX workshop will be held during the IUGG/IAHS Perugia General Assembly in 2007. The workshop will focus on the connection to the atmospheric science community and their need for parameter estimation. As always, any constructive criticism on any aspect of this paper is very welcome.

Acknowledgements Thanks go to the NOAA Office of Global Programs for providing financial support under the GEWEX Americas Prediction Project (GAPP) program that funded part of the MOPEX workshop activities and supported development of the US contribution to the MOPEX database. The financial support of Cemagref's International Affairs Department for the organization of the Paris workshop is gratefully acknowledged.

REFERENCES

Duan, Q., Gupta, H., Sorooshian, S., Rousseau, A. & Turcotte, R. (eds) (2003) *Advances in Calibration of Watershed models*. Water Science and Application, Series 6. American Geophysical Union, Washington, DC, USA.

Duan, Q., Schaake, J., Andreassian, V., Franks, S., Gupta, H.V., Gusev, Y. M., Habets, F., Hall, A., Hay, L., Hogue, T. S., Huang, M., Leavesley, G., Liang, X., Nasonova, O. N., Noilhan, J., Oudin, L., Sorooshian, S., Wagener, T. & Wood, E. F. (2005). Model Parameter Estimation Experiment (MOPEX): overview of science strategy and summary of the second and third workshop results. *J. Hydrol.* **320**(1–2), 3–17.

Franz, K., Ajami, N., Schaake, J. & Buizza, R. (2005) Hydrologic ensemble prediction experiment focuses on reliable forecasts. *Eos Trans. AGU* **86**(25), 239.

Hogue, T., Wagener, T., Schaake, J., Duan, Q., Hall, A., Gupta, H. V., Leavesley, G. & Andreassian, V. (2004) A new phase of the Model Parameter Estimation Experiment (MOPEX). *Eos Trans. AGU*, **85**(22), 217–218.

Littlewood, I. G., Croke, B. F. W., Jakeman, A. J. & Sivapalan, M. (2003) The role of "top down" modelling for Prediction in Ungauged Basins (PUB). *Hydrol. Processes* **17**, 1673–1679.

Schaake, J. C. (1981) Summary of river forecasting raingauge network density requirements. Available at http://www.nws.noaa.gov/oh/mopex/raingage%20density%20requirement.htm. (unpublished).

Schaake, J., Duan, Q., Smith, M. & Koren, V. (2000) Criteria to select basins for hydrologic model development and testing. In: *15th Conf. on Hydrology* (AMS Meeting, 9–14 January, 2000). Long Beach, California, USA.

Schaake, J. & Duan, Q. (2006) The model parameter estimation experiment (MOPEX). *J. Hydrol.* (in press).

Sivapalan, M. (2003) Prediction in ungauged basins: a grand challenge for theoretical hydrology. *Hydrol. Processes* **17**(15), 3163–3170.

Wagener, T., Sivapalan, M., McDonnell, J., Hooper, R., Lakshmi, V., Liang, X. & Kumar, P. (2004) Predictions in ungauged basins as a catalyst for multidisciplinary hydrology. *Eos Trans. AGU* **85**(44), 451.

Wagener, T., Freer, J., Zehe, E., Beven, K., Gupta, H. V. & Bardossy, A. (2006) Towards an uncertainty framework for predictions in ungauged basins: The uncertainty working group. In: *Predictions in Ungauged Basins: Promise and Progress* (ed. by M. Sivapalan, T. Wagener, S. Uhlenbrook, X. Liang, V. Lakshmi, P. Kumar, E. Zehe, & Y. Tachikawa), 454–462. IAHS Publ. 303. IAHS Press, Wallingford, UK.

Key word index

Predictions in Ungauged Basins: International Perspectives on the State of the Art and Pathways Forward

PUB

Hydrological prediction where data are available is relatively easily achieved, albeit subject to uncertainty that is often unquantified.

But, ungauged catchments, by far the majority, present major difficulties for hydrological prediction, hence the IAHS PUB initiative.

The PUB (Predictions in Ungauged Basins) initiative, launched in Brasilia in November 2002, will run for 10 years. For information see:

www.pub.iwmi.org

Edited by Stewart Franks, Murugesu Sivapalan, Kuniyoshi Takeuchi & Yasuto Tachikawa

IAHS Publication 301 (*2005*) ISBN 1-901502-38-4, 348 + xxii pp. Price £60.00

The PUB Workshop held at Perth, Western Australia, in 2004, brought together researchers and practitioners to assess current hydrological prediction capabilities, and future research directions to achieve significant improvements given the decline of hydrological gauging networks, natural variability, and long-term climate and land-use changes. This volume is the outcome of those deliberations. It combines chapters presenting innovative theoretical and practical possibilities of different approaches for prediction, with contributions describing the differing perspectives and specific needs of Australia and Japan.

The final chapter summarizes group discussions which took a common approach to tackling topics from flood and drought hydrology, to water quality and ecosystem health, to assess prediction practicalities: What are Society's requirements? What is available? How can improvements in predictive capability be achieved? It reflects an innovative and successful approach to achieving a convergence of actions out of a divergence of views and perspectives.

Stimulating reviews and perspectives on the state of the art and innovative approaches for future progress

Predictions in Ungauged Basins: Promises and Progress

Edited by Murugesu Sivapalan, Thorsten Wagener, Stefan Uhlenbrook, Erwin Zehe, Venkat Lakshmi, Xu Liang, Yasuto Tachikawa & Praveen Kumar

IAHS Publication 303 (*2006*) ISBN 1-901-502-48-1; 520 + viii pp. Price £84.00

The peer-reviewed papers were selected from a symposium held at Foz do Iguaçu, Brazil, in 2005, which coincided with the completion of Phase 1 of PUB and provided the opportunity for the global PUB community to get together to discuss the progress that had been achieved, to identify areas of weakness or lack of activity, re-iterate the plans and promises of PUB, and to make necessary modifications/refinements to the PUB implementation activities with a view to achieving faster and more demonstrable progress in the future.

The 51 papers are organized into five sections that focus on the distinct PUB themes:

1 Model Improvements Through Detailed Process Studies
2 Model Evaluation and Comparison: Uncertainty Analysis and Diagnostics
3 New Data Collection Approaches and Model Development
4 New Modelling Approaches and Methods for Testing Against Observations
5 Progress in PUB Implementation: PUB Working Groups

Please send book orders and enquiries to:

Mrs Jill Gash
IAHS Press, Centre for Ecology and Hydrology
Wallingford, Oxfordshire OX10 8BB, UK

jilly@iahs.demon.co.uk
tel.: + 44 1491 692442
fax: + 44 1491 692448/692424

See the IAHS website ***www.iahs.info*** for information about the International Association of Hydrological Sciences (IAHS), membership, meetings, and publications.